T0250131

RIVER PUBLISHERS SERIES IN INFORMATION SCIENCE AND TECHNOLOGY

Volume 2

Consulting Series Editor

KWANG-CHENG CHEN
National Taiwan University
Taiwan

Other books in this series:

Volume 1
Traffic and Performance Engineering for Heterogeneous Networks
Demetres D. Kouvatsos
February 2009
ISBN 978-87-92329-16-5

Performance Modelling and Analysis of Heterogeneous Networks

Performance Modelling and Analysis of Heterogeneous Networks

Editor

Demetres D. Kouvatsos

*PERFORM – Networks & Performance Engineering
Research Unit, University of Bradford, U.K.*

River Publishers

Aalborg

ISBN 978-87-92329-18-9 (hardback)

Published, sold and distributed by:
River Publishers
P.O. Box 1657
Algade 42
9000 Aalborg
Denmark

Tel.: +45369953197
www.riverpublishers.com

To Maria and Mihalis

Table of Contents

Preface

Over the recent years, a considerable amount of effort has been devoted, both in industry and academia, into performance modelling and evaluation of convergent multi-service heterogeneous networks, supported by internetworking and the evolution of diverse access and switching technologies, towards the design and dimensioning of the next and future generation Internets. However, there are still challenging performance analysis and applications issues to be addressed and resolved before the establishment of a global and wide-scale integrated broadband network infrastructure for the efficient support of multimedia applications.

Performance Modelling and Analysis of Heterogeneous Networks presents recent advances in networks of diverse technology reflecting the state-of-the-art research achievements in the performance modelling, analysis and applications of heterogeneous networks. The book contains 21 extended/revised research papers, which have their roots in the series of the HET-NETs International Working Conferences on the 'Performance Modelling and Evaluation of Heterogeneous Networks'. These events were staged under the auspices of the EU Networks of Excellence (NoE) Euro-NGI and Euro-FGI and are associated with the NoE Work-packages WP.SEA.6.1 and WP.SEA.6.3, respectively.

The research papers are organised into five technical parts dealing with current research themes in 'Multiservice Switching Networks', 'Wireless Ad Hoc Networks', 'Wireless Sensor Networks', 'Wireless Cellular Networks' and 'Optical Networks'.

In *Part One* 'Multi-Service Switching Networks', Stasiak and Głąbowski present a credible analytic methodology, based on the concept of the effective availability, for the approximation of the blocking probability in multi-service multi-stage switching networks with point-to-group selection and several attempts of setting up a connection. Do et al. propose a novel queueing model for the performance evaluation of multipath routing in Multi Protocol Label Switching (MPLS) networks supported by label switched paths (LSPs), which enhances subscribers' performance and QoS. Bíró devises efficient estimates for the buffered workload ratio and bandwidth requirement based on equivalent capacity approximations in buffered statistical multiplexers in the modelling framework of many sources asymptotics. Kabaciński and Żal undertake a performance investigation into the conditions for strict-sense non-blocking and rearrange-able operation for a new architecture of high-speed self-routing switching networks.

In *Part Two* 'Wireless Ad Hoc Networks', Baccelli et al. focus on the convergence of the Internet and wireless ad-hoc networks through extensions to the Open Shortest Path First (OSPF) routing protocol and propose a new mechanism based on an adaptation of OSPF for link-state database exchanges in a wireless ad-hoc domain. Coenen et al. present a new flow level model for multihop wireless ad hoc networks and investigate efficient capacity allocation between network users under different operational scenarios.

In *Part Three* 'Wireless Sensor Networks', Delye de Clauzade de Mazieux et al. address the issues of cluster creation, addressing mechanism and auto routing in wireless fixed sensor networks. A new cluster address auto-configuration mechanism is devised, based on multi-hop performance criterion, allowing auto routing via a systematic pattern to assign addresses without the use of any routing protocol. Kacimi et al. propose new protocols for the deployment of ad hoc wireless sensor networks (WSNs) for a cold supply chain monitoring, which are implemented over a ZigBee/IEEE802.15.4 protocol stack whilst the energy save is integrated to the conception of application level functionalities. Levendovszky et al. present new energy balancing protocols, based on statistical traffic characteristics, to maximise life span in wireless sensor networks (WSN). An optimal packet forwarding mechanism is developed and the concept of generalized statistical bandwidth is used to evaluate the energy demands of the network. Czachórski et al. formulate a diffusion approximation model for the estimation of the probability density function of the distribution of a packet travel time in a multihop wireless sensor network. The model includes retransmission in case of a packet loss, subject to the propagation medium whilst the distribution of relay nodes may be heterogeneous in space and the system characteristics may change over time.

In *Part Four* 'Wireless Cellular Networks', Pla et al. propose a new and efficient QoS algorithm for computing the optimal traffic load parameters of the Multiple Fractional Guard Channel (MFGC) admission policy for multi-service mobile wireless networks. Kouvatsos et al. devise an analytic methodology, based on the principle of maximum entropy (ME), for the performance modelling and analysis of a fourth generation (4G) cellular network with a novel hand off priority channel assignment scheme and a generalised partial sharing traffic handling protocol. Schefczik and Wiedemann describe a simulation modelling approach for an efficient assessment of different evolution scenarios of the UMTS architecture with regard to signaling performance. Kallos et al. construct a new model, named Wideband Threshold Model (WTM), for the analysis of Wideband Code Division Multiple Access (W-CDMA) Networks supporting elastic traffic classes and propose a credible method to determine the call blocking probabilities of mobile users. Głąbowski et al. present a methodology, based on the Kaufman–Roberts recursion, to calculate during design planning stages the uplink blocking probability in cellular systems with W-CDMA Radio interface and finite population source. García-Roger and Casares-Giner propose a new approximation method, based on an efficient recursive algorithm, for the estimation of the uplink capacity in WCDMA systems and the computation of several parameters in a multiservice scenario.

In *Part Five* 'All-Optical Networks', Nguyên and Atmaca carry out performance analysis and relative comparisons involving the new technologies of circuit emulation service (CES) and modified packet bursting (MPB) on an all-optical Ethernet ring. Chakka et al. propose an analytic methodology for the solution of a generalised Markovian queueing model, based on the transformation of the global balance equations into a computable form and describe applications into the performance evaluation of an optical burst switching node and the High Speed Downlink Packet Access protocol in wireless networks. Chaitou et al. consider a novel aggregation technique and an analytic model for the efficient support of IP packets in a slotted optical wavelength-division-multiplexing (WDM) layer with several QoS requirements. Mouchos and Kouvatsos present optical edge node simulations, implemented on the Compute Unified Device Architecture (CUDA), which can accelerate significantly by using parallel generators of optical bursts employing the Graphics Processing Unit (GPU), as a cost-effective platform, on a commercial NVidia GeForce 8800 GTX. Finally, Vardakas et al. develop teletraffic loss models, based on one-dimensional Markov chains, for calculating connection failure probabilities and call blocking probabilities in hybrid time-wavelength division multiplexing (TDM-WDM) passive optical networks with multiple service-classes, either infinite or finite traffic sources and dynamic wavelength allocation.

I would like to end this preface by expressing my thanks to the EU Networks of Excellence Euro-NGI and Euro-FGI for sponsoring in part the publication of this research book and to the members of HET-NETs Advisory Boards and Programme Committees as well as the external referees worldwide for their invaluable and timely reviews. My thanks are also due to Professor Ramjee Prasad, Director of the Center for TeleInFrastruktur (CTIF), Aalborg University, Denmark, for his encouragement and valuable advice during the preparation of this book.

Demetres D. Kouvatsos

Participants in the Peer Review Process

Samuli Aalto
Ramon Agusti
Sohair Al-Hakeem
Eitan Altman
Jorge Andres
Vladimir Anisimov
Laura Aspirot
Salam Adli Assi
Tulin Atmaca
Zlatka Avramova
Frank Ball
Simonetta Balsamo
Ivano Bartoli
Alejandro Beccera
Monique Becker
Pablo Belzarena
Andre-Luc Beylot
Andreas Binzenhoefer
Jozsef Biro
Pavel Bocharov
Miklos Boda
Sem Borst
Nizar Bouabdallah
Richard Boucherie
Christos Bouras
Onno Boxma
Chris Blondia
Alexandre Brandwajn

George Bravos
Oliver Brun
Herwig Bruneel
Alberto Cabellos-Aparicio
Patrik Carlsson
Fernando Casadevall
Vicente Casares-Giner
Hind Castel
Llorenc Cerda
Eduardo Cerqueira
Mohamad Chaitou
Ram Chakka
Meng Chen
Stefan Chevul
Tom Coenen
Doru Constantinescu
Marco Conti
Laurie Cuthbert
Tadeusz Czachorski
Alexandre Delye de Clauzade
   de Mazieux
Koen De Turck
Danny De Vleeschauwer
Stijn De Vuyst
Luc Deneire
Felicita Di Giandomenico
Manuel Dinis
Tien Do

Jose Domenech-Benlloch
Rudra Dutta
Joerg Eberspaecher
Antonio Elizondo
Khaled Elsayed
Peder Emstad
David Erman
Melike Erol
Jose Oscar Fajardo Portillo
Fatima Ferreira
Markus Fiedler
Jean-Michel Fourneau
Rod Fretwell
Wilfried Gangsterer
Peixia Gao
Ana Garcia Armada
David Garcia-Roger
Georgios Gardikis
Vincent Gauthier
Alfonso Gazo
Xavier Gelabert Doran
Leonidas Georgiadis
Bart Gijsen
Jose Gil
Cajigas Gillermo
Stefano Giordano
Jose Gonzales
Ruben Gonzalez Benitez
Annie Gravey
Klaus Hackbarth
Slawomir Hanczewski
Guenter Haring
Peter Harrison
Hassan Hassan
Dan He
Gerard Hebuterne
Bjarne Helvik
Robert Hines
Enrique Hernandez

Helmut Hlavacs
Amine Houyou
Hanen Idoudi
Ilias Iliadis
Dragos Ilie
Paola Iovanna
Andrzej Jajszczyk
Lorand Jakab
Sztrik Janos
Robert Janowski
Terje Jensen
Laszlo Jereb
Mikael Johansson
Hector Julian-Bertomeu
Athanassios Kanatas
Tamas Karasz
Johan Karlsson
Stefan Koehler
Daniel Kofman
Vangellis Kollias
Huifang Kong
Kimon Kontovasilis
Rob Kooij
Goerge Kormentzas
Ivan Kotuliak
Harilaos Koumaras
Demetres Kouvatsos
Tasos Kourtis
Udo Krieger
Koenraad Laevens
Samer Lahoud
Jaakko Lahteenmaki
Juha Leppanen
Amaia Lesta
Hanoch Levy
Wei Li
Yue Li
Fotis Liotopoulos
Renato Lo Cigno

Michael Logothetis
Carlos Lopes
Johann Lopez
Andreas Maeder
Tom Maertens
Thomas Magedanz
Sireen Malik
Lefteris Mamatas
Jose Manuel Gimenez-Guzman
Michel Marot
Alberto Martin
Jim Martin
Simon Martin
Ignacio Martinez Arrue
Jose Martinez-Bauset
Martinecz Matyas
Lewis McKenzie
Madjid Merabti
Bernard Metzler
Geyong Min
Isi Mitrani
Nicholas Mitrou
Is-Haka Mkwawa
Hala Mokhtar
Sandor Molnar
Edmundo Monteiro
Ioannis Moscholios
Harry Mouchos
Luis Munoz
Maurizio Naldi
Victor Netes
Pal Nilsson
Simon Oechsner
Sema Oktug
Mohamed Ould-Khaoua
Antonio Pacheco
Michele Pagano
Zsolt Pandi
Panagiotis Papadimitriou

Stylianos Papanastasiou
Nihal Pekergin
Izaskun Pellejero
Roger Peplow
Paulo Pereira
Gonzalo Perera
Jordi Perez-Romero
Rubem Perreira
Guido Petit
Maciej Piechowiak
Michal Pioro
Jonathan Pitts
Vicent Pla
Nineta Polemi
Daniel Popa
Adrian Popescu
Dimitris Primpas
David Remondo-Bueno
David Rincon
Roberto Sabella
Francisco Salguero
Sebastia Sallent
Werner Sandmann
Ana Sanjuan
Lambros Sarakis
Wolfgang Schott
Raffaello Secchi
Maria Simon
Swati Sinha Deb
Charalabos Skianis
Amaro Sousa
Dirk Staehle
Maciej Stasiak
Panagiotis Stathopoulos
Bart Steyaert
Zhili Sun
Kannan Sundaramoorthy
Riikka Susitaival
Janos Sztrik

Yutaka Takahashi
Sotiris Tantos
Leandros Tassiulas
Luca Tavanti
Silvia Terrasa
Geraldine Texier
David Thornley
Florence Touvet
Phuoc Tran-Gia
Chia-Sheng Tsai
Thanasis Tsokanos
Krzysztof Tworus
Rui Valadas
Rob Van der Mei
Vassilios Vassilakis
Vasos Vassiliou
Sandrine Vaton
Tereza Vazao
Speros Velentzas

Dominique Verchere
Pablo Vidales
Nguyen Viet Hung
Manolo Villen-Altamirano
Bart Vinck
Jorma Virtamo
Kostas Vlahodimitropoulos
Joris Walraevens
Xin Gang Wang
Wemke Weij
Sabine Wittevrongel
Mehti Witwit
Michael Woodward
George Xilouris
Mohammad Yaghmaee
Bo Zhou
Stefan Zoels
Piotr Zwierzykowski

PART ONE
MULTISERVICE SWITCHING NETWORKS

1

Multi-Service Switching Networks with Point-to-Group Selection and Several Attempts of Setting up a Connection

Maciej Stasiak and Mariusz Głąbowski

Chair of Computer and Communication Networks, Poznań University of Technology, ul. Polanka 3, 60-965 Poznań, Poland; e-mail: {stasiak, mglabows}@et.put.poznan.pl

Abstract

This paper presents a new analytical method of blocking probability calculation in multi-service multi-stage switching networks with point-to-group selection and several attempts of setting up a connection. The basis for the proposed calculation method is the availability distribution. In this paper, the conditions for the availability distribution are formulated. On the basis of them, approximate function of this distribution has been chosen. The relationship between the availability distribution and the number of connection-setting attempts made within the system is also presented. The moments of the random variable, calculated on the basis of the availability distribution and the occupation distribution within the outgoing group are the direct data for the probability calculations of internal and external blocking in multi-service switching networks with point-to-group selection and several attempts of setting up a connection. The results of the analytical calculations of blocking probability are compared with the results of the digital simulation of three-stage switching networks, thereby, proving the validity of the assumptions used in the model.

Keywords: Multiservice switching networks, blocking probability calculation, effective-availability method.

D. D. Kouvatsos (ed.), Performance Modelling and Analysis of Heterogeneous Networks, 3–26.

1.1 Introduction

The researches carried out recently in many research centres indicate the necessity of determining methods of analytical modelling of traffic phenomena occurring in B-ISDN networks (Broadband Integrated Services Digital Network). As a result of the rapid growth in the capacity of transmission links caused by the increasing requirements of modern bandwidth-consuming multimedia applications, the elaboration of time-effective methods of analysis and designing of switching networks has become crucial. However, determining traffic characteristics of multi-service multi-stage switching networks is a complex problem. Basic problems associated with the description of such systems, from the perspective of the traffic theory, arise from a necessity of servicing various types of traffic sources by the network [1, 2]. In principle, the classification of traffic sources in a broadband network is reduced to distinguishing the CBR (Constant Bit Rate) sources and the VBR (Variable Bit Rate) sources. To define loads introduced into networks by the VBR sources, it is proposed to determine the so-called *equivalent bandwidth* for particular classes of traffic streams generated by the sources [1, 2]. The assignment of several constant bit rates to the VBR sources enables the evaluation of traffic characteristics of switching systems in the B-ISDN network by means of multi-rate models worked out for the multi-rate circuit switching [1–6]. Accurate methods of switching networks calculation, based on multi-rate models, employ the method of statistic equilibrium. However, in spite of its great accuracy, this method cannot be used for calculations of larger systems which have practical meaning. The reason for this is an excessive number of states in which a multi-dimensional Markov process occurring within the system can take place [7]. The most effective methods of blocking probability calculation in multi-service switching networks are thus well-proven methods of the effective availability. The basic parameter of these methods is the term *effective availability* which is defined as the availability in a multi-stage switching network in which the blocking probability is equal to the blocking probability of a single-stage network (grading) with the same capacity of the outgoing group and at analogous parameters of the traffic stream offered. The methods of effective availability were introduced in [8–13] for switching networks carrying single-rate traffic, and extended in [4, 14, 15] for 3-stage switching networks carrying multi-rate traffic. In [16–18], the practical and universal formulae for calculating the effective availability were derived for arbitrary multi-stage multi-service switching networks with both point-to-point and point-to-group selection.

This group of method was subsequently extended to switching networks with BPP (Binomial–Poisson–Pascal) traffic [19, 20]. Simultaneously, non-blocking switching networks have been studied. In [21–23], non-blocking conditions for switching networks with multi-rate traffic were devised.

Even though much research has been and is still being done on the subject, no method of blocking probability calculation in multi-stage multi-service switching networks with several attempts of setting up a connection has been worked out so far. The methods which appeared in the literature are limited to switching networks with single-rate traffic [11, 24, 25]. In this paper, a new original PGBMTn method (Point-to-Group Blocking for Multi-rate Traffic, *n* attempts of setting up a connection) of blocking probability calculation in multi-stage multi-service switching networks with point-to-group selection and several attempts of setting up a connection has been proposed. The PGB-MTn method is based on the PPRD method which was proposed in [26] for blocking probability calculation in switching networks with point-to-point selection. Following the proposed conception, calculations of internal blocking probability in multi-stage multi-service switching networks are reduced to calculations of this blocking in an equivalent model of switching network carrying single-rate traffic. The basis for the proposed PGBMTn method is the so-called *effective availability distribution* which determines the probability that the availability for class *i* call takes a given value. In this paper, a function which approximates the effective availability distribution has been chosen. Additionally, it has been proved that internal blocking probability in switching network with *n* attempts of setting up a connection depends on the first *n* moments of the random variable, calculated on the basis of the effective availability distribution. This probability also depends upon the distribution of available links in an outgoing group (direction) which is approximated by the distribution of available subgroups (links) in the limited-availability group. This distribution is the basis for calculating external blocking probabilities in switching networks with multi-rate traffic.

The remaining part of this paper is organised as follows. Section 1.2 describes the algorithm of setting up connections in switching networks with point-to-group selection and several attempts of setting up a connection. Section 1.3 of this paper is devoted to the elaboration of the PGBMTn method for blocking probability calculation in a multi-stage switching networks carrying multi-rate traffic. In Section 1.4, the calculation results have been compared with the simulation results of selected switching networks. Finally, Section 1.5 concludes the paper.

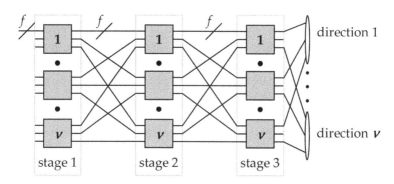

Figure 1.1 3-stage switching network.

1.2 Point-to-Group Selection with Several Attempts of Setting up a Connection

Let us consider a multi-stage switching network with multi-rate traffic. Let us assume that the system services call demands having an integer number of the so-called BBUs (Basic Bandwidth Units).[1] Let us assume that each of inter-stage links has the capacity equal to f BBUs. Furthermore, outgoing multiplexed transmission links create link groups called directions. The outgoing links can be wired to the directions in different ways. In the paper, the outgoing direction in switching networks under consideration have been created as follows: each first link of each last-stage switch belongs to the first direction and, analogously, each link n of each last-stage switch belongs to the same direction with a serial number equal to n (Figure 1.1).

The switching network is offered M independent classes of Poisson traffic streams having the intensities: $\lambda_1, \lambda_2,\ldots,\lambda_M$. A class i call requires t_i BBUs to set up a connection. The holding time for calls of particular classes has an exponential distribution with the parameters: $\mu_1, \mu_2, \ldots, \mu_M$. Thus, the mean traffic offered to the system by the class i traffic stream is equal to:

$$a_i = \lambda_i/\mu_i. \tag{1.1}$$

The switching network under consideration operates with point-to-group selection and several attempts of setting up a connection. Following the control algorithm of this kind of selection, the control device of the switching

---

[1] While constructing multi-rate models for broadband systems, it is assumed that BBU is the greatest common divisor of the equivalent bandwidths of all call streams offered to the system [1,2].

network determines the first stage switch, on the incoming link of which a class i call appears (switch α). Then, the control system finds a last-stage switch (switch β) having a free outgoing link (i.e. the link comprising of at least t_i free BBUs) in a required direction. Subsequently, the control device tries to find a connection path between switches α and β. If such a path exists, the connection between switches α and β is set up, otherwise the control system begins the second attempt to set up a connection, i.e. the control system determines another switch β having a free outgoing link in a required direction and tries to find a new connection path between switches α and β. If such a path does not exist, the third attempt of setting up a connection will be made. If the connection cannot be set up during the last allowed n attempt assumed in the control algorithm, a class i call is lost as a result of internal blocking. A call can be lost in k attempt ($1 \leqslant k \leqslant n$) if, for a given state of the switching network, $k - 1$ last-stage switches which have a free outgoing link exist (i.e. the control device would not be able to find a next switch β during the k attempt). When none of outgoing links of the demanded direction of the switching network can service the class i call (i.e. does not have t_i free BBUs) a call is lost due to the phenomenon of external blocking.

The point-to-point selection is a particular case of a point-to-group selection with several attempts of setting up a connection and can be considered as the point-to-group selection with one attempt of setting up a connection. Thus, the point-to-point selection leads to an increase in the loss probability, particularly internal blocking probability in comparison to the point-to-group selection (the point-to-group selection is defined as the point-to-group selection with ν attempts of setting up a connection, where ν is the number of links of an outgoing direction). As the number of attempts of setting up a connection increases, the internal blocking probability decreases and approaches the values obtained in the switching network with point-to-group selection [24].

With the same assumed internal blocking probability in a switching network, the point-to-point selection requires an additional number of crosspoints in comparison to the point-to-group selection. In [24], it was shown that the transition from a point-to-point selection to a point-to-group selection, in the case of switching networks servicing single-rate traffic, made it possible to save about 25% of crosspoints. On the other hand, however, a control of the switching network with the point-to-group selection is much more complex. The application of the point-to-group selection with several attempts of setting up a connection allows us to do away with this inconsistency.

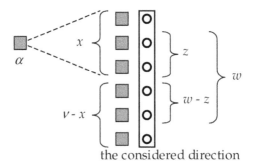

the considered direction

Figure 1.2 One of the possible states of the outgoing direction of the switching network.

1.3 PGBMTn Method

1.3.1 Basic Theoretical Considerations

Let us consider a z-stage multi-service switching network with point-to-group selection and several attempts of setting up a connection. Now, for the class i call we can determine an equivalent switching network[2] carrying a single-rate traffic [17]. Providing that the outgoing direction of the considered switching network has the capacity equal to v links (i.e. the outgoing links of the real switching network consist of v links), let us consider one of the possible states of the equivalent switching network, presented in Figure 1.2, where:

- x – links available for the switch α,
- $(v - x)$ – links non-available for the switch α,
- w – free links in a given direction,
- z – free links in the set of x available links for the switch α,
- $(w - z)$ – free links in the set of $(v - x)$ non-available links for the switch α.

The probabilities corresponding to such a state of the equivalent switching network can be specified as follows:

- $P(x)$ – the probability of the availability of x outgoing links for the switch α,
- $P(w)$ – the probability of an event in which there is w free links in a given outgoing direction,

---

[2] In an equivalent switching network each link is treated as a single-rate link with a fictitious load equal to blocking probability for a class i stream in a link of the real switching network (Section 1.3.4).

- $P(z)$ – the probability of an event in which there is z free links in the set of x available links; simultaneously, it is the probability of an event in which there is $(w - z)$ free links in a set of $(v - x)$ non-available links,
- $P_b(k)$ – the probability of the internal blocking in the equivalent switching network for k $(k \leqslant n)$ unsuccessful attempts of setting up a connection,
- $P_b(k, x)$ – the probability of the internal blocking for k $(k \leqslant n)$ unsuccessful attempts of setting up a connection providing the outgoing group has x available links for the switch α,
- $P_b(k, x, w, z)$ – the probability of the internal blocking for k $(k \leqslant n)$ unsuccessful attempts of setting up a connection providing that there is x available links for the switch α in the outgoing group, w free links and z free links in the set of x links available for the switch α,
- $B_{\text{in}}(i, n)$ – the probability of the internal blocking for class i calls in switching network with point-to-group selection and n attempts of setting up a connection. $B_{\text{in}}(i, n)$ is calculated in an equivalent switching network determined for class i calls.

The probabilities $P_b(k)$, $P_b(k, x)$, $P_b(k, x, w, z)$ determine the internal blocking probability for k unsuccessful attempts of setting up a connection, i.e. they express a probability of an event, in which k attempts of setting up a connection fail.

The algorithm of setting up a connection with the point-to-group selection and k attempts of setting up a connection is characterised by the fact that each of the following attempts can only be made when there is a subsequent free outgoing link. So the realisation of k attempts of setting up a connection depends on the existence of at least k free outgoing links, i.e. the inequality $w \geqslant k$ must be attained in a given switching network. Calculations of the probabilities $P_b(k)$, $P_b(k, x)$, $P_b(k, x, w, z)$ also require the inequality to be attained.

If we assume, that the traffic is uniformly distributed between all v outgoing links, the occupation probabilities of each link are equal and independent. Thus, the probability distribution $P(z)$ in the given state of outgoing links can be approximated by hypergeometric distribution:

$$P(z) = \frac{\binom{x}{z}\binom{v-x}{w-z}}{\binom{v}{w}}. \tag{1.2}$$

Let the control device make k attempts to set up a connection in a given state of the switching network. If each of these attempts fails to be successful, then the call will be lost and the probability of such an event can be determined as follows:

$$P_b(k, x, w, z) = \frac{\binom{w-z}{k}}{\binom{w}{k}} P(z)P(w)P(x).$$ (1.3)

The combinatorial coefficient in equation (1.3) expresses the probability of an event in which all k free links will be chosen from the set of $(w - z)$ non-available links for the switch α. The probability of the internal blocking when n attempts of setting up a connection are allowed can be expressed by the following formula:

$$B_{\text{in}}(i, n) = \frac{1}{1 - P(w = 0)} \sum_{w=1}^{\nu} \sum_{x=0}^{\nu} \sum_{z=0}^{w} \frac{\binom{w-z}{k}}{\binom{w}{k}} P(w)P(x)P(z),$$ (1.4)

where $k = w$ for $w < n$, $k = n$ for $w \geqslant n$. The maximum value of k parameter in equation (1.4) is equal to the number of free links in a demanded outgoing direction — w. This is due to the fact that the control device cannot make more attempts of setting up a connection than the number of the free outgoing links of the equivalent switching network ($k \leqslant w$).

Internal blocking probability in a switching network is calculated on the assumption that there is at least one outgoing link belonging to the considered direction and having at least t_i free BBUs. Thus, the calculations must include a transition from the distribution $P(w)$ determined in the sample space $\{w : 0, 1, 2, \ldots, \nu\}$ to a truncated distribution $P'(w)$ determined in the space $\{w : 1, 2, \ldots, \nu\}$. The relation between the distributions $P(w)$ and $P'(w)$ is as follows:

$$P'(w) = 1/[1 - P(w = 0)].$$ (1.5)

This relation is included in equation (1.4). Let us notice that equation (1.4) can be expressed as follows:

$$B_{\text{in}}(i, n) = \frac{1}{1 - P(w = 0)} \sum_{w=1}^{\nu} \sum_{x=0}^{\nu} P_b(k, x)P(w)P(x),$$ (1.6)

where

$$P_b(k, x) = \sum_{z=0}^{w} \frac{\binom{w-z}{k}}{\binom{w}{k}} P(z),$$ (1.7)

expresses internal blocking probability in switching networks with k unsuccessful attempts of setting up a connection and the availability equal to x.

Knowing that internal blocking probability in switching networks with k unsuccessful attempts of setting up a connection is equal to:

$$P_b(k) = \sum_{x=0}^{v} P_b(k, x) P(x), \qquad (1.8)$$

Equation (1.6) can be expressed as follows:

$$B_{in}(i, n) = \frac{1}{1 - P(w = 0)} \sum_{w=1}^{v} P_b(k) P(w), \qquad (1.9)$$

where $k = w$ for $w < n$, $k = n$ for $w \geqslant n$.

$P(w)$ is the distribution of free outgoing links of a given direction in an equivalent switching network. In a real switching network this is the distribution of free links for class i calls. In the proposed PGBMTn method, this distribution is approximated by the combinatorial distribution $P_K(i, w)$ of the available links in the limited-availability group[3] (Section 1.3.2). Simultaneously $P(w = 0) = P_K(i, 0)$ determines a state in which all links of a given direction are busy, that is to say, the external blocking probability $B_{ex}(i)$:

$$B_{ex}(i) = P(w = 0) = P_K(i, 0). \qquad (1.10)$$

Total blocking probability $B(i, n)$ for the class i call is a sum of external and internal blocking probabilities. Assuming the independence of internal and external blocking events, we obtain

$$B(i, n) = B_{ex}(i) + B_{in}(i, n)[1 - B_{ex}(i)]. \qquad (1.11)$$

To calculate blocking probability according to equations (1.9), (1.10) and (1.11), it is necessary to determine the combinatorial distribution $P_K(i, w)$ of available links and the value of internal blocking $P_b(k)$ in the equivalent switching network (determined for class i calls) with k unsuccessful attempts of setting up a connection. The next sections will deal with methods used to determine parameters mentioned above.

---

[3] In the proposed PGBMTn method it has been assumed that the groups of outgoing links (outgoing directions) are approximated by the model of a limited-availability group.

1.3.2 Distribution of Available Links

In the proposed method of blocking probability calculation in switching networks with point-to-group selection and several attempts of setting up a connection we assume that groups of outgoing links (outgoing directions) are approximated by the model of a limited-availability group. The limited-availability group is a model of separated transmission links [7, 17, 27] carrying a common mixture of different multi-rate traffic streams. The limited-availability group is the group divided into v identical subgroups (links), each of the capacity equal to f BBUs. Thus, the total capacity of the system is equal to $V = vf$. The system services a call — only when this call can be entirely carried by the resources of an arbitrary single subgroup. According to the approximate simple method of blocking probability calculation in the limited availability group [27], the state probabilities in this system are approximated by the generalised Kaufman–Roberts recursion [3, 15, 27]:

$$nP(n) = \sum_{i=1}^{M} a_i t_i \sigma_i (n - t_i) P(n - t_i), \qquad (1.12)$$

where $P(n)$ is the state probability, i.e. the probability of an event where there is n busy BBUs in the multi-rate system. The $\sigma_i(n)$ parameter is the probability of admission of the class i call to the service when the system is found in the state n. This probability is usually called the conditional probability of passing between the adjacent states of the process occurring in the system and can be approximated by the following formula [27]:

$$\sigma_i(n) = \frac{F(V - n, v, f, 0) - F(V - n, v, t_i - 1, 0)}{F(V - n, v, f, 0)}, \qquad (1.13)$$

where $F(x, v, f, t)$ is the number of arrangements of x free BBUs in v subgroups, calculated with the assumption that the capacity of each subgroup is equal to f BBUs (the upper limitation) and each subgroup has at least t free BBUs (the lower limitation) [27]:

$$F(x, v, f, t) = \sum_{i=0}^{\left\lfloor \frac{x - vt}{f - t + 1} \right\rfloor} (-1)^i \binom{v}{i} \binom{x - v(t - 1) - 1 - i(f - t + 1)}{v - 1}. \qquad (1.14)$$

On the basis of the occupancy distribution of the limited-availability group (formulae (1.12), (1.13)), the so-called *distribution of available subgroups* was determined [17]. This distribution determines the probability

$P_K(i, w)$ of an event in which each of arbitrarily chosen w subgroups ($w = 1, \dots, v$) can carry the class i call. In order to calculate $P_K(i, w)$ distribution, it is indispensable to know the so-called *conditional distribution of available subgroup* $P(i, w|x)$ [17]. This distribution determines the probability of an arrangement of x free BBUs, in which each of w arbitrarily chosen subgroups has at least t_i free BBUs, while in each of the remaining $(v-w)$ subgroups the number of free BBUs is lower than t_i. This means that exactly w subgroups from among the v ones in the given limited-availability group can service the class i call. Thus, this problem is reduced to the determination of the number of arrangements of u free BBUs in w subgroups at two-side limitations, where t_i is the lower and f is the upper limitation. According to equation (1.14), the number of such arrangements is $F(u, w, f, t_i)$. Each chosen arrangement of u free BBUs in w subgroups has a number of corresponding arrangements of $(x - u)$ free BBUs in $(v - w)$ subgroups at the upper limitation $(t_i - 1)$. The number of such arrangements in accordance with equation (1.14) is $F(x - u, v - w, t_i - 1, 0)$. The number of all possible arrangements of x free BBUs in v subgroups is also determined by equation (1.14) and is equal to $F(x, v, f, 0)$. Taking into consideration the possible variations in the parameter u and all the chances for choosing the w subgroups, we finally obtain a conditional probability of arrangements, as a result of which w subgroups are available for a class i call. This probability can be expressed as follows [17]:

$$P(i, w|x) = \frac{\binom{v}{w} \sum_{u=wt_i}^{\Psi} F(u, w, f, t_i) F(x - u, v - w, t_i - 1, 0)}{F(v, x, f, 0)}, \quad (1.15)$$

where $\Psi = wf$, if $x \geqslant wf$ and $\Psi = x$, if $x < wf$.

On the basis of the distribution $P(i, w|x)$ and of the theorem of total probability, the distribution of available subgroups $P_K(i, w)$ is equal to

$$P_K(i, w) = \sum_{n=0}^{V} P(n) P(i, w|V - n), \quad (1.16)$$

where $P(n)$ is the occupancy distribution in the limited-availability group (equation (1.12)).

The $P_K(i, w)$ distribution approximates the distribution of free outgoing links (for class i calls) of a given direction of a switching network with point-to-group selection and several attempts of setting up a connection.

1.3.3 Internal Blocking Probability for k Unsuccessful Attempts of Setting up a Connection

The internal blocking probability $P_b(k)$ for k unsuccessful attempts of setting up a connection is given by equation (1.8). This probability depends on conditional distribution $P_b(k, x)$ expressed by equation (1.7). Taking into consideration equation (1.2), equation (1.7) can be rewritten as

$$P_b(k, x) = \sum_{z=0}^{w} \frac{\binom{w-z}{k}}{\binom{w}{k}} \frac{\binom{x}{z}\binom{v-x}{w-z}}{\binom{v}{w}}, \qquad (1.17)$$

that can be expressed as follows:

$$P_b(k, x) = \frac{1}{\binom{w}{k}\binom{v}{w}} \sum_{z=0}^{w-k} \binom{w-z}{k}\binom{x}{z}\binom{v-x}{w-z} = \frac{1}{\binom{w}{k}\binom{v}{w}} B(x). \qquad (1.18)$$

The parameter $B(x)$ can be expressed by the following formula:

$$B(x) = \sum_{z=0}^{w-k} \binom{w-z}{k}\binom{x}{z}\binom{v-x}{w-z}$$

$$= \sum_{z=0}^{w-k} \binom{x}{z} \frac{(v-x)!}{k!(w-k-z)!(v-x-w+z)!}. \qquad (1.19)$$

Multiplying the numerator and the denominator of equation (1.19) by $(v-x-k)!$, we obtain:

$$B(x) = \frac{(v-x-k)!(v-x)!}{(v-x-k)!k!} \sum_{z=0}^{w-k} \binom{x}{z} \frac{1}{(w-k-z)!(v-x-w+z)!}$$

$$= \binom{v-x}{k} \sum_{z=0}^{w-k} \binom{(v-k)-x}{(w-k)-z}\binom{x}{z} = \binom{v-x}{k}\binom{v-k}{w-k}. \qquad (1.20)$$

When we substitute equation (1.20) for equation (1.18) we obtain the formula for internal blocking probability calculation in a switching network with k unsuccessful attempts of setting up a connection where there is x available links for the switch α in an outgoing direction:

$$P_b(k, x) = \frac{1}{\binom{w}{k}\binom{v}{w}} \binom{v-x}{k}\binom{v-k}{w-k} = \prod_{j=0}^{k-1} \left(1 - \frac{x}{v-j}\right). \qquad (1.21)$$

Consequently, including equation (1.21) in equation (1.8) the unconditional probability of internal blocking in switching network with k unsuccessful attempts of setting up a connection can be rewritten as follows:

$$P_b(k) = \frac{1}{\prod_{j=0}^{k-1}(v-j)} \sum_{x=0}^{v} \prod_{j=0}^{k-1}(v-x-j)P(x). \qquad (1.22)$$

In equation (1.22), the following expression:

$$C_k(x) = \prod_{j=0}^{k-1}(v-x-j) = a_k x^k + a_{k-1} x^{k-1} + \cdots + a_1 x + a_0 \qquad (1.23)$$

is a polynomial of degree n with real coefficients a_j, where $0 \leqslant j \leqslant k$. As the j-order moment of the random variable x is equal to

$$M_j(x) = \sum_{x=0}^{v} x^j P(x), \qquad (1.24)$$

so taking into account equations (1.23) and (1.24), equation (1.22) can be expressed as follows:

$$\begin{aligned} P_b(k) &= \frac{1}{\prod_{j=0}^{k-1}(v-j)} [a_k M_k(x) + \cdots + a_1 M_1(x) + a_0] \\ &= \frac{1}{\prod_{j=0}^{k-1}(v-j)} A_k(x), \end{aligned} \qquad (1.25)$$

where
$$A_k(x) = C_k(x)P(x) \qquad (1.26)$$

is a polynomial of the random variable of availability equal to x with the same real coefficients as in $C_k(x)$ polynomial.

The analysis of equation (1.25) shows that determination of the first k moments of the random variable of the effective availability is essential to the internal blocking probability calculation in switching networks with k unsuccessful attempts of setting up a connection. If we apply equations (1.9), (1.22) and (1.24) when only one attempt of setting up a connection is allowed in the switching network ($k = 1$, point-to-point selection), we obtain

$$B_{in}(i, 1) = P_b(1) = \frac{v - M_1(x)}{v}. \qquad (1.27)$$

In [18] it was proved that point-to-point internal blocking probability for a class i call is equal to

$$B_{\text{in}}(i, 1) = \frac{v - d_e(i)}{v}. \qquad (1.28)$$

Thus equations (1.27) and (1.28) determine the expected value of the availability distribution (the first moment). We assumed, that the expected availability is approximated by the effective availability parameter $d_e(i)$ for class i calls. The parameter $d_e(i)$ is determined on the basis of the equivalent switching network model. Calculations of higher-order moments of the random variable of availability require determination of the probability function $P(x)$ (Section 1.3.5).

1.3.4 Effective Availability Parameter

The concept of the so-called equivalent switching network [17] is the base for the effective availability calculation for class i traffic stream. Following this concept, the network with multi-rate traffic is reduced to an equivalent network carrying a single-rate traffic. Each link of an equivalent network is treated as a single-rate link with a fictitious load $b_j(i)$ equal to blocking probability for a class i stream in a link of the real switching network between section j and $j + 1$. The proposed PGBMTn method assumes that this probability can be calculated on the basis of the occupancy distribution in the full-availability group with multi-rate traffic [28–30]. Due to this assumption blocking probability for class i calls can be calculated by

$$b(i) = \sum_{n=V-t_i+1}^{V} P(n), \qquad (1.29)$$

where $P(n)$ is determined on the basis of the *Kaufman–Roberts recursion* [28, 29]:

$$nP(n) = \sum_{i=1}^{M} a_i t_i P(n - t_i). \qquad (1.30)$$

The effective availability in a real z-stage switching network is equal to the effective availability in an equivalent switching network and can be determined by the formula derived in [17]:

$$d_e(i) = [1 - \pi_z(i)]v + \pi_z(i)\eta Y_1(i) + \pi_z(i)[v - \eta Y_1(i)]b_z(i)\sigma_z(i), \quad (1.31)$$

where

- $d_e(i)$ – the effective availability for the class i traffic stream in an equivalent network,
- $\pi_r(i)$ – the probability of non-availability of a stage r switch for the class i call. $\pi_r(i)$ is the probability of an event where a class i connection path cannot be set up between a given first-stage switch and a given stage r switch. Evaluation of this parameter is based on the channel graph of the equivalent switching network and can be calculated by the Lee method [31];
- v – the number of outgoing links from the first stage switch;
- $Y_j(i)$ – the average value of the fictitious traffic served by the the stage j switch: $Y_j(i) = lb_j(i)$;
- l – the number of outgoing links from the last stage switch;
- η – a portion of the average fictitious traffic from the switch of the first stage which is carried by the direction in question. If the traffic is uniformly distributed between all h directions, we obtain: $\eta = 1/h$;
- $\sigma_z(i)$ – the so-called *secondary availability coefficient* [17] which is the probability of an event in which the connection path of the class i connection passes through directly available switches of intermediate stages:

$$\sigma_z(i) = 1 - \prod_{r=2}^{z-1} \pi_r(i). \tag{1.32}$$

A determination of the effective availability in the equivalent switching network is tantamount to a determination of the effective availability of the real network for class i calls.

1.3.5 Effective Availability Distribution

Analytical determination of availability distribution requires to solve corresponding systems of equations of the state of equilibrium. These equations are determined on the basis of a multi-dimensional Markov process which is a model for real process occurring in switching networks. However, due to an excessive number of states in which a multi-dimensional Markov process can take place, this method cannot have practical meaning. Consequently, the distribution of effective availability is approximated by a corresponding theoretical distribution. The theoretical distribution is specified in the manner which ensures both conformity of analytical calculation of blocking probability ($P_b(k)$ and $B_{in}(i, n)$) with simulation results and simplicity of $M_j(x)$ calculation process. This mainly applies to the first four moments (in practice, no more attempts of setting up a connection are made, and the value $B_{in}(i, 4)$ is already practically equal to the internal point-to-group blocking probability. Here the effective availability parameter $d_e(i)$ was selected as the expected value of the approximate distribution.

One of the simplest distributions which fulfill the condition of simplicity of higher-order moments calculations is the binomial distribution

$$P(x) = \binom{\nu}{x} p^x (1 - p)^{\nu - x},$$
(1.33)

where $P(x)$ is the probability of an event in which an effective availability in equivalent switching network is equal to x; p is the probability of the availability of an arbitrarily chosen outgoing link:

$$p = d_e(i)/\nu.$$
(1.34)

Simulation studies confirm that the effective availability distribution in an equivalent switching network can be approximated by Bernoulli distribution. The binomial distribution is characterised by a simple calculation of moments of the random variable. These parameters can be, for instance, determined by the application of an equation that binds the successive moments of the random variable with Bernoulli distribution:

$$M_{j+1} = p \sum_{r=0}^{\nu} \left[\nu \binom{j}{r} - \binom{j}{r+1} \right] M_{j-r},$$
(1.35)

where p is determined by equation (1.34). Equation (1.35) allows us to calculate all the necessary moments of the approximate availability distribution. For instance, the first four moments of the binominal distribution are determined by the following equations:

$$M_1 = p\nu = d_e(i),$$
(1.36)
$$M_2 = p[(\nu - 1)M_1 + \nu],$$
(1.37)
$$M_3 = p[(\nu - 2)M_2 + (2\nu - 1)M_1 + \nu],$$
(1.38)
$$M_4 = p[(\nu - 3)M_3 + 3(\nu - 1)M_2 + (3\nu - 1)M_1 + \nu].$$
(1.39)

1.3.6 Coefficients of the Polynomial $A_k(x)$

Considering equation (1.25) we notice that, in order to calculate the probability $P_b(k)$, we must determine coefficients of the polynomial $A_k(x)$, where

$$A_k(x) = \sum_{i=0}^{k} a_i M_i. \tag{1.40}$$

As the polynomials $A_k(x)$ i $C_k(x)$ have the same coefficients (equation (1.26)), we will consider subsequently only the polynomial $C_k(x)$. According to equation (1.23) we have

$$C_{k+1}(x) = C_k(x)(\nu - k - x). \tag{1.41}$$

Equation (1.41) can be expressed as follows:

$$(a_k x^k + a_{k-1} x^{k-1} + \cdots + a_1 x + a_0)(\nu - k - x)$$
$$= a'_{k+1} x^{k+1} + a'_k x^k + \cdots + a'_1 x + a'_0, \tag{1.42}$$

where a_j is a coefficient of a polynomial of degree k $(0 \leqslant j \leqslant k)$ and a'_j is a coefficient of a polynomial of degree $k + 1$ $(0 \leqslant j \leqslant k + 1)$. Equation (1.42) gives the direct recurrent relations between coefficients of a polynomial of degree k and coefficients of a polynomial of degree $k + 1$:

$$\begin{cases} a'_{k+1} = -a_k, \\ a'_j \;\; = -a_{j-1} + a_j(\nu - k) \quad \text{for } 0 < j < k+1, \\ a'_0 \;\; = a_0(\nu - k), \end{cases} \tag{1.43}$$

where coefficient a_0 of a polynomial of degree zero is equal to one.

On the basis of equations (1.43), (1.35) and (1.40) it can be stated that the procedure of the determination of the coefficients of the polynomial $A_k(x)$ is elementary simple. Four polynomials $A_k(x)$, for k equal 1, 2, 3, 4, exclusively, are given below. All the coefficients of the polynomials are worked out directly from equation (1.43):

$$A_1(x) \;=\; -M_1 + \nu, \tag{1.44}$$

$$A_2(x) \;=\; M_2 - (2\nu - 1)M_1 + \nu(\nu - 1), \tag{1.45}$$

$$A_3(x) \;=\; -M_3 + 3(\nu - 1)M_2 - [3(\nu - 1)^2 - 1]M_1$$
$$+ \nu(\nu - 1)(\nu - 2), \tag{1.46}$$

$$A_4(x) \;=\; M_4 - (4\nu - 6)M_3 + [6(\nu - 1)(\nu - 2) - 1]M_2 +$$
$$- (4\nu^3 - 18\nu^2 + 22\nu - 6)M_1 + \nu(\nu - 1)(\nu - 2)(\nu - 3). \tag{1.47}$$

After a determination of the polynomial $A_k(x)$, we can calculate the value of the internal blocking probability in an equivalent switching network (determined for class i calls) with k unsuccessful attempts of setting up a connection (equation (1.25)). For example, we obtain the following formula for the successive values of the parameter $k = 1, 2, 3, 4$:

$$P_b(1) = \frac{v - M_1}{v}, \tag{1.48}$$

$$P_b(2) = 1 - \frac{2v - 1}{v(v - 1)}M_1 + \frac{1}{v(v - 1)}M_2, \tag{1.49}$$

$$P_b(3) = 1 - \frac{3(v - 1)^2 - 1}{v(v - 1)(v - 2)}M_1 + \frac{3(v - 1)}{v(v - 1)(v - 2)}M_2$$
$$- \frac{1}{v(v - 1)(v - 2)}M_3, \tag{1.50}$$

$$P_b(4) = 1 - \frac{4v^3 - 18v^2 + 22v - 6}{v(v - 1)(v - 2)(v - 3)}M_1 + \frac{6(v - 1)(v - 2) - 1}{v(v - 1)(v - 2)(v - 3)}M_2 +$$
$$- \frac{4v - 6}{v(v - 1)(v - 2)(v - 3)}M_3 + \frac{1}{v(v - 1)(v - 2)(v - 3)}M_4. \tag{1.51}$$

1.3.7 Algorithm of Blocking Probability Calculation

Total blocking probability for class i calls in multi-service multi-stage switching networks with point-to-group selection and several attempts of setting up a connection can be calculated according to the following algorithm:

1. The considered network is reduced to an equivalent switching network in which the equivalent availability parameter $d_e(i)$ is determined (equation (1.31)).
2. Knowing the probability $p = d_e(i)/v$, all necessary moments of availability distribution are calculated (equation (1.35)).
3. Distribution $P_k(i, w)$ of available links of outgoing directions is approximated by the distribution of available subgroups in the limited-availability group (equations (1.12)–(1.16)). Probability $P_k(i, 0)$ determines external blocking for class i calls in the considered switching network.
4. On the basis of the distribution $P_k(i, s)$ and the blocking probability $P_b(k)$ for k unsuccessful attempts of setting up a connection (equa-

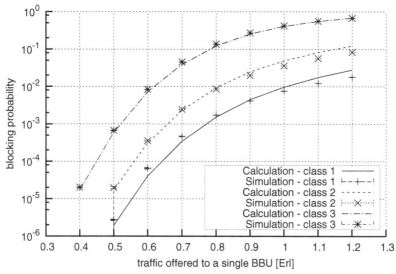

Figure 1.3 Point-to-group blocking probability for two attempts of setting up a connection; switching network structure: $v = 4$, $f = 30$, $V = 120$; traffic structure: $t_1 = 1$, $t_2 = 2$, $t_3 = 6$.

tion (1.25)), the internal blocking probability $B_{in}(i, n)$ for n attempts (assumed in control device) is calculated (equation (1.9)).

5. Finally, the total blocking probability $B(i, n)$ is determined (equation (1.11)).

1.4 Calculation and Simulation Results

In order to confirm the adopted assumptions in the PGBMTn method, the results of analytical calculations were compared with the results of digital simulation of a multi-service switching network. The researches were carried out for a switching network consisting of the switches of $v \times v$ links (Figure 1.1). The outgoing links create v identical directions, each with a capacity equal to v links. The total capacity of the direction expressed in bandwidth units is equal to $V = vf$. Additionally, it was assumed that traffic offered to each direction had the same value.

In Figures 1.3–1.6, the calculation results of blocking probability are compared with the simulation results for three switching networks with the following parameters:

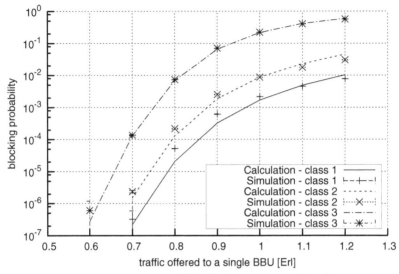

Figure 1.4 Point-to-group blocking probability for three attempts of setting up a connection; switching network structure: $\nu = 4$, $f = 155$, $V = 620$; traffic structure: $t_1 = 1$, $t_2 = 2$, $t_3 = 10$.

1. switching network structure: $\nu = 4$, $f = 30$, $V = 120$; traffic structure: $t_1 = 1$, $t_2 = 2$, $t_3 = 6$;
2. switching network structure: $\nu = 4$, $f = 155$, $V = 620$; traffic structure: $t_1 = 1$, $t_2 = 2$, $t_3 = 10$;
3. switching network structure: $\nu = 8$, $f = 63$, $V = 504$; traffic structure: $t_1 = 1$, $t_2 = 2$, $t_3 = 4$, $t_4 = 10$.

System no. 1 and no. 2 were offered three classes of multi-rate traffic in the following proportions: $a_1 t_1 : a_2 t_2 : a_3 t_3 = 1 : 1 : 1$ whereas system no. 3 was offered four classes of multi-rate traffic in the proportions: $a_1 t_1 : a_2 t_2 : a_3 t_3 : a_4 t_4 = 1 : 1 : 1 : 1$.

The analytical and simulation results are presented in Figures 1.3–1.5 in relation to the value of total traffic $a = \sum_{i=1}^{M}(a_i t_i)/V$ offered to a single BBU. Figure 1.3 shows the results of point-to-group blocking probability in the switching network no. 1 with two attempts of setting up a connection, whereas Figure 1.4 shows the results obtained in the switching network no. 2 with three attempts of setting up a connection. The results of point-to-group blocking probability obtained in the switching network no. 3 with four attempts of setting up a connection are presented in Figure 1.5. It was

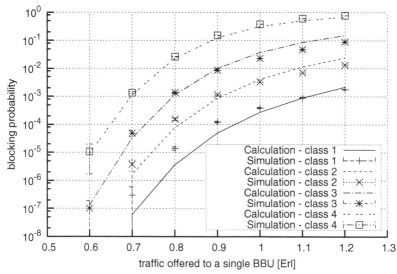

Figure 1.5 Point-to-group blocking probability for four attempts of setting up a connection; switching network structure: $\nu = 8$, $f = 63$, $V = 504$; traffic structure: $t_1 = 1$, $t_2 = 2$, $t_3 = 4$, $t_4 = 10$.

assumed that the switching networks work under random hunting strategy of connection paths. The results of the simulation are shown in the charts in the form of marks with 95% confidence intervals that have been calculated after the t-Student distribution for the five series with 1,000,000 calls of this traffic class that generates the lowest number of calls.

To determine the influence of the point-to-group selection with several attempts of setting up a connection on traffic characteristics of switching networks, Figure 1.6 shows the percentage changes in the point-to-group blocking in relation to the number of the attempts of setting up a connection (switching system no. 1). It can noticed that for $k > 1$ the results of blocking probability are lower than the results obtained in switching network with one attempt of setting up a connection (point-to-point selection). This phenomenon results from the increasing value of the availability of an outgoing direction caused by the increasing number of attempts of setting up a connection. On the basis of the obtained results it can be stated that an application of more than three attempts is ineffective.

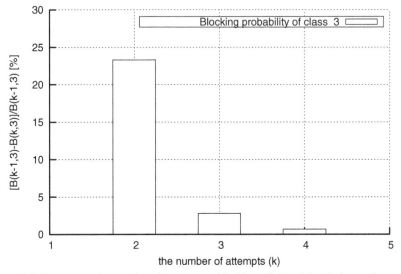

Figure 1.6 Percentage changes in point-to-group blocking of class 3 in relation to the number of attempts of setting up a connection; switching network structure: $v = 4$, $f = 30$, $V = 120$; traffic structure: $a = 0.7$ Erl., $t_1 = 1$, $t_2 = 2$, $t_3 = 6$.

1.5 Conclusions

In the paper an analytical method of point-to-group blocking probability calculation – the PGBMTn method – has been derived for the switching networks with several attempts of setting up a connection. This method is based on the concept of effective availability. The calculation results have been compared with the simulation results of multi-service switching networks. The simulation study has confirmed fair accuracy of the proposed analytical method. The obtained accuracy is comparable to that in the switching network with single-rate traffic [11, 25]. The simulation research carried out by the authors implies that the proposed method ensures fair accuracy of calculations for switching networks with any number of attempts of setting up a connection, for various network structures and for various traffic structures. Calculations made according to the proposed formulae are not complicated and are easily programmable. The proposed PGBMTn method can be used to determine the most effective number of attempts of setting up a connection in multi-service switching networks in relation to the structure of a switching network and the structure of traffic.

References

[1] J. W. Roberts (Ed.), *Performance Evaluation and Design of Multiservice Networks*, Final Report COST 224, Commission of the European Communities, Brussels, 1992.

[2] J. W. Roberts, V. Mocci and I. Virtamo (Eds.), *Broadband Network Teletraffic*, Final Report of Action COST 242, Commission of the European Communities, Springer Verlag, Berlin, 1996.

[3] J. W. Roberts, Teletraffic models for the Telcom 1 integrated services network, in *Proceedings of 10th International Teletraffic Congress*, Montreal, Canada, p. 1.1.2, 1983.

[4] G. J. Fitzpatrick, M. E. Beshai and E. A. Munter, Analysis of large-scale three-stage networks serving multirate traffic, in *Proceedings of 13th International Teletraffic Congress*, Copenhagen, Denmark, vol. 14, pp. 905–910, 1991.

[5] M. I. Reiman and J. A. Schmitt, Performance models of multirate traffic in various network implementations, in *Proceedings of 14th International Teletraffic Congress*, J. Labetoulle and J. W. Roberts (Eds.), Antibes Juan-les-Pins, France, Elsevier Science, vol. 1b, pp. 1217–1227, 1994.

[6] I. Moscholios, M. Logothetis and G. Kokkinakis, Connection dependent threshold model: A generalization of the Erlang multiple rate loss model, *Journal of Performance Evaluation*, vol. 48, nos. 1–4, pp. 177–200, 2002.

[7] J. Conradt and A. Buchheister, Considerations on loss probability of multi-slot connections, in *Proceedings of 11th International Teletraffic Congress*, Kyoto, Japan, pp. 4.4B–2.1, 1985.

[8] N. Binida and W. Wend, Die Effektive Erreichbarkeit für Abnehmerbundel hinter Zwischenleitungsanungen, *NTZ*, vol. 11, no. 12, pp. 579–585, 1959.

[9] A. D. Charkiewicz, An approximate method for calculating the number of junctions in a crossbar system exchange, *Elektrosvyaz*, vol. 2, pp. 55–63, 1959.

[10] A. Lotze, A. Röder and G. Thierer, PPL – A reliable method for the calculation of point-to-point loss in link systems, in *Proceedings of 8th International Teletraffic Congress*, Melbourne, Australia, pp. 547/1–547/14, 1976.

[11] A. Lotze, A. Röder and G. Thierer, Point-to-point loss in case of multiple marking attempts, in *Proceedings of 8th International Teletraffic Congress*, Melbourne, Australia, vol. supplement to paper 547/1–547/14, 1976.

[12] K. Rothmaier and R. Scheller, Design of economic PCM arrays with a prescribed grade of service, *IEEE Transactions on Communications*, vol. 29, no. 7, pp. 925–935, 1981.

[13] V. A. Ershov, Some further studies on effective accessibility: Fundamentals of teletraffic theory, in *Proceedings of 3rd International Seminar on Teletraffic Theory*, Moscow, pp. 193–196, 1984.

[14] M. Butto and G. Colombo, Analytic models for switching networks with wideband traffic, in *Proceedings of 12th International Teletraffic Congress*, Torino, Italy, North Holland-Elsevier Science Publishers, p. 5.1A.4.1, 1988.

[15] M. E. Beshai and D. R. Manfield, Multichannel services performance of switching networks, in *Proceedings of 12th International Teletraffic Congress*, Torino, Italy, North Holland-Elsevier Science Publishers, p. 5.1A.7, 1988.

[16] M. Stasiak, An approximate model of a switching network carrying mixture of different multichannel traffic streams, *IEEE Transactions on Communications*, vol. 41, no. 6, pp. 836–840, 1993.

[17] M. Stasiak, Combinatorial considerations for switching systems carrying multi-channel traffic streams, *Annales des Télécommunications*, vol. 51, nos. 11–12, pp. 611–625, 1996.

[18] M. Stasiak and M. Głąbowski, Point-to-point blocking probability in switching networks with reservation, in *Proceedings of 16th International Teletraffic Congress*, Edinburgh, UK, June, Elsevier Science, vol. 3A, pp. 519–528, 1999.

[19] M. Głąbowski, Point-to-point and point-to-group blocking probability in multi-service switching networks with BPP traffic, *Electronics and Telecommunications Quarterly*, vol. 53, no. 4, pp. 339–360, 2007.

[20] M. Głąbowski, Blocking probability in multi-service switching networks with finite source population, in *Proceedings of The 14th IEEE International Conference On Telecommunications*, Penang, Malaysia, May 2007.

[21] G. Danilewicz and W. Kabaciński, Wide-sense nonblocking multicast ATM switching networks, *Journal of Performance Evaluation*, vol. 41, no. 2–3, pp. 165–177, 2000.

[22] R. Melen and J. S. Turner, Nonblocking multirate networks, *SIAM Journal on Computing*, vol. 18, no. 2, pp. 301–313, 1989.

[23] Y. Yang and M. Masson, Nonblocking broadcast switching network, *IEEE Transactions on Communications*, vol. 40, no. 9, pp. 1005–1015, 1991.

[24] A. Lotze, A. Röder and G. Thierer, Point-to-point selection versus point-to-group selection in link systems, in *Proceedings of 8th International Teletraffic Congress*, Melbourne, Australia, pp. 541/1–541/5, 1976.

[25] E. B. Ershova and V. A. Ershov, Digital systems for information distribution, *Radio and Communications, Moscow*, pp. 89–148, 1983 [in Russian].

[26] M. Głąbowski and M. Stasiak, Point-to-point blocking probability in switching networks with reservation, *Annales des Télécommunications*, vol. 57, no. 7–8, pp. 798–831, 2002.

[27] M. Stasiak, Blocking probability in a limited-availability group carrying mixture of different multichannel traffic streams, *Annales des Télécommunications*, vol. 48, nos. 1–2, pp. 71–76, 1993.

[28] J. S. Kaufman, Blocking in a shared resource environment, *IEEE Transactions on Communications*, vol. 29, no. 10, pp. 1474–1481, 1981.

[29] J. W. Roberts, A service system with heterogeneous user requirements – Application to mutli-service telecommunications systems, in *Proceedings of Performance of Data Communications Systems and Their Applications*, G. Pujolle (Ed.), North Holland, Amsterdam, pp. 423–431, 1981.

[30] K. W. Ross, *Multiservice Loss Models for Broadband Telecommunication Network*, Springer Verlag, London, 1995.

[31] C. Lee, Analysis of switching networks, *Bell Systems Technical Jornal*, vol. 34, no. 6, 1955.

2

A Performance Model of MPLS Multipath Routing with Failures and Repairs of the LSPs

Tien Van Do[1], Dénes Papp[1], Ram Chakka[2] and Mai X. T. Truong[1]

[1]*Department of Telecommunications, Budapest University of Technology and Economics, H-1117 Budapest, Hungary; e-mail: do@hit.bme.hu*
[2]*Meerut Institute of Engineering and Technology, Meerut 250005, India; e-mail: ramchakka@yahoo.com*

Abstract

Efficient operation of multipath routing is considered as one of the most important aspects for the success of Multiprotocol Label Switching (MPLS) traffic engineering. Multipath routing has an advantage over the traffic routing based on single shortest paths because single path routing may lead to unbalanced traffic situations and degraded QoS (Quality of Service).

This paper proposes a novel analytical queueing model for the performance evaluation of multipath routing in MPLS networks. This paper applies an extension of a generalized Markovian queue, called the *HetSigma* queue, to model the operation of ingress nodes and derives the performance measures. That extended queue is essentially the MM $\sum_{k=1}^{K}$ CPP$_k$/GE/c/L queue with heterogeneous servers, and with server-breakdowns and repairs according to an appropriate repair strategy. The model-validation is accomplished by comparing the numerical results of the model with those of the simulation of the system. The model is fast with good accuracy, hence useful to network designers in an number of aspects. Moreover, a comparison of some repair strategies is presented with the use of the proposed model.

Keywords: MPLS, multipath routing, Hetsigma queue, performability, reliability.

D. D. Kouvatsos (ed.), Performance Modelling and Analysis of Heterogeneous Networks, 27–43.

2.1 Introduction

MPLS (Multiprotocol Label Switching) is the recent development of IETF (Internet Engineering Task Force) in response to the explosive growth of the Internet [10]. One of the aims of the MPLS development is to expand the granularity of administrative traffic control (i.e., Traffic Engineering) [3].

Efficient operation of multipath routing is considered as one of the essential aspects in MPLS traffic engineering [3]. Multipath routing has an advantage over traffic routing based on single shortest path because single path routing may lead to unbalanced traffic situations and degraded QoS (Quality of Service) in some cases (even if the network is not overloaded). In MPLS networks multiple LSPs (label switched path) can be established between MPLS ingress and egress nodes to enhance the network performance and meet the QoS (Quality of Service) requirements.

Previous works on the performance of multipath routing were only done with simulations [11, 13, 14]. Recent works [7, 9, 12, 15] on MPLS multipath routing have also used simulation tools to show the advantage of the multipath algorithms over single-path constraint-based routing algorithms. Moreover, the mentioned works considered the impact of multipath routing in a network-scale, but did not deal with the operation of network nodes. Cidon et al. [8] have proposed an analytical solution for the performance analysis of multipath routing. However, this work considers the system only at the call level.

This paper proposes a new analytical performance model at the packet level for an MPLS ingress node participating in state-dependent multipath routing. We apply an extended version of the *HetSigma* queue [4] to evaluate the performance of ingress nodes, taking into account the failures and the repairs of the MPLS paths. That would essentially be the MM $\sum_{k=1}^{K}$ CPP$_k$/GE/c/L queue with server breakdowns and repairs according to an appropriate strategy.

The rest of the paper is organized as follows. We describe the operation of MPLS multipath routing in Section 2.2. From the system operation, the inherent queueing model is elucidated. Then, in Section 2.3, we propose an extension of the *HetSigma* queue, i.e. the MM $\sum_{k=1}^{K}$ CPP$_k$/GE/c/L queue [4] with heterogenous-servers, and with server breakdowns and repairs, to model an ingress-egress node pair. The steady-state solution approach is explained in Section 2.4. Relevant application, numerical results and model-validation are presented in Section 2.5. The paper concludes with Section 2.6.

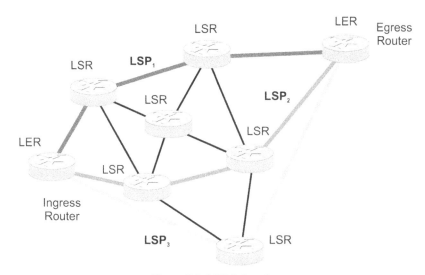

Figure 2.1 MPLS domain.

2.2 Operation of MPLS Multipath Routing

An example of an MPLS domain with routers and links is illustrated in Figure 2.1. Traffic demands traversing the MPLS domain are conveyed along pipes, or in the MPLS terminology, Label Switched Paths (LSPs). When a packet arrives at the ingress router called Label Edge Router (LER) of the MPLS domain, a short fixed length label is appended to it. The packet will be assigned to a specific LSP. The criteria for the assignment are the destination IP address of the incoming packet, and some additional considerations and constraints concerning the current resource availability in the domain. Afterwards, the packet is forwarded along the specified LSP in the core of the MPLS domain. At each core router called Label Switched Router (LSR), the label is simply swapped instead of interrogating IP header, significantly increasing packet forwarding efficiency.

Traffic Engineering (TE) in MPLS is introduced to avoid or at least decrease the congestions and ensure efficient resource allocation in the network. Congestion problems may arise from situations when the network could not convey the offered load, or when resource allocation is not efficient enough and some links are overutilized. The first type of congestion can be avoided through congestion control techniques, while the second one can be minimized by applying load balancing mechanism.

The constraint based routing can also be supported besides the shortest path routing because the shortest path routing can result in a situation where some links will be overwhelmed, while others remain underutilized, which may lead to unbalanced traffic situations. On the overloaded links a congestion may happen, therefore it degrades the QoS (Quality of Service) experienced by end users. In order to avoid the disadvantage, for example, an explicit LSP for a given traffic flow from the ingress LER to the egress LER can be established and maintained based on the operation of constraint based routing (CBR) algorithms and of signaling protocols (e.g. Resource Reservation Protocol – Traffic Engineering and RSVP-TE). These two components allow MPLS to decide the LSP, based not only on the link metric (as OSPF does) but also on the currently available resources along the links. By doing this way, traffic may be routed not exactly along the shortest path but along the most adequate path that has enough resources to meet a given target QoS (e.g. sufficient bandwidth, low delay). Moreover, traffic may also be split and routed simultaneously along several LSPs. All of these features enable MPLS traffic engineering to evenly distribute traffic inside the domain.

2.3 The Proposed Model

In what follows, we describe the proposed model, as illustrated in Figure 2.2.

2.3.1 Traffic

Recent studies have shown that the traffic in today's telecommunications systems often exhibits burstiness – i.e. packets arriving together in batches – and correlation among the inter-arrival times. Therefore, the packet arrival process to the ingress node for a specific ingress-egress node-pair is assumed, in this paper, to be the recently proposed (MM $\sum_{k=1}^{K}$ CPP$_k$) process [4], which can indeed accommodate traffic-burstiness and correlations among inter-arrival times. Here, CPP stands for compound Poisson process with geometrically distributed batch sizes.

The arrival process is thus inherently modulated by a continuous-time, discrete-state irreducible Markov process X with N_1 states or phases of modulation. Let Q_X be the generator matrix of this modulating process, given

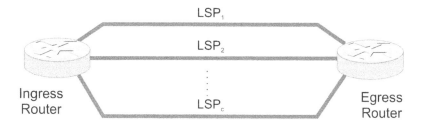

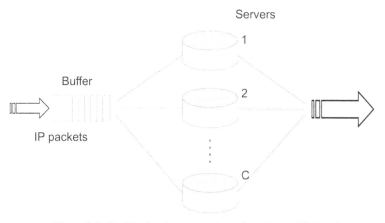

Figure 2.2 Model of an ingress node performing load balancing.

by

$$
Q_X = \begin{bmatrix}
-q_1 & q_{1,2} & \cdots & q_{1,N_1} \\
q_{2,1} & -q_2 & \cdots & q_{2,N_1} \\
\vdots & \vdots & \ddots & \vdots \\
q_{N_1,1} & q_{N_1,2} & \cdots & -q_{N_1}
\end{bmatrix},
$$

where $q_{l,k}$ $(l \neq k)$ is the instantaneous transition rate from phase l to phase k, and

$$
q_l = \sum_{j=1}^{N_1} q_{l,j}, \quad q_{l,l} = 0 \quad (l = 1, 2, \ldots, N_1).
$$

Ω_X is the transition rate matrix of process X. We can write

$$
\Omega_X(l, k) = Q_X(l, k), \quad (l \neq k);
$$

$$
\Omega_X(l, l) = 0, \quad (0 \leq l \leq N_1). \tag{2.1}
$$

Let the random variable $I_1(t)$ $(1 \leq I_1(t) \leq N_1)$ represent the phase of the modulating process X at any time t.

In each of the modulating phases, the arrivals are the superposition of K independent CPP arrival streams of packets. The parameters of the GE (generalized exponential) inter-arrival time distribution of the kth $(1 \leq k \leq K)$ customer arrival stream in phase l $(l = 1, 2, \ldots, N_1)$ are $(\sigma_{l,k}, \theta_{l,k})$. That is, the inter-arrival time probability distribution function is $1 - (1 - \theta_{l,k})e^{-\sigma_{l,k}t}$, in phase l, for the kth stream of packets. Thus, each of the K arrival *point-processes* is batch-Poisson, with batches having geometrically distributed size.

2.3.2 Multiple LSPs

Several paths can be defined and determined between a given IE (ingress-egress) node pair in the MPLS network according to some predefined criteria (e.g., paths with disjoint edges) for a single service class. Assume that there are c paths with different bandwidths to be established between the IE node pair for a specific service class.

We model the c paths in the system as the c heterogeneous servers in parallel in the queueing model. The GE-distributed service time parameters of the n th server $(n = 1, 2 \ldots, c)$ are denoted by (μ_n, ϕ_n) when the server is functional. L is the queueing capacity, in all phases, including the customers in service, if any.

2.3.3 Failures and Repairs of the LSPs

Networks are prone to failures because of different reasons (e.g., unreliable equipment, software bugs or cable cut). Such faults and failures may affect the operation of LSPs and cause packet losses. In case of failures, the load balancing mechanism can move packets that are queued for the affected LSPs to the unaffected LSPs.

When the failed link is repaired, the repaired LSP can be used again. Repair strategy defines the order of repairing failed links, when they are more than one. A number of alternative repair-strategies exist when there are failures of links. Since repair strategies can influence the performance of the network, choosing an adequate or nearly optimal repair strategy can be crucial since that can enhance the performance to meet QoS constraints. Repair strategies can be preemptive or non-preemptive. Though we have used only the latter, we believe preemptive-priority based repair strategies can be better than non-preemptive ones, in cases where the repair time distributions are exponential or they are done in phases of exponential times.

The simplest and trivially arising repair policies are Last-Come-First-Serve (LCFS) and First-Come-First-Serve (FCFS), in which the order of repairs depends on the order of link failures. As the names indicate, always the latest and the earliest link failure will be repaired respectively in these two strategies.

The key element regarding the strategies based on link-repair priorities is that links with higher priority should be set into repair sooner, even if these failures have occurred later. The priority list of links should be constructed in a greedy way, with a view to repair first the link which would result the biggest gain (that is, having highest sensitivity).

A system configuration is defined as a specific failure scenario in which both failed as well as functional LSPs exist. Note that an LSP consists of several links which may independently fail.

We assume the system states due to failures and repairs of the LSPs (which would correspond to all possible multi-server configurations with functional servers as well as failed servers in the corresponding queueing model) can be described by a continuous time, discrete state Markov process (called Z). This can be so when exponential or phase-type failures and repairs are assumed. Let Q_Z denote the generator matrix of process Z. Let N_2 be the number of the states of process Z and hence the total number of server configurations in the model. We introduce $I_2(t)$ – an integer-valued random variable – to describe the server configuration of the model (which corresponds to the system state) at time t. We define the following function:

$$\gamma(I_2(t), n) = \begin{cases} 1 & \text{the } n\text{th server is functional,} \\ 0 & \text{otherwise.} \end{cases} \tag{2.2}$$

In case of limited repairmen or repair resources, the applied repair strategy for the failed links can be a crucial factor from the viewpoint of network performance. In case of more than one failed links, the repair strategy defines the order of the repairing of failed links. Strategies can be categorized according to, not only the order of repair but also the preemptive-ness. If it is allowed to interrupt an ongoing repair task because of a new failure, the strategy is preemptive, otherwise it is non-preemptive. Different repair strategies are discussed in [6] for different contexts. The ideas presented therein can be pursued here to determine N_2 and Q_Z.

2.4 Steady-State Solution Methodology

Thus, a suitable correspondence is now established between the system operation and the queueing problem that is formulated. We shall show below that the above explained queueing problem can be represented as a useful extension of the *HetSigma* queue [4] that was introduced recently by some of the authors as a generalized Markovian queue for telecom applications [5]. The *HetSigma* queue is essentially the MM $\sum_{k=1}^{K}$ CPP$_k$/GE/c/L G queue with heterogeneous servers. Though only one variant was considered for analysis in [4], the vast flexibility for extending the model to suit different applications is also explained. The steady-state solution of the *HetSigma* queue was presented in [5], which is done by transforming the balance equations into the computable QBD-M (Quasi Simultaneous-Multiple Births and Deaths) form. That solution was shown to be computationally highly efficient.

2.4.1 Notations

The extension of the *HetSigma* model that is considered suitable to the present case has no negative customers and has, additionally, independent failures of servers and their repairs according to an appropriate repair strategy. In this section, we present how we use this *HetSigma* queue variant for the steady-state solution of the system we have considered in this paper:

- In the resulting *HetSigma* queue with server breakdowns and repairs, both the arrival and the service processes can be thought of being modulated by the same continuous time, irreducible Markov process, with N states where $N = N_1 \cdot N_2$. That generator matrix of the modulating process is denoted by Q where

$$Q = Q_Z \oplus Q_X. \tag{2.3}$$

The state $I(t)$ $(1 \leq I(t) \leq N; N = N_1 * N_2)$ of the joint modulating process is constructed by lexicographically sorting the two variables $(I_1(t), I_2(t))$ as illustrated in Table 2.1, which is also reflected in Q. From Table 2.1, we can write the following equations:

$$I(t) = I_1(t) + (I_2(t) - 1)N_1,$$

$$I_2(t) = f_2(I(t)) = \left\lfloor \frac{I(t) - 1}{N_1} \right\rfloor + 1,$$

$$I_1(t) = f_1(I(t)) = (I(t) - 1) \bmod N_1 + 1. \tag{2.4}$$

Table 2.1 Order of the phase variable.

I	(I_1, I_2)
1	$(1, 1)$
2	$(2, 1)$
$\vdots$	$\vdots$
N_1	$(N_1, 1)$
$N_1 + 1$	$(1, 2)$
$\vdots$	$\vdots$
$2N_1$	$(N_1, 2)$
$\vdots$	$\vdots$
$N_1 N_2 - N_1 + 1$	$(1, N_2)$
$\vdots$	$\vdots$
$N_1 N_2$	(N_1, N_2)

- Consequently, the parameters of the arrival process are mapped as follows. The parameters of the GE inter-arrival time distribution of the kth $(1 \le k \le K)$ customer arrival stream in phase i $(i = 1, \ldots, N)$ of the *HetSigma* queue are $(\sigma_{f_1(i),k}, \theta_{f_1(i),k})$.
- The service time parameters of the nth server $(n = 1, 2, \ldots, c)$ in phase i $(1 \le i \le N)$, denoted by $(\mu_{i,n}, \phi_{i,n})$ can be determined as

$$\mu_{i,n} = \gamma(f_2(i), n)\mu_n, \qquad (2.5)$$

$$\phi_{i,n} = \gamma(f_2(i), n)\phi_n. \qquad (2.6)$$

2.4.2 Markov Process Representation of the Queue

Let $J(t)$ $(0 \le J(t) \le L)$ be the number of customers (packets) in the system, including those receiving service. Due to the Markovian property of the distributions that are involved, the dynamics of the queue can be represented as a continuous-time, discrete-state Markov process $\bar{Y} = \{[I(t), J(t)]; t \ge 0\}$ evolving on a finite (semi-infinite and known as Y, if L is unbounded) rectangular lattice strip. For simplicity of understanding we may represent $I(t)$ on the X-axis and $J(t)$ on the Y-axis. Assuming a similar server-allocation policy as in [4], the possible transitions that underlie this Markov process are:

- $A_j(i, k)$ – purely lateral transition rate – from state (i, j) to state (k, j), for all $L \geq j \geq 0$ and $1 \leq i, k \leq N$ ($i \neq k$), caused by either a phase transition in the modulating Markov process X or a change in the multi-server configuration. $A_j(i, k) = Q(i, k)$ for $i \neq k$, and $A_j(i, i) = 0$.

- $B_{i,j,j+s}$ – s-step upward transition rate – from state (i, j) to state $(i, j + s)$ ($1 \leq s \leq L - j$ and $1 \leq i \leq N$), caused by a new batch arrival of size s. For a given j, s can be seen as bounded when L is finite and unbounded when L is infinite. When $j + s$ exceeds L, the queue would become full and $j + s - L$ requests would be lost.

$$B_{i,j-s,j} = (1 - \theta_{f_1(i)})\theta_{f_1(i)}^{s-1}\sigma_{f_1(i)}$$

$$(\forall i;\ 0 \leq j - s \leq L - 2;\ j - s < j < L)$$

$$B_{i,j,L} = \sum_{s=L-j}^{\infty} (1 - \theta_{f_1(i)})\theta_{f_1(i)}^{s-1}\sigma_{f_1(i)} = \theta_{f_1(i)}^{L-j-1}\sigma_{f_1(i)}$$

$$(\forall i;\ j \leq L - 1).$$

- $C_{i,j+s,j}$ – s-step downward transition rate – from state $(i, j + s)$ to state (i, j) due to the completion of a batch of size s of packets.

$$C_{i,j+s,j} = \sum_{n=1}^{c} \mu_{i,n}(1 - \phi_{i,n})\phi_{i,n}^{s-1}$$

$$(\forall i;\ c + 1 \leq j \leq L - 1;\ 1 \leq s \leq L - j)$$

$$= \sum_{n=1}^{c} \mu_{i,n}(1 - \phi_{i,n})\phi_{i,n}^{s-1} \quad (\forall i;\ j = c;\ 1 \leq s \leq L - c)$$

$$= \sum_{n=1}^{c} \phi_{i,n}^{s-1}\mu_{i,n} \quad (\forall i;\ j = c - 1;\ 1 \leq s \leq L - c + 1)$$

$$= 0 \quad (\forall i;\ c \geq 2;\ 0 \leq j \leq c - 2;\ s \geq 2)$$

$$= \sum_{n=1}^{j+1} \mu_{i,n} \quad (\forall i;\ c \geq 2;\ 0 \leq j \leq c - 2;\ s = 1).$$

Define

$$B_{j-s,j} = \text{Diag}\,[B_{1,j-s,j}, B_{2,j-s,j}, \ldots, B_{N,j-s,j}] \quad (j - s < j \leq L);$$

$$B_s = B_{j-s,j} \quad (j \le L - 1)$$

$$\Sigma_k = \text{Diag}\,[\sigma_{1,k}, \sigma_{2,k}, \ldots, \sigma_{N,k}] \quad (k = 1, 2, \ldots, K); \quad \Sigma = \sum_{k=1}^{K} \Sigma_k;$$

$$M_n = \text{Diag}\,[\mu_{1,n}, \mu_{2,n}, \ldots, \mu_{N,n}] \quad (n = 1, 2, \ldots, c);$$

$$C_j = \sum_{n=1}^{j} M_n \quad (1 \le j \le c);$$

$$= \sum_{n=1}^{c} M_n = C \quad (j \ge c);$$

$$C_{j+s,j} = \text{Diag}\,[C_{1,j+s,j}, C_{2,j+s,j}, \ldots, C_{N,j+s,j}].$$

Also, let $\{p_{i,j}\}$ be the steady-state probabilities defined as

$$p_{i,j} = \lim_{t \to \infty} \text{Prob}\,(I(t) = i, J(t) = j),$$

and let $\mathbf{v}_j = (p_{1,j}, \ldots, p_{N,j})$.

Then, the steady-state balance equations are

(1) For the Lth row or level:

$$\sum_{s=1}^{L} \mathbf{v}_{L-s} B_{L-s,L} + \mathbf{v}_L\,[Q - C] = 0; \tag{2.7}$$

(2) For the jth row or level:

$$\sum_{s=1}^{j} \mathbf{v}_{j-s} B_s + \mathbf{v}_j\,[Q - \Sigma - C_j] + \sum_{s=1}^{L-j} \mathbf{v}_{j+s} C_{j+s,j} = 0$$

$$(0 \le j \le L - 1); \tag{2.8}$$

(3) Normalization

$$\sum_{j=0}^{L} \mathbf{v}_j \mathbf{e}_N = 1. \tag{2.9}$$

where $\mathbf{e}_N$ is a column vector of size N with all ones.

The steady-state solution of $\bar{Y}$ and Y are very complicated because of the unbounded upward and downward transitions. It is shown in [4, 5] that these balance equations can be transformed to a QBD-M computable form. A highly computationally efficient solution-methodology has been brought out for such systems in [5]. That approach has been used in this paper to solve the present queue, which is essentially an extension of the *HetSigma* queue dealt in [4, 5], for steady-state performance and reliability measures.

2.5 Case Study, Numerical Results and Model-Validation

2.5.1 A Case Study

In order to validate the accuracy of the proposed model, a case study is carried out to determine the performance of a multipath routing scenario depicted in Figure 2.3, which involves the European Optical Network topology [1]. It contains 19 nodes and 78 optical links.

The traffic to be carried between the ingress node and egress node follows the ON-OFF process with two states (*ON-OFF*). The ON and OFF periods are exponentially distributed with mean 0.4 and 0.5 respectively. The distribution of inter-arrival times in state *ON* is generalized exponential with parameters ($\sigma = 150.6698990$, $\theta = 0.526471$). In state *OFF* no packets arrive. These parameters of the arrival stream are derived from the inter-arrival times of the recorded samples of the real Bellcore traffic trace BC-pAug89 [2].

Assume that there are three LSPs established between nodes 11 and 4 of the IP/MPLS network as seen in Figure 2.3. The first LSP is routed through node 10 and 5. The second LSP goes through node 17 and 16, while the third one is through node 9. Thus, we have, in the model, $c = 3$ servers with GE exponential distribution parameters ($\mu_i, \phi_i; i = 1, 2, 3$): $\mu_1 = 96$, $\mu_2 = 128$ $\mu_3 = 160$; $\phi_1 = \phi_2 = \phi_3 = 0.109929$. Note that these parameters are obtained, based on the service capacity of the LSPs and the packet lengths from the trace.

A specific LSP fails when a link through which the LSP is routed becomes inoperative (e.g., cable cut). Normally, the failure rate of each link can be assumed to depend on its length. The failure rate of each link is depicted in Figure 2.3. Three repair strategies are considered, which are Last-Come-First-Served Preemptive Resume (LCFS-PR), First-Come-First-Serve (FCFS) and a strategy based on static-repair priority (we assume that only one failed link can be repaired at any time). In the first two strategies, the order of repairing depends on the order of link failures. As the names show, always the last and the first link-failure will be repaired, respectively in LCFS-PR and FCFS. Over against this approach the repair strategy based on priority does not

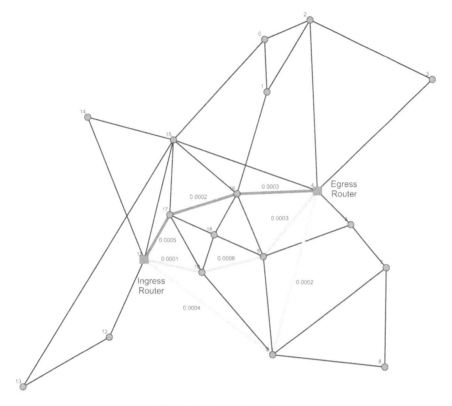

Figure 2.3 Network topology.

depend on the order of link-failures. Instead, the order of repairing will be based on a priority list. The priority list of links should be constructed on a greedy way by repairing always the link which will get the biggest gain. It can be shown easily, that it is worth repairing link with smaller failure rate first, if all link have the same repair rate.

2.5.2 Numerical Results and Validation by Simulation

In Figures 2.4 and 2.5, the variations of the average queue length and the packet loss probability with respect to the buffer size, are presented (the repair rate is fixed in such a way to have the 'availability of the connectivity' ensured by the LSPs between the ingress and egress node to be 99.9%), respectively. It can be observed that the analytical model gives acceptable results when compared with the simulated results. Note that we simulate a

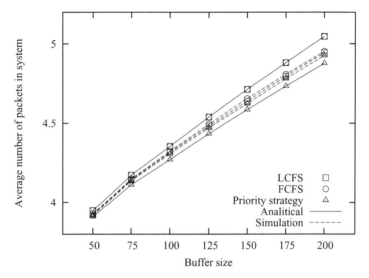

Figure 2.4 Queue length versus the buffer size.

scenario where traffic is transferred in three LSPs between the ingress and egress nodes in Figure 2.3. The link failures and repairs according to different repair strategies are also included in the simulation model. It is worth emphasizing that the analytical results can be computed very quickly, which is very crucial and attractive for the consideration in a network planning, dimensioning and design process.

In what follows, we examine the performance of the system (typically, mean queue length and packet-loss probabilities) versus the repair rate of optical links. Such a quantitative analysis should be able to help answer the question whether it is worth investing more efforts and costs in faster repairing of the links.

In order to ensure that the availability of service between the ingress and egress node to lie between 99.9 and 99.999%, the repair rate should be in the interval [0.0397703, 0.405355]. We investigate the effect of the repair rate of link within that range, on the performance. The expected queue length was reduced only by a small extent by the increase of repair rate (see Figure 2.6). The packet-loss probability varied nearly in inverse-proportion to the repair rate (see Figure 2.7). However, accurate estimation of the returns on investment that is used to increase repair rate can be made, only if the cost-patterns involved in increasing the repair rate are known. It may also be observed, in the considered range of experimentation, that the various repair

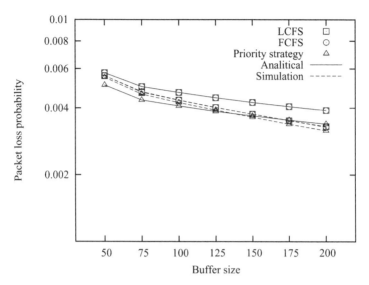

Figure 2.5 Packet loss versus the buffer size.

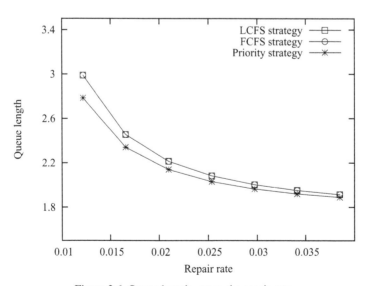

Figure 2.6 Queue length versus the repair rate.

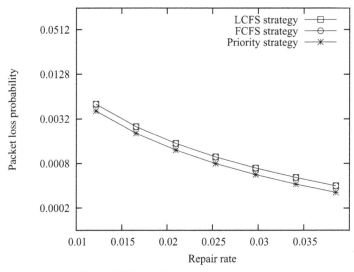

Figure 2.7 Packet loss versus the repair rate.

strategies considered do not have significant impact on the performance and the availability of the service in the network.

2.6 Conclusions

We have presented a new analytical queueing model for the multipath routing in MPLS networks. This model is indeed an extension of the *HetSigma* queue, namely, the MM $\sum_{k=1}^{K}$ CPP$_k$/GE/c/L queue with heterogeneous servers and with server-breakdowns and repairs according to specified repair strategy. The model is also able to take into account the reliability aspects of networks along with performance. A solution methodology similar to the one developed in [4, 5] has been used to solve the model for the steady-state performance measures.

We have presented a case study for the performance evaluation of an ingress-egress node pair using multiple LSPs in the European Optical Network topology. The numerical results of the model have been compared with the results of actual simulation and thus the model is validated. It is clearly argued that such a fast performance model is ideal and very useful in network design, dimensioning, planning and possible optimization studies.

References

[1] Network Research Topologies: The European optical network (EON), http://www.optical-network.com/topology.php.

[2] The Internet traffic archive, http://ita.ee.lbl.gov/index.html.

[3] D. Awduche, J. Malcolm, J. Agogbua, M. O'Dell and J. McManus, Requirements for traffic engineering over mpls. Technical Report RFC2702, Internet Engineering Task Force, 1999.

[4] R. Chakka and T. V. Do, Some new Markovian models for traffic and performance analysis in telecommunication networks, in *Proceedings of the Second International Working Conference on Performance Modelling and Evaluation of Heterogeneous Networks (HET-NETs 04)*, D. D. Kouvatsos (Ed.), Ilkley, UK, July, pp. T6/1–31, 2004.

[5] R. Chakka and T. V. Do, The MM $\sum_{k=1}^{K}$ CPP$_k$/GE/c/L G-queue with heterogeneous servers: Steady state solution and an application to performance evaluation, *Performance Evaluation*, vol. 64, pp. 191–209, March 2007.

[6] R. Chakka, O. Gemikonakli and P. Basappa, Multiserver systems with time or operation dependent breakdowns, in *Proceedings of International Symposium on Performance Evaluation of Computer and Telecommunications Systems, SPECTS*, San Diego, USA, pp. 266–277, 2002.

[7] H. Y. Cho, J. Y. Lee and B. C. Kim, Multi-path constraint-based routing algorithms for MPLS traffic engineering, in *Proceedings of IEEE International Conference on Communications, ICC'03*, vol. 3, pp. 1963–1967, May 2003.

[8] I. Cidon, R. Rom and Y. Shawitt, Analysis of multi-path routing. *IEEE Transaction on Networking*, vol. 7, pp. 885–896, 1999.

[9] S. K. J. Song and M. Lee, Dynamic load distribution in MPLS networks, in *Information Networking*, Lecture Notes in Computer Science, vol. 2662, Springer, pp. 989–999, 2003.

[10] E. Rosen, A. Viswanathan and R. Callon, Multiprotocol label switching architecture, Technical Report RFC3031, Internet Engineering Task Force, 2001.

[11] G. M. Schneider and T. Nemeth, A simulation study of the OSPF-OMP routing algorithm, *Computer Networks*, vol. 39, pp. 457–468, 2002.

[12] Y. Seok, Y. Lee, N. Choi and Y. Choi, Fault-tolerant multipath traffic engineering for MPLS networks, in *Proceedings of Communications, Internet, and Information Technology (CIIT 2003)*, 2003.

[13] C. Villamizar, MPLS optimized multipath, Internet-draft, February 1999.

[14] C. Villamizar, OSPF optimized multipath, Internet-draft, August 1999.

[15] Z. Zhao, Y. Shu, L. Zhang, H. Wang and O. W. W. Yang, Flow-level multipath load balancing in MPLS network, *Proceedings of 2004 IEEE International Conference on Communications*, vol. 2, pp. 1222–1226, June 2004.

3

Novel Equivalent Capacity Approximations through Asymptotic Analysis: A Review

József Bíró

Department of Telecommunications and Media Informatics, Budapest University of Technology and Economics, Magyar tudósok körútja 2, H-1117 Budapest, Hungary; e-mail: biro@tmit.bme.hu

Abstract

This paper is concerned with novel equivalent capacity estimators in the large deviation modeling framework for buffered communication link with fixed transmission capacity. Direct formulae for the estimation of the traffic bandwidth demand (the so-called equivalent capacity) have been presented that are suitable for satisfying prescribed QoS level on either the buffer overflow probability or the expected loss ratio due to buffer overflow. The equivalent capacity estimators have been derived and discussed in accordance with the common modeling framework of logarithmic and exact many sources asymptotics.

Keywords: Equivalent capacity, large deviations, many sources asymptotics.

3.1 Introduction

Bandwidth requirement[1] estimation is a key function in networks intending to provide quality of service (QoS) to their users. Network devices in QoS-capable networks must be able to control the amount of traffic they handle.

---

[1] The notions "bandwidth requirement" and "equivalent capacity" have been used interchangeably in this article, in accordance with the relevant literature.

D. D. Kouvatsos (ed.), Performance Modelling and Analysis of Heterogeneous Networks, 45–72.

This is generally performed by using some form of admission control. There are two commonly used methods for determining whether a new connection can be allowed to enter the system: in the first one an estimate of the buffer overflow probability (or loss ratio) is computed based on the properties of the new and the already active flows in the system, while the second method computes the bandwidth requirement of the existing traffic flows. When using the first method for admission control decisions, the devices check the computed overflow probability against the target overflow probability. If the second method is used, the bandwidth requirement of the existing flows is increased by the predicted bandwidth usage of the new flow and the result is compared to the capacity of the system.

Often, the second method is preferred over the first, mainly because it results in a quantity – the bandwidth requirement – that is more tractable and more useful than the estimate of the overflow probability. The on-line estimation of the bandwidth requirement of the traffic enables the network operator to track the amount of allocated (and free) capacity in the network. Furthermore, the impact of network management actions (e.g. directing more traffic on the link) on the resource status of the network can be more easily assessed. The overflow probability on the other hand is a less straightforward quantity that depends on the parameters of the queueing system in a more complex way, thus changes in them imply a less tractable and computationally more complex update procedure.

For buffered resources, the theory of large deviations was shown to be a very capable method for calculating the bandwidth requirement of traffic flows. There are two asymptotics that can be used for this purpose: the large buffer asymptotics and the many sources asymptotics. The large buffer asymptotics provide a rate function describing the decay rate of the tail of the probability of buffer overflow when the size of the buffer gets very large. The many sources asymptotics also offer a rate function but with the assumption that the number of traffic flows in the system gets very large, while the traffic mix, per-source buffer space and system per-source capacity are held constant. Both asymptotics discussed so far provide an overflow-probability type quantity.

Using the large buffer asymptotics it is easy to switch from the overflow probability representation to the bandwidth requirement representation. However, algorithms relying on this asymptotics [8] do not account for the gain arising from the statistical multiplexing of many traffic flows. In recent years, the second asymptotic regime, the many sources asymptotics (and its Bahadur–Rao improvement) have been described and investigated in [3–5, 10]. In the native form, the many sources asymptotics provide a rate function that can be used to estimate the probability of overflow. The computation of this rate function involves two optimisations in two variables. Yet, if it is the bandwidth requirement that is of interest, another optimisation has to be performed that requires the recomputation of the two original optimisations in each step. Despite of its complexity, this bandwidth requirement estimate is appealing because it incorporates the statistical properties of the traffic along with its QoS requirements and it also embraces the statistical multiplexing gain that occurs on the multiplexing link. However, the use of this estimator in real-time applications is not feasible because of its computational complexity.

This chapter introduces new methods for computing the bandwidth requirement of traffic flows that is based on the many sources asymptotics as well. Instead of the three embedded optimisations that previous approaches required, these comprise only of two optimisations that directly result in an estimate of the bandwidth requirement. The method is favorable to on-line measurement-based application, since the admission decision step is simplified and the more involving computations can be done in the background. It is shown that the new and the old methods for obtaining the bandwidth requirement are equivalent in case of overflow probability type approach, and are close to each other in case of loss ratio type methods. Similar results are performed for the buffer requirement estimates.

3.2 Overflow-Probability Based Admission Criteria

This section presents an overview on the many sources asymptotics. Next, a collection of admission control methods are reviewed, all of which build on the asymptotic property of the overflow probability.

3.2.1 Many Sources Asymptotics

The asymptotic regime described by the many sources asymptotics can be used to form an estimate of the probability of buffer overflow in the system as follows. Let us consider a buffered communication link with transmission capacity C, buffer size B, which carries N independent flows multiplexed in the system. N is viewed as a scaling factor, i.e. we can identify a per-source transmission capacity $c = C/N$ and a per-source buffer size $b = B/N$. Further, let the stochastic process $X[0, t)$ denote the total amount of work arriving at the system during the time interval $[0, t)$. Let us assume that $X[0, t)$ has stationary increments.

Conclusions on the behavior of this system can be derived by investigating a queueing system of infinite buffer size that is served by a finite capacity server with service rate $C = cN$. In order to account for the finite buffer size $B = bN$ of the real system, the probability of buffer overflow in the original system can be deduced from the proportion of time over which the queue length, $Q(C, N)$, is above the finite level B. In this system, where the system parameters (cN, bN) and the workload $(X[0, t))$ are scaled by the number of sources, an asymptotic equality can be obtained in N for the probability of overflow:

$$\lim_{N \to \infty} \frac{1}{N} \log P\{Q(cN, N) > bN\} = \sup_{t>0} \inf_{s>0} \left\{ \frac{\Lambda(s, t)}{N} - s(b + ct) \right\} \overset{\text{def}}{=} -I. \tag{3.1}$$

Here $\Lambda(s, t)$ (the so-called cumulant generating function) is defined as

$$\Lambda(s, t) \overset{\text{def}}{=} \frac{1}{st} \log E\left[e^{sX[0,t)} \right] \tag{3.2}$$

and I is called the asymptotic rate function, which depends on the per-source system parameters and on the scaled workload process. This result was proven for discrete time in [4] and for continuous time in [5]. Equation (3.2) practically means that for N large, the probability of overflow can be approximated as $P\{Q(C, N) > B\} \approx e^{-NI}$, where $-NI$ can be computed from (3.1) as

$$-NI = \sup_{t>0} \inf_{s>0} \left\{ \Lambda(s, t) - s(B + Ct) \right\}. \tag{3.3}$$

The approximation above can also be reasoned in a less rigorous, but brief and intuitive manner as follows [10]. The Chernoff bound can be used to approximate the probability that the workload $X[0, t)$ exceeds Ct, the offered service in $[0, t)$ and in addition it fills up the buffer space B: $P\{X[0, t) > B + Ct\} \approx \inf_{s>0} \exp\{\Lambda(s, t) - s(B + Ct)\}$. The steady state queue length distribution can be described by $Q = \sup_{t>0}\{X[0, t) - Ct\}$ provided that the $X[0, t)$ process has stationary increments. This way, the probability of the queue length exceeding the buffer level B is $P\{Q > B\} \approx P\left\{\sup_{t>0}\{X[0, t) - Ct\} > B\right\} \approx \sup_{t>0} P\{X[0, t) > B + Ct\} \approx e^{-NI}$.

For the sake of simplifying further discussions, let us define the function $J(s, t) \overset{\text{def}}{=} \Lambda(s, t) - s(B + Ct)$. In (3.3), the evaluation of $\sup_{t>0} \inf_{s>0} J(s, t)$ is computationally complex as a double optimisation has to be performed after the computation or estimation of the effective bandwidth of $X[0, t)$. Since the optimisations are embedded, first the optimal (minimal) s has to be found which still depends on t. Placing this optimal s into $J(s, t)$, the task is its maximisation with respect to t. For a more formal and concise discussion the following notation is introduced:

$$s^*(t) \stackrel{\text{def}}{=} \arg\inf_{s>0} J(s, t), \quad t^* \stackrel{\text{def}}{=} \arg\sup_{t>0} J(s^*(t), t) \quad \text{and} \quad s^* \stackrel{\text{def}}{=} s^*(t^*).$$

$$(3.4)$$

Now, the extremising pair of $J(s, t)$ is (s^*, t^*) and thus $-NI = J(s^*, t^*)$. The extremising values t^* and s^* are commonly termed as the critical time and space scales, respectively. The intuitive explanation of the critical time scale is that it is the most probable time interval after which overflows occur in the multiplexing system (i.e. the most likely length of the busy period prior to overflow). Although many other busy periods may contribute to the total overflow, large deviation theory takes into account only the most probable one, which is the most dominant in the asymptotic sense. The rationale behind the critical space parameter is that it captures the statistical behaviour of the workload process, that is the amount of achievable statistical multiplexing gain and the burstiness. Critical space values close to 0 describe a source (or an aggregate) that can benefit from statistical multiplexing, while larger values infer a higher bandwidth requirement. Finally, it is also worth noting that s^* and t^* always depend on the system parameters C, B and the statistical properties of $X[0, t)$.

In practical applications there is a QoS requirement, which is often specified as a constraint for the probability of buffer overflow ($e^{-\gamma}$). In order to admit a source the following criterion has to be satisfied:

$$P\{Q(C, N) > B\} \approx e^{-NI} \le e^{-\gamma} \quad \text{or} \quad \sup_{t>0}\inf_{s>0} J(s, t) \le -\gamma. \quad (3.5)$$

3.2.2 Equivalent Admission Criteria

The inequalities in (3.5) define an admission rule that uses the method of the many sources asymptotics in the native form. In this original form, the probability of buffer overflow is estimated using $X[0, t)$, B and C as the input quantities, whilst the target overflow probability is used as the performance criterion.

It is possible to set up two other criteria that can be used for admission control decisions. As it was mentioned in the introduction, it is often preferable to express the bandwidth requirement of the traffic and compare this quantity to the server capacity. In order to form an estimate of the bandwidth requirement of the traffic, another optimisation has to be performed. For this, the server capacity has to be treated as a free variable and given the workload process, the buffer size and the QoS requirement, the smallest server capacity has to be identified for which the system still satisfies the performance criterion put forward in (3.5). The resulting quantity

$$C_{equ} \stackrel{\text{def}}{=} \inf \left\{ C : \sup_{t>0} \inf_{s>0} J(s, t) \leq -\gamma \right\} \tag{3.6}$$

is termed in the rest of the chapter as the equivalent capacity.[2] Then the admission criterion can be written as

$$C_{equ} \leq C. \tag{3.7}$$

A similar, but less frequently used criterion can be defined that allows admission decisions to be made based on the available buffer space. In this case, the buffer requirement of the traffic is determined using a similar triple optimisation as in (3.6), but this time taking $X[0, t)$, C and the QoS requirement as the input quantities and B as the performance constraint:

$$B_{req} \stackrel{\text{def}}{=} \inf \left\{ B : \sup_{t>0} \inf_{s>0} J(s, t) \leq -\gamma \right\} \quad \text{and} \quad B_{req} \leq B. \tag{3.8}$$

Figure 3.1 presents a summary of the three methods with respect to their input parameters and the quantity they use as a constraint in the decision criterion. The methods are equivalent in the sense that in a given context they arrive at the same decision. When it comes to numerical evaluation, the first (original) method with the double optimisation is, however, significantly less demanding than the others involving three embedded optimisations.

3.3 The Improved Bandwidth Requirement Estimator

This section introduces an alternative method for computing the equivalent capacity. The advantage of this new method is that its computational complexity is reduced to a double optimisation, resulting in a similar formula to the one used in the rate-function based estimation of the buffer overflow

---

[2] Following the terminology of previous works, the term effective bandwidth is reserved for $\alpha(s, t)$, which is not directly associated with the minimal service rate required to meet the QoS target.

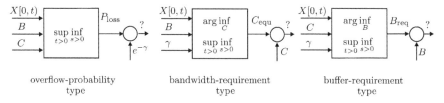

Figure 3.1 Admission decision methods.

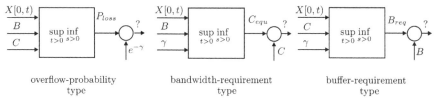

Figure 3.2 Admission decision methods with improved estimators.

probability. It is shown that the estimation of the equivalent capacity using the proposed method arrives at the same decision as the method in (3.6) and (3.7). The proposed method infer a new optimisation function resulting in an alternative set of space and time scales. The equivalence of the respective methods for estimating the buffer requirement can be proven in an identical manner (see the Appendix).

3.3.1 Alternative Definition of the Equivalent Capacity

Let us introduce $K(s, t)$ as

$$K(s, t) \stackrel{\text{def}}{=} \frac{\Lambda(s, t) + \gamma}{st} - \frac{B}{t}, \tag{3.9}$$

which is obtained from the isolation of C from $J(s, t) = -\gamma$. Namely, $K(s, t) = C$ holds after the rearrangement. By defining a new double optimisation

$$\widetilde{C}_{\text{equ}} \stackrel{\text{def}}{=} \sup_{t>0} \inf_{s>0} K(s, t), \tag{3.10}$$

similarly to (3.3), the extremisers are attained in the form of

$$s^{\dagger}(t) \stackrel{\text{def}}{=} \arg\inf_{s>0} K(s, t), \quad t^{\dagger} \stackrel{\text{def}}{=} \arg\sup_{t>0} K(s^{\dagger}(t), t) \quad \text{and} \quad s^{\dagger} \stackrel{\text{def}}{=} s^{\dagger}(t^{\dagger}). \tag{3.11}$$

The extremising pair of the double optimisation in (3.10) is then $(s^\dagger, t^\dagger)$ and these are the alternative space and time scales, respectively.

It can be proven that $\tilde{C}_{equ} = C_{equ}$ holds. In other words, we need only two optimisations instead of three to arrive at the equivalent capacity C_{equ}. This is shown in the next subsection using the subsequent theorem.

Theorem 1 ([1]). *The following two strict inequalities are equivalent:*

$$J(s^*, t^*) < -\gamma \iff K(s^\dagger, t^\dagger) < C, \qquad (3.12)$$

furthermore the equations

$$J(s^*, t^*) = -\gamma, \qquad (3.13)$$

$$K(s^\dagger, t^\dagger) = C \qquad (3.14)$$

are equivalent as well and consequently the two strict inequalities below also imply each other:

$$J(s^*, t^*) > -\gamma, \qquad (3.15)$$

$$K(s^\dagger, t^\dagger) > C. \qquad (3.16)$$

For the proof of this theorem, see [1].

Theorem 1 can now be used to prove that the equivalent capacities defined by (3.6) and (3.10) are equal.

Corollary 1. *The equivalent capacity defined by the double optimisation in (3.10) equals the one defined by the triple optimisation in (3.6):* $K(s^\dagger, t^\dagger) = \tilde{C}_{equ} = C_{equ}$.

For the proof of this corollary, see again [1].

The respective optimiser pairs $(s^*(B, C), t^*(B, C))$ and $(s^\dagger(B, \gamma), t^\dagger(B, \gamma))$ do not coincide in general, they are not even comparable as such, since they depend on a different set of variables. Nevertheless, on the boundary of the acceptance region $(J(s^*, t^*) = -\gamma \iff (C_{equ} =) K(s^\dagger, t^\dagger) = C)$ the same parameter values are the optimisers of the two problems:

Proposition 1. *If one of the double optimisations (3.3) and (3.10) has a unique extremising pair and $J(s^*, t^*) = -\gamma$ (or $K(s^\dagger, t^\dagger) = C$), then the two extremiser pairs coincide, $t^* = t^\dagger$ and $s^* = s^\dagger$.*

3.4 Comparison of the Methods for fBm Traffic

This section presents a comparison of the three admission control methods discussed in Section 3.2.2 using the new formulae developed in Section 3.3.1 and in the Appendix. The traffic case used is fractional Brownian motion (fBm), which involves closed-form formulae due to its Gaussian nature.

3.4.1 Key Formulae for Fractional Brownian Motion Traffic

The stochastic process $\{Z_t, t \in \mathbb{R}\}$ is called normalised fractional Brownian motion with self-similarity (Hurst-) parameter $H \in (0, 1)$ if it has stationary increments and continuous paths, $Z_0 = 0$, $E[Z_t] = 0$, $\text{Var}[Z_t] = |t|^{2H}$ and if Z_t is a Gaussian process. Let us define the process $X[0, t) \overset{\text{def}}{=} mt + Z_t$, for $t > 0$. It is known as fractional Brownian traffic and can be interpreted as the amount of traffic offered to the multiplexer in the time interval $[0, t)$. This is a so-called self-similar model, which has been suggested for the description of Internet traffic aggregates [6, 11].

Using this model the CGF can be written as

$$\Lambda(s, t) = stm + \frac{s^2\sigma^2 t^{2H}}{2}$$

and accordingly

$$J(s, t) = st\, m + \frac{s^2\sigma^2 t^{2H}}{2} - s(B + Ct).$$

The extremisers for $J(s, t)$ and $-NI$ can be found in Table 3.1,[3] where $\kappa(H) \overset{\text{def}}{=} H^H(1 - H)^{1-H}$.

The equivalent capacity can be evaluated in two ways, either using the definition in (3.6) or the method proposed in this chapter (3.10). $\widetilde{C}_{\text{equ}}$ requires the direct evaluation of $K(s, t)$ (3.9) at the alternative critical space and time scales $(s^\dagger, t^\dagger)$ (3.11), i.e. "only" a double optimisation is necessary. For fBm traffic

$$K(s, t) = m + \frac{1}{2}\sigma^2 st^{2H-1} + \frac{\gamma}{st} - \frac{B}{t},$$

its extremisers and $\widetilde{C}_{\text{equ}}$ are listed in Table 3.1. If C_{equ} is calculated in the conventional way (3.6), the third optimisation (with respect to C) can be

---

[3] An identical expression for the approximation of the overflow probability was obtained in [11] with a different approach.

Table 3.1 Comparison of the three admission control methods for fBm traffic.

	$f(s,t)$		
	$J(s,t)$	$K(s,t)$	$L(s,t)$
$s^{opt}(t) =$ $\arg\inf_{s>0} f(s,t)$	$s^*(t) =$ $\frac{t^{-2H}(B+(C-m)t)}{\sigma^2}$	$s^\dagger(t) =$ $\frac{\sqrt{2\gamma}t^{-H}}{\sigma}$	$s'(t) =$ $\frac{\sqrt{2\gamma}t^{-H}}{\sigma}$
$t^{opt} =$ $\arg\sup_{t>0} f(s^{opt}(t), t)$	$t^* =$ $\frac{H}{1-H}\frac{B}{C-m}$	$t^\dagger =$ $2^{-\frac{1}{2H}}\left(\frac{B}{(1-H)\sqrt{\gamma}\sigma}\right)^{\frac{1}{H}}$	$t' =$ $\left(\frac{H\sqrt{2\gamma}\sigma}{C-m}\right)^{\frac{1}{1-H}}$
$s^{opt} = s^{opt}(t^{opt})$	$s^* = \frac{1-H}{\kappa(H)^2}\cdot$ $\frac{(C-m)^{2H}B^{1-2H}}{\sigma^2}$	$s^\dagger =$ $\frac{2(1-H)\gamma}{B}$	$s' = \left(\frac{C-m}{H}\right)^{\frac{H}{1-H}}\cdot$ $\left(\sqrt{2\gamma}\sigma\right)^{\frac{1}{1-H}}2\gamma$
$\sup_{t>0}\inf_{s>0} f(s,t)$	$-NI =$ $-\frac{(C-m)^{2H}B^{2-2H}}{2\kappa(H)^2\sigma^2}$	$\tilde{C}_{equ} = m + H\cdot$ $\left(2\gamma\sigma^2\right)^{\frac{1}{2H}}\left(\frac{1-H}{B}\right)^{\frac{1-H}{H}}$	$\tilde{B}_{req} = \left(\frac{H}{C-m}\right)^{\frac{H}{1-H}}\cdot$ $(1-H)\left(\sqrt{2\gamma}\sigma\right)^{\frac{1}{1-H}}$

exchanged for solving $-NI = -\gamma$ for $C = C_{equ}$ (as seen in the proof of Corollary 1).[4] It can be checked that $C_{equ} = \tilde{C}_{equ}$ as expected (using the definition of $\kappa(H)$). In a similar way, s', t' and $\tilde{B}_{req}$ (see the Appendix) can be computed (see Table 3.1) and it also turns out that $\tilde{B}_{req} = B_{req}$.

Confirming the statements in the previous section, it is apparent from Table 3.1 that the critical space scales s^*, $s^\dagger$ and s' are usually different and depend on different parameter sets. For given B, C, γ, m, H and σ, the corresponding scales match only when equalities (3.13) and (3.14) hold. An interesting consequence of this fact is that the solution of $t^*(B, C, m, H) = t^\dagger(B, \gamma, \sigma, H)$ for C results in the equivalent capacity C_{equ} and its solution for γ is NI. Similar statements are valid for the space scales as well.

3.4.2 A Numerical Example

In this subsection a numerical example is presented to demonstrate the results of the previous subsection. Let us take the fBm model of one of the

---

[4] This simplification can be done only because $-NI$ is an explicit function of C in the fBm case (C can be isolated from the equation). In most other cases the third optimisation must be done through several double optimisations of $J(s,t)$ for different values of C in order to locate $C = C_{equ}$ for which $-NI = -\gamma$.

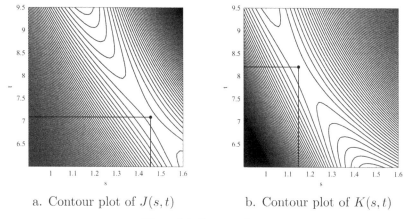

a. Contour plot of $J(s,t)$ b. Contour plot of $K(s,t)$

Figure 3.3 Contour plots.

Bellcore Ethernet data traces [6]: $m_1 = 138\,135$ byte/s, $\sigma_1 = 89\,668$ byte/s^H, $H = 0.81$. Assume that $N = 100$ of such sources are multiplexed into a buffer. Hence, the model parameters of the fBm model for the aggregate traffic workload become: $m = 13.8135$ Mbyte/s, $\sigma = 0.89668$ Mbyte/s^H, $H = 0.81$. The buffer size is chosen to be $B = 5.3$ Mbyte, the service rate is $C = 16$ Mbyte/s and let the constraint for the overflow be $e^{-16} \approx 10^{-7}$ ($\gamma = 16$). For these system parameters, the extremiser pair is $(s^*, t^*) = (1.453, 7.091)$ and therefore $-NI = -20.26$. Clearly, $-NI < -\gamma$, i.e. the QoS requirement is fulfilled. The alternative critical scales and the equivalent capacity are obtained as $(s^\dagger, t^\dagger) = (1.147, 8.203) \neq (s^*, t^*)$ and $C_{\text{equ}} = \widetilde{C}_{\text{equ}} = 16.568$ Mbyte/s, thus $C_{\text{equ}} < C$ holds and there is 0.432 Mbyte/s of free service capacity.

3.4.3 Example on the Computational Gain

The indirect method of evaluating the equivalent capacity requires a triple optimisation (3.6) in case of the many sources asymptotics. When it comes to numerical evaluation, this is significantly more demanding than the double optimisation found in the direct method (3.10), hindering the deployment of the solution in real systems.

The optimisation in (3.6) with respect to C generally takes numerous steps. Consequently, the extremising (s, t) pair of $J(s, t)$ needs to be computed several times (for each intermediate value of C). In contrast, the direct method involves only one double optimisation for locating the optimal (s, t)

pair of $K(s,t)$, which corresponds to one step from the perspective of the indirect method.

The rest of this section demonstrates the degree of this discrepancy in computational complexity through a numerical example and also sheds some light on its implications on the implementation aspects of the direct and indirect methods.

In the following, a queueing system is investigated that is fed by the superposition of several Markov on-off sources. For this kind of traffic the cumulant generating function can be computed analytically. If the arrival process $X[0,t)$ is described by a finite Markov chain with transition rate matrix $\mathbf{Q}$ and instantaneous arrival rates h_i (valid in state i), then its CGF can be expressed as [7]:

$$\Lambda(s,t) = \log E\left[\boldsymbol{\pi}\, e^{[\mathbf{Q}+\mathbf{H}s]t}\mathbf{1}\right], \tag{3.17}$$

where $\boldsymbol{\pi}$ is the vector of the steady-state distribution of the Markov chain ($\boldsymbol{\pi}\mathbf{Q} = \mathbf{0}$, $\boldsymbol{\pi}\mathbf{1} = 1$) and $\mathbf{H} = \mathrm{diag}(h_i)_i$ ($\mathbf{1}$ is a column vector of 1-s).

Let us consider N independent and identical Markov modulated on-off sources generating packets at the peak rate p in the on state and remaining silent in the off state. Let λ denote the transition rate from the off state to the on state and let μ be the transition rate from the on state to the off state. The arrival process generated by these sources can be described by a corresponding finite Markov chain with the equilibrium vector $\boldsymbol{\pi}$:

$$\pi_i = \binom{N}{i}\left(\frac{\lambda}{\lambda+\mu}\right)^i\left(1-\frac{\lambda}{\lambda+\mu}\right)^{N-i}, \tag{3.18}$$

where $\pi_i = P$ (i sources being in 'on' state) and by the instantaneous arrival rate matrix $\mathbf{H}$:

$$\mathbf{H} = \mathrm{diag}(ip)_{i=0,\dots,N}. \tag{3.19}$$

With the analytical expression for $\alpha(s,t)$ at hand the rate function surface $J(s,t)$ and the equivalent capacity surface $K(s,t)$ can be evaluated and plotted in arbitrary system configurations.

The parameters of the system under study are the following: the traffic is made up by $N = 15$ Markov modulated on-off flows with per-flow peak rate $p = 4$ Mbps, $\lambda = 10$ 1/s and $\mu = 90$ 1/s. The system is equipped with a buffer of size $B = 52500$ byte. The target overflow probability is set to $\varepsilon = e^{-\gamma} = 10^{-4}$.

Figure 3.4 illustrates the trail of the saddle point of $J(s,t)$ projected over the (s,t) plane for several values of C. In an imaginary implementation of

(3.6) one would start the search for the bandwidth requirement by optimising the $J(s, t)$ function with a service rate that is definitely above the equivalent capacity of the traffic. Provided that no a priori information is available on the traffic, this could be the service rate of the physical link. Subsequent iterations would employ smaller and smaller values of C until the best approximation of the equivalent capacity is reached, for which the overflow probability estimated by $J(s, t)$ still satisfies the predefined target value. The size of the search region shall be chosen in such a way that the (s, t) pair marking the saddle point of the $J(s, t)$ surface lies within the search region for all intermediate C values. In the course of the search the optimal (s, t) pair steers through a broad range of values in both variables (Figure 3.4), which implies that the search region must be large.

In the same set-up, but this time with the direct method, the identification of the equivalent capacity would take one step, as settled in (3.10). As a practical consequence, the range of s and t values over which the search for the saddle point of the $K(s, t)$ surface is performed can be narrowed down. In a real implementation of the C_{equ} estimator this brings along an improvement in the achievable precision of the saddle-point locator algorithm, and consequently in the accuracy of the equivalent capacity estimate. This is because the $K(s, t)$ function is typically evaluated (measured) over a grid made up by a fixed number of (s, t) points and the search for the supinf point of this sampled surface is performed in discrete steps. Hence, in case of a narrower search range the spacing of the grid becomes considerably finer, leading to a more precise estimate of C_{equ}.

In summary, the implementation of the optimisation in the direct method can work with a finer granularity than its indirect counterpart, which is forced to use a coarse-grained grid to ensure the wide s and t ranges necessary for enclosing the saddle points of the $J(s, t)$ surface for all intermediate values of C in (3.6). In addition to the reduction in computational complexity the direct method thus offers an increased precision for the equivalent capacity estimate.

In the last step of the search the values of the extremising parameters for the two optimisation regions would tally, as the saddle point of the $J(s, t)$ and the $K(s, t)$ functions are co-located when the indirect expression (3.6) is evaluated for C_{equ}. Figures 3.5 and 3.6 depict fragments of the $J(s, t)$ surface for $C = C_{equ}$ and the $K(s, t)$ surface in which the saddle points can be found. The exact location of the supinf point is marked by a dot on both surfaces. The extremising values in both cases are $s_{B,\gamma} = 17.15$ 1/Mbit, $t_{B,\gamma} = 0.0685$ s and the equivalent capacity of the traffic is $C_{equ} = 12.87$ Mbps.

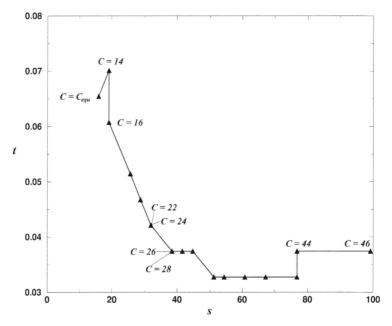

Figure 3.4 Optimising (s, t) values of $J(s, t)$ as a function of C.

This simple analytical example thus demonstrates that the direct and the indirect methods for obtaining the equivalent capacity lead to the same solution. Note, however, that in terms of computational complexity the direct method is an order of magnitude smaller than its indirect counterpart.

In the previous section [1] a new method has been introduced for computing the equivalent capacity of traffic flows that is based on the overflow probability estimation in the many sources (logarithmic) asymptotics. Instead of the three embedded optimisations that previous approaches required, it comprises only of two optimisations that directly result in an estimate of the bandwidth requirement.

In [4] buffer overflow asymptotics has been analyzed under the many flows *exact* asymptotics. In [9] a refined approximation for the buffered workload loss ratio based on many sources *exact* asymptotics has been introduced and analyzed. These approaches has also been compared to other methods and turned out to be a viable computational method for obtaining accurate enough estimation of buffer overflow probability and workload loss ratio. Equivalent bandwidth and buffer requirement can also be defined in this case similarly to (3.6), (3.8), when the QoS constraint is imposed directly

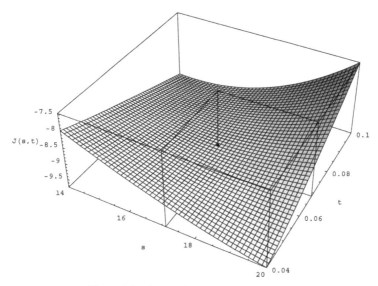

Figure 3.5 The rate function surface $J(s, t)$.

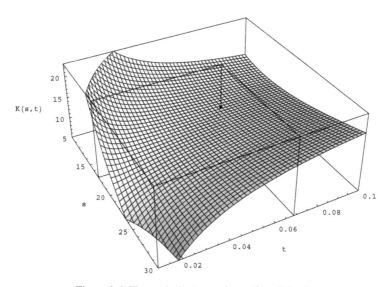

Figure 3.6 The equivalent capacity surface $K(s, t)$.

on these refined estimates. In connection with this, the problem of having computationally more feasible resource estimators quite naturally emerges.

3.5 Refined Approximation for the Equivalent Capacity

Based on the work of Bahadur and Rao on the deviation from the sample mean [2] a refined (*exact*) asymptotic approximation has been developed in [9] for the buffer overflow probability and workload loss ratio under the many sources asymptotics regime. The following theorem presents this result:

Theorem 2. *Assume that there exists a unique* $t^* < \infty$ *such that:*

$$J(s^*(t^*), t^*) = \max_{t>0} J(s^*(t), t) < 0 . \tag{3.20}$$

Then, as $N \to \infty$ *the overflow probability and the workload loss ratio LR is given by*

$$P(Q > B) = \frac{e^{-NI}}{s^* \sqrt{2\pi \Lambda''(s^*, t^*)}} \left(1 + O\left(\frac{1}{N}\right)\right), \tag{3.21}$$

$$LR = \frac{e^{-NI}}{Ms^{*2} \sqrt{2\pi \Lambda''(s^*, t^*)}} \left(1 + O\left(\frac{1}{N}\right)\right), \tag{3.22}$$

where

$$M \overset{\text{def}}{=} E[X(0, t)/t] ,$$

$$\Lambda''(s, t) \overset{\text{def}}{=} \frac{\partial^2 \Lambda(s,t)}{\partial s^2} , \quad s^*(t) \overset{\text{def}}{=} \arg\inf_s J(s, t)$$

$$t^* \overset{\text{def}}{=} \arg\sup_t J(s^*(t), t) , \quad s^* \overset{\text{def}}{=} s^*(t^*) .$$

A practical consequence of this theorem is that for finite system parameters B, C and N the overflow probability and the loss ratio can be approximated as

$$P(Q > B) \approx \frac{1}{s^* \sqrt{2\pi \Lambda''(s^*, t^*)}} e^{-NI}, \tag{3.23}$$

$$LR \approx \frac{1}{Ms^{*2} \sqrt{2\pi \Lambda''(s^*, t^*)}} e^{-NI} . \tag{3.24}$$

The advantage of these formulae to those based on logarithmic asymptotics, that the possible $o(N)$ error term in the exponent is eliminated, instead, a more attractive $O(1/N)$ term is present. However, the serious drawback of the direct application of them is that the second derivative of the cumulant generating function ($\Lambda''(s)$) would also be needed for the computation. Nevertheless, the use of the second derivative can be eliminated by applying a second-order approximation of $\Lambda(s, t)$ around s^*, t^* [10]:

$$s^{*2}\Lambda''(s^*, t^*) \approx -2(\Lambda(s^*, t^*) - s^*(B + Ct^*) = 2NI. \qquad (3.25)$$

Due to this reasonable approximation

$$\frac{1}{Ms^{*2}\sqrt{2\pi\Lambda''(s^*)}} \approx \frac{1}{Ms^*\sqrt{4\pi NI}} \qquad (3.26)$$

holds, and then the right hand side of (3.23) and (3.24) can be approximated as

$$P(Q > B) \approx \exp\left\{-NI - \frac{1}{2}\log 4\pi NI\right\}, \qquad (3.27)$$

$$LR \approx \exp\left\{-NI - \frac{1}{2}\log 4\pi NI - \log s^*M\right\} \overset{\text{def}}{=} \widetilde{LR}. \qquad (3.28)$$

Now the equivalent capacities can be defined, similarly to (3.6) when a QoS constraint is imposed on the overflow probability and *LR*:

$$C_{\text{equ,ov}} \overset{\text{def}}{=} \inf\left\{C : -NI - \frac{1}{2}\log 4\pi NI \leq -\gamma\right\}, \qquad (3.29)$$

$$C_{\text{equ,lr}} \overset{\text{def}}{=} \inf\{C : \widetilde{LR} \leq e^{-\gamma}\}. \qquad (3.30)$$

At this point, the problem is phrased again: Is it possible to give computationally more feasible estimates for $C_{\text{equ,lr}}$, similarly to those in (3.6) and (3.8), i.e. which would require a double optimisation instead of a triple one?

First, the overflow probability type equivalent capacity estimate is presented.

Theorem 3.

$$C_{\text{equ,ov}} \approx \sup_{t>0} \inf_{s>0}\left\{\frac{\Lambda(s, t) + \gamma\left(1 - \frac{\log 4\pi\gamma}{1+2\gamma}\right)}{st} - \frac{B}{t}\right\} \overset{\text{def}}{=} \widetilde{C}_{\text{equ,ov}}. \qquad (3.31)$$

Proof. Let $NI = \gamma + \varepsilon$. In this way, towards finding the equivalent capacity the following equation based on (3.29) should be solved for ε

$$\gamma + \varepsilon + \frac{1}{2} \log 4\pi (\gamma + \varepsilon) = \gamma . \tag{3.32}$$

Because it is essentially a non-algebraic equation, an approximate solution is to be find, which is possible with using the first two terms in the Taylor series expansion of $\log 4\pi (\gamma + \varepsilon)$ around $4\pi \gamma$. That is,

$$\log 4\pi (\gamma + \varepsilon) \approx \log 4\pi \gamma + \frac{\varepsilon}{\gamma},$$

hence, the approximate solution for ε is

$$\varepsilon \approx -\frac{\gamma \log 4\pi \gamma}{1 + 2\gamma} \stackrel{\text{def}}{=} \tilde{\varepsilon} . \tag{3.33}$$

From this we have the equation

$$NI = \gamma - \frac{\gamma \log 4\pi \gamma}{1 + 2\gamma} \tag{3.34}$$

for determining the equivalent capacity which is an approximate to $C_{\text{equ,ov}}$. Also from Corollary 1 the solution of the equation above for C is exactly the right hand side of (3.31). $\qquad \square$

The immediate consequence of this theorem and the equivalence between B_{req} and $\tilde{B}_{\text{req}}$, that the buffer requirement estimator for this case is as follows:

$$B_{\text{req,ov}} \stackrel{\text{def}}{=} \inf\{B : -NI - \frac{1}{2} \log 4\pi NI \leq -\gamma\}$$

$$\approx \sup_{t>0} \inf_{s>0} \left\{ \frac{\Lambda(s, t) + \gamma \left(1 - \frac{\log 4\pi \gamma}{1+2\gamma}\right)}{s} - Ct \right\} . \tag{3.35}$$

Note that taking into account the refined estimate of the buffer overflow probability (3.23) in the improved equivalent capacity and buffer requirement formula is manifested in a modified QoS constraint $\gamma - \frac{\gamma \log 4\pi \gamma}{1+2\gamma}$ which is always smaller (less stringent) then the original γ. It also means that $\tilde{C}_{\text{equ,ov}}$ is always smaller than $\tilde{C}_{\text{equ}}$ (3.10). Because both of them are approximations, it does not follow from this that $\tilde{C}_{\text{equ,ov}}$ would be better approximation than

$\tilde{C}_{equ}$, however, it is expected from the different types of asymptotics. In the previous subsection it has been performed that $\tilde{C}_{equ} = C_{equ}$. Here, one can say only that $\tilde{C}_{equ,ov} \approx C_{equ,ov}$. The accuracy of this approximation is quite acceptable for large range of γ, this can be illustrated by the very same example with fBm traffic used in Section 2.4.2. For that example the computations are straightforward using Table 3.1. The results are $\tilde{C}_{equ,ov} = 16.2857$ while $C_{equ,ov} = 16.2864$, which is really a negligible difference.

It is considerably more involved to find a loss ratio type equivalent capacity formula than $\tilde{C}_{equ,ov}$ in the previous case. The reason for this is that the presence of the logarithm of the asymptotic rate function and also the logarithm of the optimising parameter s^* in the estimate as additive terms besides the rate function. To answer the underlying question, the following theorem is presented:

Theorem 4. *If the assumption of Theorem 2 holds, then a reasonable approximation for $C_{equ,lr}$ is as follows:*

$$C_{equ,lr} \approx \sup_{t>0} \inf_{s>0} \left\{ \frac{\Lambda(s,t) + \gamma + \tilde{\varepsilon}(\gamma, B, M)}{st} - \frac{B}{t} \right\} \stackrel{def}{=} \tilde{C}_{equ,lr} \qquad (3.36)$$

where

$$\tilde{\varepsilon}(\gamma, B, M) = \frac{2\gamma \log \left(e^{-\frac{1}{2}-\gamma} + \frac{B+2B\gamma}{4M\sqrt{\pi}\gamma^{3/2}} \right)}{1+2\gamma}. \qquad (3.37)$$

Before proving this theorem two lemmas have been presented.

Lemma 1. *Assume there exists a function $\varepsilon(\gamma, B, M)$ for which the following equivalence holds:*

$$\widetilde{LR} = e^{-\gamma} \Leftrightarrow NI = \gamma + \varepsilon(\gamma, B, M) \qquad (3.38)$$

where $\varepsilon(\gamma, B, M)$ may depend on γ, B and M, but not on C. Then $C_{equ,lr}$ defined in (3.30) can be expressed as

$$C_{equ,lr} = \sup_{t>0} \inf_{s>0} \left\{ \frac{\Lambda(s,t) + \gamma + \varepsilon(\gamma, B, M)}{st} - \frac{B}{t} \right\} \qquad (3.39)$$

Proof. (Lemma 1) Because $C_{equ,lr}$ is a solution of the equation $\widetilde{LR} = e^{-\gamma}$ with respect to C as a variable (and with fixing the other system parameters), it is also a solution of $NI = \gamma + \varepsilon(\gamma, B, M)$. Based on the similar equivalence $\tilde{C}_{equ} = C_{equ}$, proven in [1]) the statement of the lemma follows. $\qquad\square$

Before proving the main theorem an important property of s^* is also highlighted:

Lemma 2.

$$s^* = \frac{\partial NI}{\partial B} \, . \tag{3.40}$$

Proof of Lemma 2. By the definition of $-NI$ we have

$$\frac{\partial NI}{\partial B} = \frac{\partial (s^*(B + Ct^*) - \Lambda(s^*, t^*))}{\partial B} \, . \tag{3.41}$$

Note that s^* and t^* depend on B, hence, the right hand side of the equation above can be formed as

$$\frac{\partial NI}{\partial B} = (B + Ct^*) \frac{\partial s^*}{\partial B} + s^* \left(1 + C \frac{\partial t^*}{\partial B} \right) - \frac{\partial \Lambda(s, t)}{\partial B} \bigg|_{s=s^*, t=t^*} \, . \tag{3.42}$$

The last term in the equation above can be written as

$$\frac{\partial \Lambda(s, t)}{\partial B} \bigg|_{s=s^*, t=t^*} = \frac{\partial \Lambda(s, t)}{\partial s} \bigg|_{s=s^*, t=t^*} \frac{\partial s^*}{\partial B} + \frac{\partial \Lambda(s, t)}{\partial t} \bigg|_{s=s^*, t=t^*} \frac{\partial t^*}{\partial B} \, . \tag{3.43}$$

The extremising pair of s^*, t^* satisfies the following equations:

$$\frac{\partial \Lambda(s, t)}{\partial s} \bigg|_{s=s^*, t=t^*} = B + Ct^* \, , \qquad \frac{\partial \Lambda(s, t)}{\partial t} \bigg|_{s=s^*, t=t^*} = s^* C \, . \tag{3.44}$$

Combining this with the equations (3.42) and (3.43) the statement of the lemma follows. □

Proof of Theorem 4. What is left to show in this proof the existence of $\varepsilon(\gamma, M, B)$ and that $\tilde{\varepsilon}(\gamma, B, M)$ in (3.37) is a reasonable approximation for it.

Based on Lemma 2 the expression of $-\log \widetilde{LR}$ in (3.28) can be rewritten as

$$-\log \widetilde{LR} = NI + \frac{1}{2} \log 4\pi NI + \log M + \log \frac{\partial NI}{\partial B} \, . \tag{3.45}$$

Therefore, for satisfying the equivalence in (3.38) such ε has to be find which satisfies the following differential equation:

$$\varepsilon + \frac{1}{2} \log 4\pi (\gamma + \varepsilon) + \log M + \log \frac{\partial \varepsilon}{\partial B} = 0 \, . \tag{3.46}$$

For positive M and B there always exists such ε function which may depend on γ, B and M, but not on C. Unfortunately, there is no closed form $\varepsilon(\gamma, M, B)$ solution of this differential equation, hence, an approximate closed-form solution has to be find. To this end, let us consider the modified (approximate) differential equation

$$\varepsilon + \frac{1}{2}\log 4\pi\gamma + \frac{1}{2}\frac{\varepsilon}{\gamma} + \log M + \log\frac{\partial\varepsilon}{\partial B} = 0 \tag{3.47}$$

which is obtained from (3.46) by applying the first-order approximation $\log 4\pi(\gamma + \varepsilon) \approx \log 4\pi\gamma + \varepsilon/\gamma$. A general solution of (3.47) is as follows:

$$\tilde{\varepsilon}(\gamma, B, M, K) = \frac{2\gamma\log\left(-\frac{(1+2\gamma)(-B+K)}{4M\sqrt{\pi}\gamma^{3/2}}\right)}{1 + 2\gamma}, \tag{3.48}$$

where K is a constant to be determined by some appropriate boundary condition. The boundary condition comes from the extreme case of $B = 0, C = 0$, i.e.

$$-NI = \sup_{t>0}\inf_{s>0}(\Lambda(s, t) - s(B + Ct))|_{B=0, C=0} = 0 . \tag{3.49}$$

From this and (3.38) one can deduce $\tilde{\varepsilon}(\gamma, B = 0, M, K) = -\gamma$ and determine K from it. Substituting this constant back into the right hand side of (3.48) the formula of $\tilde{\varepsilon}(\gamma, B, M)$ in (3.37) appears. Based on Lemma 1 the statement of the theorem also follows. $\qquad\square$

3.5.1 On the Accuracy of the Formulae

First we briefly discuss the asymptotical accuracy of the large deviation (many sources asymptotic) based equivalent capacity estimators defined in equation (3.29). The important consequence of the large deviation result in (3.1) to the exact value of the buffer overflow probability is that (for large number of sources)

$$P\{Q(C, N) > B\} = \exp(-NI + o(N)) . \tag{3.50}$$

It immediately follows for the corresponding exact equivalent capacity value that

$$C_{\text{exact}} = \sup_{t}\inf_{s}\left\{\frac{\Lambda(s, t) + \gamma + o(N)}{s} - \frac{B}{t}\right\} \tag{3.51}$$

which differs from $\widetilde{C}_{\text{equ,ov}}$ (see equation (3.31)) in the prescribed value of the QoS represented by γ. Hence, the problem with the result based on logarithmic asymptotics is that the function denoted by $o(N)$ might even be increasing with N causing significant bias in the equivalent capacity estimators based on logarithmic asymptotics.

As opposed to the logarithmic asymptotics, we have used an exact asymptotical result for the buffer workload loss ratio (3.22) which implies that

$$C_{\text{exact,lr}} = \inf\left\{C : \log \widetilde{LR} < -\gamma - O\left(\log\left(1 + \frac{1}{N}\right)\right)\right\} \qquad (3.52)$$

because

$$\log\left(1 + O\left(\frac{1}{N}\right)\right) = O\left(\log\left(1 + \frac{1}{N}\right)\right). \qquad (3.53)$$

It can be seen that the equivalent capacity estimator $C_{\text{equ,lr}}$ defined in (3.30) differs from the exact value through a vanishing (as the number of sources increases) additive term $O(\log(1 + 1/N))$ in the prescribed QoS level γ. This clearly clarifies the benefit in terms of accuracy of our equivalent capacity estimators based on exact asymptotics as $\widetilde{C}_{\text{equ,lr}}$.

Next we shed considerable light on the difference between $C_{\text{equ,lr}}$ and $\widetilde{C}_{\text{equ,lr}}$. There are two sources of the inaccuracy imposed on the equivalent capacity estimators $\widetilde{C}_{\text{equ,lr}}$ (3.36) compared to the original approximation $C_{\text{equ,lr}}$ (3.30). One of them originates from the approximation of the pre-factor in equation (3.26). As it can be seen this approximation is based on the second-order approximation of $\Lambda(s, t)$ around s^*, and t^* as follows by the direct application of Taylor's theorem:

$$0 = \Lambda(0, t^*) = \Lambda(s^*, t^*) + (0 - s^*)\frac{\partial \Lambda(s, t)}{\partial s}\Big|_{s=s^*, t=t^*}$$

$$+ \frac{(0 - s^*)^2}{2}\frac{\partial^2 \Lambda(s, t)}{\partial s^2}\Big|_{s=s^*, t=t^*} + r, \qquad (3.54)$$

where the term r representing the error in the approximation is

$$r = -\frac{s^{*3}}{3}\frac{\partial^3 \Lambda(s, t)}{\partial s^3}\Big|_{0 < s < s^*, t=t^*}. \qquad (3.55)$$

The pre-factor can be expressed in this way as

$$\frac{1}{Ms^{*2}\sqrt{2\pi \Lambda''(s^*)}} = \frac{1}{Ms^*\sqrt{4\pi NI + 2\pi r}}. \qquad (3.56)$$

The error term r is exactly zero, i.e. $r = 0$ when the process $X(0, t)$ is Gaussian, that is

$$\Lambda(s, t) = Mst + \frac{1}{2}s^2 v(t), \tag{3.57}$$

where $v(t)$ is the variance function of the process, $v(t) = EX^2(0, t) - (EX(0, t))^2$. In case of the many sources asymptotic regime it is a reasonable assumption that the aggregate arrival process is not far from a Gaussian one, hence, $2\pi r$ is negligible. This assumption can analytically be supported by several functional Central Limit Theorems for stochastic processes. We refer the interested readers to [12] for details.

The other source of the difference between $C_{equ,lr}$ and $\widetilde{C}_{equ,lr}$ is the introduction of an approximate (and in this case solvable in closed form) differential equation (3.47) instead of using the original one (3.46). The order of the error in approximating the term $\log(4\pi(\gamma + \varepsilon))$ by $\log 4\pi\gamma + \frac{\varepsilon}{\gamma}$ is $o(\varepsilon^2)$. In our numerical investigations this error turned out to be negligible in most cases, which is also reflected in the examples in the next section.

3.5.2 Numerical Examples

In this section numerical examples are presented illustrating the extensive numerical analysis on the accuracy of the resource estimates proposed. Results for the loss ratio oriented approximations are performed, due to the similar nature of approximation, the observations are also valid for the overflow type equivalent capacity estimators. As we have seen, the reason that $C_{equ,lr}$ and $\widetilde{C}_{equ,lr}$ does not exactly coincide is the difference between the two differential equations (3.46), (3.47). The relative difference can be quantified by the term

$$\text{diff}\left(\gamma, \frac{B}{M}\right) = 1 - \frac{\log 4\pi\gamma + \frac{1}{\gamma}\tilde{\varepsilon}(\gamma, B, M)}{\log 4\pi(\gamma + \tilde{\varepsilon}(\gamma, B, M))}. \tag{3.58}$$

Note that this quantity depends only on γ and the ratio B/M (just as $\tilde{\varepsilon}$). Based on our extensive numerical analysis the following general consequences can be drawn. For a wide range of γ and B/M ($\gamma \in (2, 25)$, $B/M \in (1/10, 100)$) the relative differences are within 0.05 (5%), in several cases within 0.01 (1%). Reasonable values of γ are between 2 and 25 (which corresponds to about the region 10^{-1}–10^{-10} of QoS constraint of the workload loss ratio). For increasing γ (more stringent QoS constraint) and/or increasing ratio of B/M the function of relative difference are usually decreasing. The figure below show the relative difference in the function of γ for $B/M = 50, 10, 1, 1/2$.

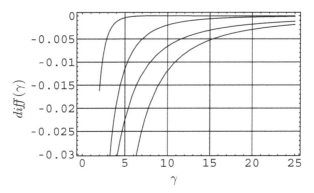

Figure 3.7 The relative difference in the function of γ.

These observations on the relative difference between the differential equations empower us to investigate the difference between the equivalent capacity estimators, however, the impacts of diff$(\gamma, B/M)$ on the accuracy of $\widetilde{C}_{\text{equ,lr}}$ seems to be complex and hardly quantifiable.

In what follows we demonstrate through numerical examples that the direct computation of the equivalent capacity by $\widetilde{C}_{\text{equ,lr}}$ presented in Theorem 4 is an accurate enough approximation of $C_{\text{equ,lr}}$. For this purpose the widely-known and used fractional Brownian motion (fBm) traffic model is used as input process to the queue.

Let us take the fBm model of one of the Bellcore Ethernet data traces [6]: $m_1 = 138\,135$ byte/s, $\sigma_1 = 89\,668$ byte/s^H, $H = 0.81$. Assume that $N = 100$ of such sources are multiplexed into a buffer. Hence, the model parameters of the fBm model for the aggregate traffic workload become: $M = 13.8135$ Mbyte/s, $\sigma = 0.89668$ Mbyte/s^H, $H = 0.81$. The buffer size is chosen to be $B = 6.9, 13.8, 138$ Mbyte corresponding to the three cases of $B/M = 1/2, 1, 10$. Figure 3.8 shows the relative difference between the equivalent capacity estimates. The thinner the line is the higher the ratio of B/M is.

We also present a figure on the relative difference, for which the ratio B/M is set to $1/2$ ($B = 6.9$, $M = 13.8135$, $\sigma = 0.89668$) and the Hurst parameter is varied from 0.6 to 0.9 (0.6, 0.7, 0.8, 0.9). The thicker lines corresponds to higher Hurst parameter.

These drawings illustrates the following common observations based on our thorough numerical analysis.

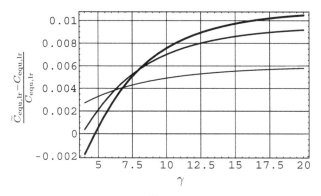

Figure 3.8 The relative difference between $\widetilde{C}_{equ,lr}$ and $C_{equ,lr}$ in the function of γ and B/M.

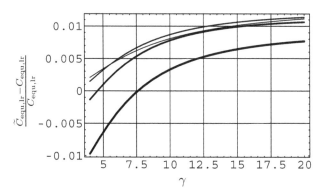

Figure 3.9 The relative difference between $\widetilde{C}_{equ,lr}$ and $C_{equ,lr}$ in the function of γ and H.

Observations:

- In case of higher γ (more stringent QoS constraint) and smaller buffer sizes (smaller B/M ratio) the accuracy of $\widetilde{C}_{equ,lr}$ is usually smaller.
- Nevertheless, in the region of smaller γ parameters, the decrease of the buffer size can improve the accuracy of the estimate (as illustrated by Figure 3.8).
- When the Hurst parameter is in the interval $(0.5, 0.8)$ the relative error is not greatly influenced by H. Nevertheless, for H values higher than 0.8, the relative difference quickly decreases with increasing H, the estimate becomes less accurate for smaller γ and more accurate for larger γ.
- In the wide region of γ and B/M of interest ($\gamma \in (2, 25)$, $B/M \in (0.1, 100)$) the accuracy of the estimate $\widetilde{C}_{equ,lr}$ is always within 0.03

(the absolute value of the relative difference is under 0.03, i.e. 3%), for significant part of this parameter region the relative difference is even below 1% (see Figures 3.8 and 3.9).

Note that the applicability of our novel equivalent capacity estimator relies on the applicability and accuracy of the refined workload loss ratio estimator in the many sources asymptotics framework. The formula of $\widetilde{LR}$ defined in (3.28) turned out to be powerful [10], especially for Gaussian sources (like fBm traffic).

3.6 Conclusions

This paper has introduced new methods for the computation of the equivalent capacity (and the buffer requirement) of traffic flows that is based on the many sources asymptotics. In contrast to the method directly building on the asymptotic rate function, the new methods involves only two embedded optimisations instead of three, thus it may significantly reduce the computational complexity of the task. It has been shown that the two methods are equivalent in case of buffer overflow probability, and have acceptable difference in case of buffer workload loss probability.

The presented method of deriving the equivalent capacity leads to an alternative domain of time and space scales. In a given system the optimisation defining the equivalent capacity estimate $(C_{\text{equ}} =) \widetilde{C}_{\text{equ}}$ (3.10) yields different optimal parameter values than that defining the estimate of the overflow probability e^{-NI} (3.3) or the workload loss ratio. Consequently, the substitution of the extremisers of $J(s, t)$ into $K(s, t)$ (3.9) does not lead to a correct estimate of the equivalent capacity (as was proposed in some papers [3,4]). The only exception is the boundary of the admission region, where the two extremising pairs coincide.

In terms of applicability, it can be shown that the method of the equivalent capacity computation is more appropriate for real-time operation than those based on the asymptotic rate function, especially if the workload process is measured on-line (measurement-based admission control). Recall the admission methods defined by (3.5) and (3.6), (3.7). In practice these admission rules are performed at the arrival of a new flow. The effective bandwidth estimate has to be adjusted in order to take the new flow into account. For example, let us assume that the new flow is described by its peak rate only. Then $\Lambda^+(s, t) = \Lambda(s, t) + stp$ is a conservative adjustment. With the rate-function based admission method, the double optimisation has to be re-evaluated in order to update the estimate of the overflow probability:

$$-NI^+ = \sup_{t>0} \inf_{s>0} \left\{ \Lambda^+(s, t) - s(B + Ct) \right\}.$$

The decision criterion remains the same in this case.

Using the equivalent-capacity based admission criterion is more convenient. Here, the estimation of the equivalent capacity of the existing flows can be maintained in the background, i.e. the estimate of C_{equ} can be recomputed based on periodic measurements. At the arrival of a new flow, the $C_{equ} + p \leq C$ criterion has to be checked, which differs from (3.7) only in a correction term that is the peak rate of the new flow. Hence, the timing-sensitive operation (the admission decision) involves only a simple addition and a comparison, while the time-consuming double optimisation can be performed in the background, with more relaxed timing requirements.

The proposed method thus enables the deployment of the many sources asymptotics in practice not only through the reduction of its complexity, but also through shifting the computations away from the critical decision instant.

The bandwidth requirement estimates have also been investigated in the framework of many sources *exact* asymptotics, in which the overflow probability and the buffer workload loss ratio themselves (not the logarithm of them) converges to some functions. In accordance with the exact asymptotics refined estimates for the buffer overflow probability and the workload loss ratio has been considered in this context, and the bandwidth requirements have been identified, when a QoS constraint imposed on either the overflow probability or the workload loss ratio. The significance of these estimates lies in that, the advantageous properties of $\widetilde{C}_{equ}$ can be inherited in terms of quite acceptable accuracy from engineering point of view. In this case, efficient decision rules for admission control could also be designed using these improved and refined equivalent capacity formulae.

References

[1] G. Seres, Á. Szlávik, J. Zátonyi and J. Bíró, Towards efficient decision rules for admission control based on the many sources asymptotic, *Performance Evaluation*, vol. 53, nos. 3–4, pp. 145–296, August 2003.

[2] R. R. Bahadur and R. Rao, On deviations of the sample mean, *Ann. Math. Statis*, vol. 31, no. 27, pp. 1015–1027, 1960.

[3] C. Courcoubetis, V. A. Siris and G. D. Stamoulis, Application of the many sources asymptotic and effective bandwidths to traffic engineering, *Telecommunication Systems*, vol. 12, pp. 167–191, 1999.

[4] C. Courcoubetis and R. Weber, Buffer overflow asymptotics for a buffer handling many traffic sources, *Journal of Applied Probability*, vol. 33, pp. 886–903, 1996.

[5] N. G. Duffield, Economies of scale for long-range dependent traffic in short buffers, *Telecommunication Systems*, vol. 7, pp. 267–280, 1997.

[6] R. J. Gibbens and Y. C. Teh, Critical time and space scales for statistical multiplexing in multiservice networks, in *Proceedings of International Teletraffic Congress (ITC'16)*, Edinburgh, Scottland, pp. 87–96, 1999.

[7] G. Kesidis, J. Walrand and C.-S. Chang, Effective bandwidths for multiclass Markov fluids and other ATM sources, *IEEE/ACM Transactions on Networking*, vol. 1, no. 4, pp. 424–428, 1993.

[8] J. T. Lewis, R. Russell, F. Toomey, S. Crosby, I. Leslie and B. McGurk, Statistical properties of a near-optimal measurement-based CAC algorithm, in *Proceedings of IEEE ATM*, Lisbon, Portugal, pp. 103–112, June 1997.

[9] N. Likhanov and R. R. Mazumdar, Cell loss asymptotics in buffers fed with a large number of independent stationary sources, in *Proceedings of the Conference on Computer Communications (IEEE Infocom)*, San Francisco, USA, pp. 339–346, March/April 1998.

[10] M. Montgomery and G. de Veciana, On the relevance of time scales in performance oriented traffic characterizations, in *Proceedings of the Conference on Computer Communications (IEEE Infocom)*, San Francisco, USA, vol. 2, pp. 513–520, March 1996.

[11] I. Norros, A storage model with self-similar input, *Queueing Systems*, vol. 16, nos. 3/4, pp. 387–396, 1994.

[12] W. Whitt, *Stochastic-Process Limits*, Springer-Verlag, 2001.

4

Performance Evaluation of $\log_2 \sqrt{N}$ Switching Networks

Wojciech Kabaciński and Mariusz Żal

Institute of Electronics and Telecommunications, Poznań University of Technology, ul. Piotrowo 3A, 60-965 Poznań, Poland; e-mail: {kabacins, mzal}@et.put.poznan.pl

Abstract

In this paper we present a new architecture of self-routing switching networks called multi-$\log_2 \sqrt{N}$. This architecture was considered by us in several earlier papers. We proved condition for strict-sense nonblocking and rearrangeable operation of this switching network. Strict-sense nonblocking multi-$\log_2 \sqrt{N}$ switching networks for $N < 128$ require less switching elements than nonblocking multi-$\log_2 N$ switching networks proposed by Lea [1]. Rearrangeable multi-$\log_2 \sqrt{N}$ switching networks for some value of N requires less switching elements than nonblocking multi-$\log_2 N$ switching networks. In this paper we will consider a performance of the proposed architecture when a single plane is used (i.e. blocking one), and compare it with standard $\log_2 N$ switching networks of the same capacity, as well as with multi-$\log_2 N$ switching networks of the same cost.

Keywords: Banyan type switching network, self-routing switching network, packet switching, packet blocking probability.

4.1 Introduction

In future telecommunication networks switches of great capacity will be needed. This capacity refers to both the total throughput of a switch and the

D. D. Kouvatsos (ed.), Performance Modelling and Analysis of Heterogeneous Networks, 73–90.

number of ports served by the switch. When the number of ports and their bit transfer rate increase, then the less time a control unit has to determine how to route packets through a switching network of a switch. Therefore, switching networks with self-routing capability are particularly attractive candidates for packet switching in broadband networks. The most popular architecture of such self-routing switching networks is a switching network composed of $d \times d$ switches arranged in $\log_d N$ stages [2]. In most popular solutions $d = 2$. An example is a banyan network, other equivalent architectures, however, like baseline, omega, or perfect shuffle networks are also well known [3].

The main drawback of self-routing multi-stage switching networks is that not all possible permutations can be realized by them, or in other words, they are blocking. Nonblocking operation of such networks may be obtained by vertically stacking several copies of such networks. Switching networks obtained in this way are called multi-$\log_2 N$, or $\log_2(N, m, p)$ when more general architecture is considered [4] (p is the number of networks connected in parallel). Multi-$\log_2 N$ switching networks are composed of n stages, where $n = \log_2 N$, while multi-$\log_2(N, m, p)$ ones consist of $n + m$ stages. Another architecture of strict-sense nonblocking switching networks composed of 2×2 switches was proposed by Richards [5]. In general, these switching networks consist of s stages, the minimum number of elements, however, is obtained for $s = n + m$, and is the same as in $\log_2(N, m, p)$ switching networks [3].

Recently, we have proposed another self-routing architecture composed of only $s = \lceil \log_2 \sqrt{N} \rceil$ stages [6]. Similarly as banyan or baseline structures, this is a blocking one, and nonblocking operation is obtained by connecting multiple copies of such network in parallel. In the paper cited we analyzed a nonblocking operation of this new architecture. We have shown that for $N < 128$ the proposed structure requires less switching elements than nonblocking multi-$\log_2 N$ switching networks of the same capacity. In this paper we will consider the performance of the proposed architecture when a single plane is used (i.e. blocking one), and compare it with standard $\log_2 N$ switching networks of the same capacity, as well as with multi-$\log_2 N$ switching networks of the same cost.

The rest of the paper is organized as follows. In Section 4.2 the switching element and the switching network architecture proposed in [6] are given for convenience. In Section 4.3 throughput and blocking probability of the considered architecture are calculated. In Section 4.4 performance of this architecture is then compared with $\log_2 N$ and multi-$\log_2 N$ switching net-

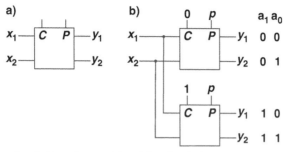

Figure 4.1 A 2 × 2 switching element (a) and 2 × 4 switch composed of 2 × 2 switches (b).

works analyzed earlier by Kruskal and Snir [7]. Finally, in Section 4.5 some conclusions are given.

4.2 Architecture of $\log_2 \sqrt{N}$ Switching Networks

4.2.1 A Basic Switching Element and Its Operation

In many switching networks a 2 × 2 switching element is used to build a switching network of greater capacity. One of popular architectures is a cross-bar structure; its capacity however, is limited due to the number of switching elements needed (N^2) and implementation complexity. In our architecture we also use such element as a basic building block [6], [8]. In some networks, called $\log_2 N$, this element poses self-routing capability, and the whole switching network is called a self-routing switching network. The self-routing capability of a 2 × 2 element means that the first bit of a transmitted packet determines to which output the packet should be directed. In our solution a switching element is modified so that two additional inputs C and P are added, and two address bits are used to control a switching element. These modifications allow connecting switches in parallel. In this way we obtain 2 × 4 switch with self-routing capability. Both architectures (2 × 2 and 2 × 4) are shown in Figure 4.1, while a switching table of a 2 × 4 switch is given in Table 4.1. Outputs are addressed by two bits, a_1a_0, where a_1 is the most significant bit. When a packet from input x_i, $i = 1, 2$, is directed to output y_j, $j = 1, 2$, the first address bit a_1 is compared with bit C. If $C = a_1$, a packet is directed through this switch to an output depending on address bit a_0 (0 to an upper and 1 to a lower output). When $a_1 \neq C$, a packet is not allowed to go through the switch. Bit P is used to give a priority to one of two inputs in case packets from two inputs are to be directed to the same output at the same time.

Table 4.1 Switching table of 2×2 switch (x denotes 'don't care' inputs).

x_1		x_2		P	y_1	y_2
a_1	a_0	a_1	a_0			
c	0	c	1	x	x_1	x_2
c	1	c	0	x	x_2	x_1
$\bar{c}$	x	c	0	x	x_2	–
$\bar{c}$	x	c	1	x	–	x_2
c	0	$\bar{c}$	x	x	x_1	–
c	1	$\bar{c}$	x	x	–	x_1
$\bar{c}$	x	$\bar{c}$	x	x	–	–
c	0	c	0	0	x_1	–
c	0	c	0	1	x_2	–
c	1	c	1	0	–	x_1
c	1	c	1	1	–	x_2

4.2.2 The $\log_2 \sqrt{N}$ Switching Network

The switching network was proposed by us in [6]; here we will give its description for convenience. The switching network is composed of switching elements presented in Figure 4.1. Let us assume that the switching network should have a capacity of $N \times N$, where $N = 2^n$. Such switching network is composed of $s = \lceil \log_2 \sqrt{N} \rceil$ stages of switches and, therefore, we called it a $\log_2 \sqrt{N}$ switching network. Inputs and outputs are numbered 0, 1, ..., $N - 1$, from top to bottom, while stages are numbered 1, 2 ..., s, from left to right. The structure of the switching network differs slightly, depending on n being odd or even.

For n odd, the first stage of the switching network is composed of 2×2 switches, while the remaining stages are composed of 2×4 switching elements, as shown in Figure 4.1b. The first stage contains $N/2$ switches. The next stage contains N 2×4 switching elements. In the following stages the number of switching elements in a given stage is two times the number of switching elements in the previous stage. In general, in stage i we have x_i 2×4 switching elements numbered from 0 to $x_i - 1$, where $x_i = 2^{n+i-2-(n \bmod 2)}$ and $2 \leq i \leq s$.

For n even, all stages of the switching network contains 2×4 switching elements. The first stage is composed of $N/2$ such elements. The next stage contains N such elements, and so forth. In this case, in stage i we have x_i 2×4 switching elements numbered from 0 to $x_i^2 - 1$, where $x_i = 2^{n+i-2-(n \bmod 2)}$ and $1 \leq i \leq s$.

In the last section there are $2^{n+s-(n \bmod 2)-2}$ 2×4 switches. All $2^{n+s-(n \bmod 2)}$ outputs of switches in section s are connected with outputs of $\log_2 \sqrt{N}$ switching network through N $2^{n-s} \times 1$ multiplexers. These multiplexers are placed after stage s.

To obtain a switching network, outputs of switches of the previous stage are to be connected with inputs of switches in the next stage. We will show connection patterns in each stage by means of permutations. Let outputs of switches in stage i and inputs of switches in stage $i+1$ be numbered $0, 1, \ldots,$ $R - 1$, from top to bottom, where $R = 2^r, r = i + n - (n \bmod 2)$. We will use binary representation of these numbers. Let $L^i_{(l,m)}$ denote an interstage link between output l in stage i and input m in stage $i + 1$, and be defined in the following way:

Definition 1. *Let $l = l_{r-1} \ldots l_2 l_1 l_0$ and $m = m_{r-1} \ldots m_2 m_1 m_0$ be binary representations of output l in stage i, and input m in stage $i + 1$, respectively. Output l is connected with input m, where $m_{r-1} \ldots m_0 = P_{(l_{r-1} \ldots l_0)}$, and*

$$
P_{(l_{r-1} \ldots l_0)} = \begin{cases} l_{r-1} \ldots l_{s-i+1} l_0 l_{s-i} l_{s-i-1} \ldots l_1 & \text{for } i = 1 \text{ and odd } n, \\ l_{r-1} \ldots l_{s-i+2} l_1 l_0 l_{s-i+1} l_{s-i} l_{s-i-1} \ldots l_2 & \text{for other } i \text{ and } n. \end{cases}
$$

$$(4.1)$$

Definition 1 describes a connection pattern between outputs in stage i and inputs in stage $i + 1$. We will now define a connection pattern between the last section of switches and multiplexers. Let outputs of switches in stage s and inputs of multiplexers be numbered $0, 1, \ldots, O - 1$, from top to bottom, where $O = 2^o$, $o = 2n - s$. We will use binary representation of these numbers. Let $L^{MUX}_{(l,m)}$ denotes an interstage link between output l in stage s and input m in stage of multiplexers, and be defined in the following way:

Definition 2. *Let $l = l_{o-1} \ldots l_2 l_1 l_0$ and $m = m_{o-1} \ldots m_2 m_1 m_0$ be binary representation of output l in stage s and input m in stage of multiplexers, respectively. Output l is connected with input m, where $m_{o-1} \ldots m_0 = Q_{(l_{o-1} \ldots l_0)}$, and*

$$
Q_{(l_{o-1} \ldots l_0)} = l_{n-1} l_{n-2} \ldots l_2 l_1 l_0 l_{o-1} \ldots l_{o-n+s}.
$$

$$(4.2)$$

In [3] permutations describing connection patterns between switches in $\log_2 N$ switching networks were defined. These permutations are as follows:

- $\delta_h(a_{n-1} \ldots a_0) = a_{n-1} \ldots a_{h+1} a_{h-1} \ldots a_0 a_h \ (0 \le h \le n-1)$,
- $\delta_h^{-1}(a_{n-1} \ldots a_0) = a_{n-1} \ldots a_{h+1} a_0 a_h \ldots a_1 \ (0 \le h \le n-1)$,
- $\beta_h(a_{n-1} \ldots a_0) = a_{n-1} \ldots a_{h+1} a_0 a_{h-1} \ldots a_1 a_h \ (0 \le h \le n-1)$,
- $j(a_{n-1} \ldots a_0) = a_{n-1} \ldots a_0$.

None of these permutations can be used in $\log_2 \sqrt{N}$ switching networks. We must define two permutations denoted by $\kappa_{w,v}$ and τ_v:

$$\kappa_{w,v}(a_{t-1} \ldots a_1 a_0)$$

$$= a_{t-1} \ldots a_{w+v} a_{v-1} a_{v-2} \ldots a_1 a_0 a_{w+v-1} a_{w+v-2} \ldots a_v$$

$$(0 \le w + v \le t - 1), \tag{4.3}$$

$$\tau_v(a_{t-1} a_{t-2} \ldots a_1 a_0) = a_{v-1} a_{v-2} \ldots a_2 a_1 a_0 a_{t-1} a_{t-2} \ldots a_{v+1} a_v$$

$$(0 \leqslant v \leqslant t - 1). \tag{4.4}$$

Using permutation $\kappa_{w,v}$ and τ_v, permutations (4.1) and (4.2) can be written as follows:

$$P_{(l_{r-1} \ldots l_0)} = \begin{cases} \kappa_{s-i,1} \text{ for } i = 1 \text{ and even } n, \\ \kappa_{s-i,2} \text{ for other } i \text{ and } n, \end{cases} \tag{4.5}$$

$$Q_{(m_{r-1} \ldots m_0)} = \tau_n. \tag{4.6}$$

The architecture of $\log_2 \sqrt{N}$ switching network is modular. Such a switching network can be obtained recursively from networks of lower capacity. Parts of smaller networks can be used to construct a switching network of greater capacity. Rules of extending $\log_2 \sqrt{N}$ switching networks were presented in [6]. Examples of 16×16 and 32×32 switching networks are shown in Figures 4.2 and 4.3, respectively.

4.3 Performance Evaluation

4.3.1 Preliminaries

Let $p(1, i)$ denote the probability that a packet will appear at an output of a switch in stage i. In similar way, $p(0, i)$ denotes the probability that an output of a switch in stage i is not requested by any packet.

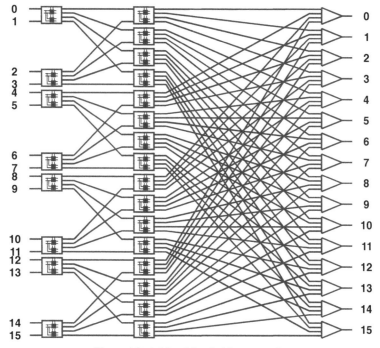

Figure 4.2 A 16×16 switching network.

The probability that a packet will be at an input of a switching network or at an output of the last section of switches will be denoted by $p(1,0)$ and $p(1,s)$, respectively. We assume that a switching network is synchronous, i.e. packets are placed in time slots. Event of a packet appearing in any input of a switching network is independent of any other input. When two packets are to be sent to the same output of a switch, only one packet is sent, and the second packet is lost.

We will consider two types of traffic: uniform and permutation. In uniform traffic, an output to which packet is directed is a random variable with a uniform distribution. In this case, it is possible that two or more packets want to reach the same output. In permutation traffic, not more than one packet may be directed to any output. For performance evaluation of $\log_2 \sqrt{N}$ switching networks we will use terminology used for instance in [3, 9, 10].

Throughput of a switching network is the normalized amount of traffic carried by the switch expressed as the utilization factor of its input links. It is defined as the probability that a packet received on an input link will appear

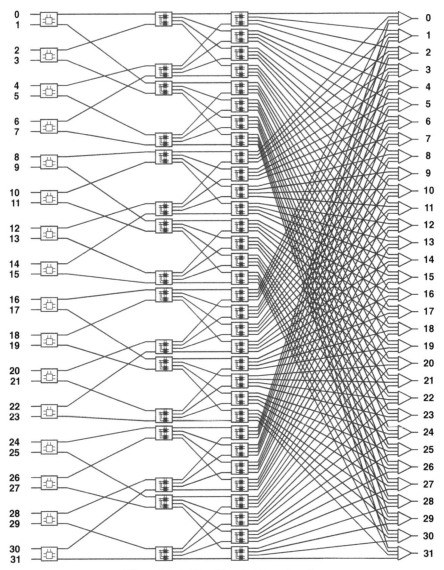

Figure 4.3 A 32 × 32 switching network.

at an addressed switch output and is denoted by ρ [3]:

$$\rho = p(1, n). \tag{4.7}$$

Throughput of a switching network depends on probability $p(1, 0)$ that a packet will be received on an input link.

Blocking probability is defined as the probability that a packet received on an input link will be lost and will not appear at the addressed switching network output. It is denoted by π and is given by

$$\pi = 1 - \frac{\rho}{p(1, 0)}. \tag{4.8}$$

4.3.2 Uniform Traffic – Model 1

The probability that a packet will appear at an output of stage $i + 1$, composed of $k \times k$ switches, was given in [7]. Since $\log_2 \sqrt{N}$ switching networks are composed of 2×2 and 2×4 switches, we extend this probability to $k \times l$ switches. Let us consider the first stage switch. Let $p(1, 0)$ be the probability that a packet arrives at an input of a switch. The probability that a packet will appear at a given output of the switch is $p(1, 0)/l$. The probability that a packet will not appear at a given output is $1 - (p(1, 0)/l)$. Since the probability of directing packet to an output is equal for all outputs, the probability that there will be no packets at a given output is given by

$$p(0, 1) = \left(1 - \frac{p(1, 0)}{l}\right)^k. \tag{4.9}$$

Finally, the probability that there will be at least one packet directed to a given output of a switch is given by

$$p(1, 1) = 1 - \left(1 - \frac{p(1, 0)}{l}\right)^k. \tag{4.10}$$

Since events in next stages are independent, equation (4.10) can be generalized for any stage of a switching network.

$$p(1, i + 1) = 1 - \left(1 - \frac{p(1, i)}{l}\right)^k. \tag{4.11}$$

Probability $p(1, s)$ determines the probability that a packet will arrive at an output of a switch in stage s. Outputs of stage s are connected with switching network outputs through $2^{n-s} \times 1$ multiplexers. Therefore, the probability that a packet will arrive at an output of a $\log_2 \sqrt{N}$ switching network is

$$\rho = 1 - (1 - p(1, s))^{2^{n-s}}. \tag{4.12}$$

4.3.3 Uniform Traffic – Model 2

The model presented in the previous section is a simple analytical model. This model may be used to calculate quantitative measures of a network performance. But to calculate the distribution function of blocking probability or throughput another analytical model will be used. Because the considered switching network is a synchronous one without buffers, switching is a memoryless process (events in time $t(i)$ do not influence events in time $t(i + 1)$) and Bernoulli process may be used as approximated model of probability that a packets will appear at switch inputs.

Let $p(1, 0)$ be the probability that a packet arrives at an input of $k \times l$ switch. The probability that a packets arrive at k inputs of this switch in one time slot is binomially distributed:

$$Pr[A = a] = \binom{k}{a} p(1, 0)^k (1 - p(1, 0))^{k-a}. \tag{4.13}$$

Like in the previous section, the calculated probability that a packet arrives at a particular output of a switch in the first stage will be generalized for stage i of the switching network, where $1 \le i \le s$.

In uniform traffic, the packet which appears at an input of the switch is directed to an output which is a random variable with uniform distribution. Because the number of outputs of $\log_2 \sqrt{N}$ switching network accessible from an input of $k \times l$ switch is equal for all outputs, the packet selects one of l outputs with probability $1/l$. Suppose that a requests arrive at inputs of a switch. The probability that b of these packets will request one of l outputs is equal $(1/l)^b$. The remaining $a - b$ packets select other outputs with probability $(1 - (1/l))^{a-b}$. The number of ways of selecting b packets from a packets arriving at inputs of a switch is a choose b. Hence, the probability that one or more packets select the same output, given that a packets arrive at inputs of switch in one time slot, is given by

$$Pr[1 \le B \le a] = \sum_{b=1}^{a} \left[\binom{a}{b} \left(\frac{1}{l}\right)^b \left(1 - \frac{1}{l}\right)^{a-b} \right]. \tag{4.14}$$

The probability that a packet arrives at an output of a switch in the first stage, is given by

$$p(1, 1) = \sum_{a=1}^{k} \left\{ \binom{k}{a} p(1, 0)^a (1 - p(1, 0))^{k-a} \right.$$

$$\left. \times \sum_{b=1}^{a} \left[\binom{a}{b} \left(\frac{1}{l}\right)^b \left(1 - \frac{1}{l}\right)^{a-b} \right] \right\}. \tag{4.15}$$

Since events in next stages are independent, equation (4.15) can be generalized for any stage of a switching network. We have

$$p(1, i + 1) = \sum_{a=1}^{k} \left\{ \binom{k}{a} p(1, i)^a (1 - p(1, i))^{k-a} \right.$$

$$\left. \times \sum_{b=1}^{a} \left[\binom{a}{b} \left(\frac{1}{l}\right)^b \left(1 - \frac{1}{l}\right)^{a-b} \right] \right\}. \tag{4.16}$$

Probability $p(1, s)$ determines the probability that a packet will arrive at the output of a switch in stage s. Outputs of stage s are connected with switching network outputs through $2^{n-s} \times 1$ multiplexers. From (4.16), for $k = 2^{n-s}, l = 1$ and $i = s$ we get

$$\rho = p(1, s + 1) = 1 - (1 - p(1, s))^{2^{n-s}}. \tag{4.17}$$

4.3.4 Permutation Traffic

Let us consider $k \times l$ switch. The number of outputs of $\log_2 \sqrt{N}$ switching network accessible from an output of $k \times l$ switch in stage i is equal l^{s-i}. From an input of $k \times l$ switch we have access to l^{s-i+1} outputs of $\log_2 \sqrt{N}$ switching network. Let us define function $p\_w(i, d, e, l)$ as the probability that the packet, which arrived at an input of a switch in stage i will be directed to a particular output of this switch, given that d packets have already selected that output, and that e packets have already selected other outputs. This probability is given by

$$p\_w(i, d, e, l) = \begin{cases} \frac{k^{s-i} - d}{k^{s-i+1} - d - e} & \text{for } d \leq k^{s-i}, d + e < k^{s-i+1}, \\ 0 & \text{in other cases.} \end{cases} \tag{4.18}$$

In a similar manner, let us define function $p\_nw(i, d, e, l)$ as the probability that a packet, which arrived at an input of a switch in stage i will not be directed to a particular output of this switch, given that d packets have already selected that output, and that e packets have already selected other outputs. This probability is given by

$$p\_nw(i, d, e, l) = \begin{cases} \frac{(k-1)k^{s-i}-e}{k^{s-i+1}-d-e} & \text{for } e < (k-1)k^{s-i}, d+e < k^{s-i+1}, \\ 0 & \text{in other cases.} \end{cases}$$

(4.19)

Suppose that a requests arrive at inputs of a switch. The probability that a packets arrive at k inputs of this switch in one time slot is given by (4.13). The probability that b of these packets select a particular output is given by

$$\prod_{d=0}^{b-1} p\_w(i, d, e, l).$$

The probability that $a - b$ packets do not select that particular output, given that b packets have already selected it, is given by

$$\prod_{e=0}^{a-b-1} p\_nw(i, d, e, l).$$

The number of ways of selecting b packets from a packets arriving at inputs of a switch is a choose b. Finally, the probability that there will be at least one packet directed to a given output of a switch is given by

$$Pr[1 \leqslant B \leqslant a] = \sum_{b=1}^{a} \left[\binom{a}{b} \prod_{d=0}^{b-1} p\_w(i, d, 0, l) \prod_{e=0}^{a-b-1} p\_nw(i, b, e, l) \right].$$

(4.20)

The probability that a packet arrives at an output of a switch in the first stage is given by

$$p(1, 1) = \sum_{a=1}^{k} \left\{ \binom{k}{a} p(1, 0)^a (1 - p(1, 0))^{k-a} \right.$$

$$\times \left. \sum_{b=1}^{a} \left[\binom{a}{b} \prod_{d=0}^{b-1} p\_w(1, d, 0, l) \prod_{e=0}^{a-b-1} p\_nw(1, b, e, l) \right] \right\}. \quad (4.21)$$

Since events in next stages are independent, equation (4.22) can be generalized for any stage of a switching network. We have

$$p(1, i+1) = \sum_{a=1}^{k} \left\{ \binom{k}{a} p(1, i)^a (1 - p(1, i))^{k-a} \right.$$

$$\left. \times \sum_{b=1}^{a} \left[\binom{a}{b} \prod_{d=0}^{b-1} p\_w(i+1, d, 0, l) \prod_{e=0}^{a-b-1} p\_nw(i+1, b, e, l) \right] \right\}.$$

$$(4.22)$$

Probability $p(1, s)$ determines the probability that a packet will arrive at an output of a switch in stage s. Outputs of stage s are connected with switching network outputs through $2^{n-s} \times 1$ multiplexers. Therefore, the probability that a packet will arrive at the output of a $\log_2 \sqrt{N}$ switching network is

$$\rho = 1 - (1 - p(1, s))^{2^{n-s}}. \qquad (4.23)$$

4.3.5 Numerical Results

Throughputs of crossbar, $\log_2 N$ and $\log_2 \sqrt{N}$ switching networks under uniform traffic are compared in Figure 4.4. It can be seen that $\log_2 \sqrt{N}$ switching networks always perform better than $\log_2 N$ switching networks, and that crossbar performs better than $\log_2 \sqrt{N}$, except for $N = 4$. In case of $N = 4$ $\log_2 \sqrt{N}$ switching network consists of two 2×4 switches arranged in one stage. This switching network has the same properties as the crossbar switching network. Differences between values computed from (4.11) and (4.16) are very small.

In Figure 4.5 a throughput calculated by an analytical model is compared with simulation results. The results of the simulation are shown in the charts in the form of marks with 99.5 confidence intervals, that have been calculated after the t-Student distribution for the ten series with $p(1, 0) = 1.0$ and $20000 \times N$ time slots. Figure 4.6 compares throughputs of crossbar, $\log_2 N$ and $\log_2 \sqrt{N}$ switching networks under permutation traffic.

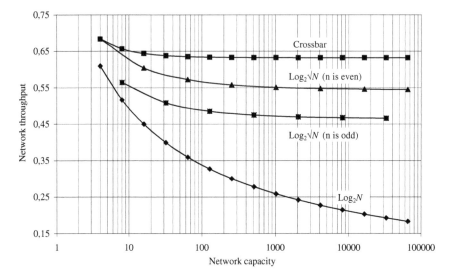

Figure 4.4 Throughputs of crossbar, $\log_2 N$ and $\log_2 \sqrt{N}$ under uniform traffic.

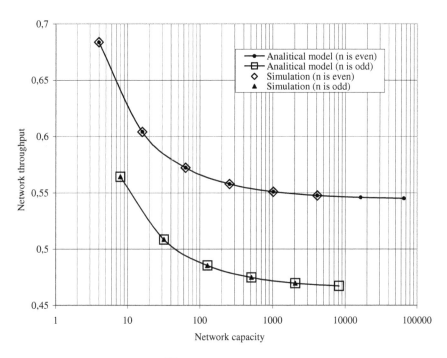

Figure 4.5 Throughputs of $\log_2 \sqrt{N}$ switching network for simulation and analytical model.

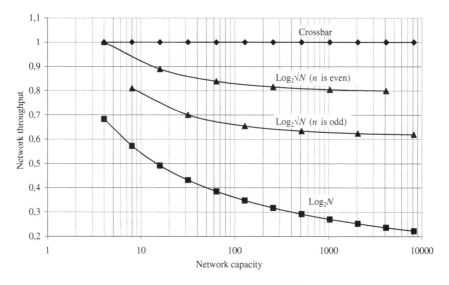

Figure 4.6 Throughputs of crossbar, $\log_2 N$ and $\log_2 \sqrt{N}$ under permutation traffic.

4.4 Cost and Performance Comparison

In the previous section performance evaluation for $\log_2 \sqrt{N}$ and $\log_2 N$ switching networks composed of one plane was done. In this section switching networks composed of similar number of switches will be considered. To compare blocking probability, we will used one plane of $\log_2 \sqrt{N}$ switching network and p copies of $\log_2 N$ switching network called planes. $\text{Log}_2 N$ switching networks composed of p planes are called multi-$\log_2 N$ switching networks, and may be obtained by stacking p planes vertically.

Let $C_{\log_2 N}$ denote the number of 2×2 switches in one plane of $\log_2 N$ switching networks. This number is given by

$$C_{\log_2 N} = n \cdot \frac{N}{2}. \tag{4.24}$$

Let $C_{\log_2 \sqrt{N}}$ denote the number of 2×2 switches in one plane of $\log_2 \sqrt{N}$ switching networks. A cost function, where the cost is given by the number of 2×2 switches in a switching network, was derived in [11], and is given by

$$C_{\log_2 \sqrt{N}} = \begin{cases} 1/2N(2^s - 1) = 1/2N(\sqrt{N} - 1) & \text{for odd } n, \\ N(2^s - 1) = N(\sqrt{N} - 1) & \text{for even } n. \end{cases} \tag{4.25}$$

Because the number $C_{\log_2 N}$ is not a multiple of the number $C_{\log_2 \sqrt{N}}$, we will consider multi-$\log_2 N$ switching net-works with total number of switches not smaller than the number of switches in one plane of a $\log_2 \sqrt{N}$ switching network of the same capacity. The number of planes is given by

$$p = \left\lceil \frac{C_{\log_2 \sqrt{N}}}{C_{\log_2 N}} \right\rceil, \tag{4.26}$$

where

$$\lceil x \rceil = \min\{y \mid y \geqslant x, \ \text{integer } y\}.$$

Kruskal and Snir [7] analyzed multi-$\log_2 N$ switching networks under uniform traffic. Let $p'(1, 0)$ be probability that a packet arrives at an input of multi-$\log_2 N$ switching network. This packet is randomly sent to one of the p planes. Events occurring at inputs of each plane are statistically dependent; if a request is sent to one particular plane, then it will not be sent to any other copy. The probability that a packet arrives at an input of one particular plane is

$$p(1, 0) = p'(1, 0)/p.$$

The probability that a packet will arrive at an output of a switch in the last stage of the plane may be computed from

$$p(1, i + 1) = 1 - \left(1 - \frac{p(1, i)}{2}\right)^2. \tag{4.27}$$

The probability that an output of a switch in the last stage of $\log_2 N$ is not requested by any packet is given by:

$$p(0, n) = 1 - p(1, n).$$

Finally, the blocking probability in multi-$\log_2 N$ composed of p copies of $\log_2 N$ switching network is given by

$$\pi = 1 - \frac{1 - p(0, n)^p}{p'(1, 0)}. \tag{4.28}$$

Figure 4.7 compares the blocking probability of $\log_2 N$ and $\log_2 \sqrt{N}$ switching networks under permutation traffic. Total number of switches in multi-$\log_2 N$ switching networks is equal or greater than total number of switches in $\log_2 \sqrt{N}$ switching networks. It can be seen that the blocking probability of multi-$\log_2 N$ switching network for practical capacity is greater

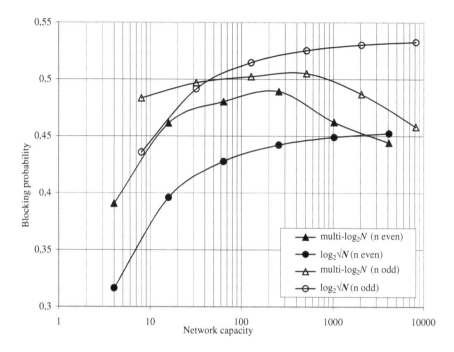

Figure 4.7 The blocking probability of $\log_2\sqrt{N}$ and multi-$\log_2 N$ composed of similar number of switches.

than the blocking probability of $\log_2\sqrt{N}$ switching networks. The results shown in Figure 4.7 remains similar also when cost of mux (demux) are taken into account in both kinds of switching networks.

4.5 Conclusions

In this paper we proposed a new architecture of high-speed self-routing switching networks composed of 2×2 switches. This architecture requires only $\log_2\sqrt{N}$ stages. We have presented analytic models for the blocking probability of $\log_2\sqrt{N}$ switching networks under assumption of uniform and permutation traffic. We have shown that $\log_2\sqrt{N}$ switching networks always have lower blocking probability than $\log_2 N$ switching networks. We also compared $\log_2\sqrt{N}$ switching networks and multi-$\log_2 N$ switching networks, with total number of switches not smaller than the number of switches in one plane of $\log_2\sqrt{N}$ switching network of the same capacity. The blocking probability of $\log_2\sqrt{N}$ switching networks for some values of N is smaller than the blocking probability of multi-$\log_2 N$ switching networks. It is because $\log_2 \sqrt{N}$ switching networks have lower internal blocking probability for smaller N. When N grows the cost of $\log_2 \sqrt{N}$ switching fabrics grows quicker than $\log_2 N$ switching fabric. Therefore, for greater N, more planes can be used in multi-$\log_2 N$ switching fabrics of the some cost as $\log_2 \sqrt{N}$ switching fabrics. More planes results in lower internal blocking probability in multi-$\log_2 N$ switching fabrics.

References

[1] C.-T. Lea, Multi-$\log_2 N$ networks and their applications in high-speed electronic and photonic switching systems, *IEEE Transactions on Communications*, vol. 38, no. 10, pp. 1740–1749, 1990.

[2] C.-T. Lea, Multirate $\log_d(N, e, p)$ strictly nonblocking networks, in *Proceedings of IEEE Globecom'94*, San Francisco, CA, pp. 319–323, 1994.

[3] A. Pattavina, *Switching Theory – Architectures and Performance in Broadband ATM Networks*, John Wiley & Sons, England, 1998.

[4] D.-J. Shyy and C.-T. Lea, $\text{Log}_2(N, m, p)$ strictly nonblocking networks, *IEEE Transactions on Communications*, vol. 39, no. 10, pp. 1502–1510, 1991.

[5] G. W. Richards, Theoretical aspects of multi-stage networks for broadband networks, Tutorial presentation at INFOCOM 93, San Francisco, April–May 1993.

[6] W. Kabaciński and M. Żal, Multi-$\log_2 \sqrt{N}$ switching networks for high-speed switching, in *Proceedings of IEEE International Conference on Communications (IEEE ICC 2002)*, New York, vol. 4, pp. 2174–2178, 2002.

[7] C. P. Kruskal and M. Snir, The performance of multistage interconnection networks for multiprocessors, *IEEE Transactions on Computers*, vol. 32, no. 12, pp. 1091–1098, 1983.

[8] W. Kabaciński and M. Żal, A new control algorithm for rearrangeable multi-$\log_2 N$ switching networks, in *Proceedings of International Conference on Telecommunications (ICT 2001)*, Bucharest, Romania, vol. 1, pp. 501–506, 2001.

[9] J. H. Patel, Performance of processor-memory interconnection for multiprocessors, *IEEE Transactions on Computers*, vol. 30, no. 10, pp. 771–780, 1981.

[10] T. H. Szymanski and V. C. Hamacher, On the permutation capability of multistage interconnection networks, *IEEE Transactions on Computers*, vol. 36, no. 7, pp. 810–822, 1987.

[11] W. Kabaciński and M. Żal, Rearrangeable multi-$\log_2 \sqrt{N}$ switching networks, in *Proceedings of International Conference on Telecommunications (ICT 2002)*, Beijing, China, pp. 1002–1006, 2002.

PART TWO
WIRELESS AD HOC NETWORKS

5

Database Exchanges for Ad-hoc Networks Using Proactive Link State Protocols

E. Baccelli[1], P. Jacquet[1] and T. Clausen[2]

[1]*Project Hipercom, INRIA Rocquencourt, BP 105, 78153 Le Chesnay Cedex, France; e-mail: {emmanuel.baccelli, philippe.jacquet}@inria.fr*
[2]*LIX, Ecole Polytechnique, 91128 Palaiseau Cedex, France; e-mail: thomas@thomasclausen.org*

Abstract

The OSPF routing protocol is currently the predominant IGP in use on the fixed Internet of today. This routing protocol scales "world wide", under the assumptions of links being relatively stable, network density being rather low (relatively few adjacencies per router) and mobility being present at the edges of the networks only. Recently, work has begun towards extending the domain of OSPF to also include ad-hoc networks – i.e. dense networks, in which links are short-lived and most nodes are mobile.

In this paper, we focus on the convergence of the Internet and ad-hoc networks, through extensions to the OSPF routing protocol. Based on WOSPF, a merger of the ad-hoc routing protocol OLSR and OSPF, we examine the feature of OSPF database exchange and reliable synchronisation in the context of ad-hoc networking. We find that the mechanisms, in the form present in OSPF, are not suitable for the ad-hoc domain. We propose an alternative mechanism for link-state database exchanges in wireless ad-hoc networks, aiming at furthering an adaptation of OSPF to be useful also on ad-hoc networks, and evaluate our alternative against the mechanism found in OSPF.

Our proposed mechanism is specified with the following applications in mind: (i) Reliable diffusion of link-state information replacing OSPF ac-

D. D. Kouvatsos (ed.), Performance Modelling and Analysis of Heterogeneous Networks, 93–111.

knowledgements with a mechanism suitable for mobile wireless networks; (ii) Reduced overhead for performing OSPF style database exchanges in a mobile wireless network; (iii) Reduced initialisation time when new nodes are emerging in the network; (iv) Reduced overhead and reduced convergence time when several wireless OSPF ad hoc network clouds merge.

Keywords: OSPF, OLSR, ad-hoc networking, routing, database exchange.

5.1 Introduction

Wireless ad-hoc networks are characterised by being networks of autonomous and mobile nodes, communicating over a wireless medium whereby they form an arbitrary, dynamic and random graph of wireless links. When the network size grows to the point where direct links do no longer exist between all node-pairs, ensuring connectivity in such a network becomes the task of *routing*. Routing in wireless ad-hoc networks brings a host of challenges not present in traditional wired networks, including "hidden nodes", low and commonly shared bandwidth, limited resources in the nodes (processing and battery-power), a high degree of network dynamics, etc.

The possible use of OSPF [2] as a routing protocol in such wireless ad hoc networks has lately been the subject of several different efforts. OLSR [4], a link-state protocol developed within the IETF specifically for routing in wireless ad-hoc networks, is in its essential functioning very close to that of OSPF, yet is without several key features of OSPF – notably the ability to perform routing in a heterogeneous environment such as wired and wireless ad-hoc routing.

There is indeed a need for a generic wired/wireless IP routing solution. Due to its widespread use on wired networks, as well as its likeness to OLSR, OSPF seems like a designated candidate. However, OSPF in its basic form is not at all tailored for mobile wireless environments and features several problems when run in these [6, 7].

A solution for making OSPF operate efficiently on wireless ad hoc networks is Wireless-OSPF (WOSPF), proposed in [1], where a new type of OSPF interface is specifically defined for manet interfaces. This interface type operates through employing the ad-hoc network specific optimisations of OLSR (i.e. periodic unreliable message transmission and optimised flooding through multipoint relays [8]) while maintaining OSPF messages (i.e. Link-State Advertisements, LSA) for diffusing topological information. However, Ahrenholz et al. [1] propose only a partial adaptation of OSPF for wireless ad-hoc networks: adjacencies are not formed on wireless ad-hoc network interfaces, which implies that the usual OSPF database exchange and reliable synchronisation mechanisms are not in action on these interfaces.

The idea behind the periodic unreliable flooding of topology information is, that since the topology of the network is thought to be changing frequently, LSAs (in OSPF) and TCs (similar messages in OLSR, called Topology Control messages) are transmitted periodically and frequently to reflect these changes. Consequently, loss of a single LSA or TC is relatively unimportant since the information contained within the message will be repeated shortly. This approach may not work well if LSA or TC periods are not roughly homogeneous and short: in a heterogeneous wired/wireless network, the LSAs generated by usual wired nodes running OSPF will have long periods (up to 1 hour) while LSAs generated by wireless nodes (running WOSPF) will typically have a period of (often much) less than a minute. In this case, of course, the short period argument fails, at least for the LSAs with a long period, and there is a definite need for a mechanism to device mechanisms for conducting the usual OSPF database exchange and reliable synchronisation in a wireless ad-hoc network.

In this paper we propose a mechanism, adapted for the low-bandwidth high-dynamics conditions of wireless ad-hoc networks, for conducting efficient database synchronisation in WOSPF. We qualify the performance of the proposed mechanism and compared to the performance of the original mechanism of OSPF. We furthermore discuss a selection of applicability scenarios for the mechanism, including reliable diffusion of link-state information through, reduced overhead for performing OSPF-style database exchanges in a wireless ad-hoc network, reduced initialisation time when new nodes are emerging in the network and reduced overhead and reduced convergence time when several network clouds merge.

5.1.1 Outline of Paper

The remainder of this paper is organised as follows: in Section 5.2, a brief description of the usual OSPF database exchange and reliable synchronisation mechanisms is reviewed, and briefly discussed in the context of wireless ad-hoc networks.

Section 5.3 describes a mechanism for conducting database exchange and reliable synchronisation, specifically adapted to and described in the context of WOSPF [1]. This mechanism respects the fact that not all LSAs carry information which is long-lived enough to justify the efforts of maintaining consistency, while it still provides an efficient mechanism for allowing nodes to maintain consistency when needed.

Section 5.4 evaluates the performance of the proposed signature exchange mechanism in comparison with the performance of the native OSPF signature exchange mechanism. The following section, Section 5.5, discusses the applicability of the proposed database exchange mechanism, and Section 5.6 concludes this paper.

5.2 Database Exchange in OSPF

The objective of the OSPF routing protocol is to provide, in each node, sufficient topological information about the network to be able to compute (using some metric) a suitable path between any source and destination in the network.

OSPF [2] employs two independent mechanisms for maintaining globally consistent topology information in the node: (i) reliable transport of LSA messages and (ii) database exchanges between pairs of routers.

5.2.1 Reliable Transmission

OSPF employs positive acknowledgements (ACK) on delivery with retransmissions, i.e. an ACK is a retransmission repressing message. In mostly static point-to-point-like network topologies (e.g. fixed wired networks), ACKs and retransmissions occur over a single link in the network. More importantly, an ACK transmitted by the recipient of an LSA message will be received by a node which is directly able to interpret the ACK message, i.e., the recipient of an ACK will be the node which sent the LSA to which the ACK corresponds.

In wireless ad-hoc networks, the network topology may be assumed to be changing frequently (node mobility). Interfaces are typically wireless (hence subject to fading), of broadcast nature. Any transmission may thus interfere with all the neighbours of the node originating the transmission. An ACK, which can be interpreted by the node which relayed the to the ACK corresponding LSA, will thus be interfering with all the nodes in the neighbourhood. If, due to node mobility or fading radio links, a node does not receive an expected ACK, unnecessary retransmissions will occur, consuming precious bandwidth. In other words, reliable topology information diffusion through ACKs imposes the assumption that the network conditions are such that an ACK that is sent can be received by the intended node. This does not hold for a wireless ad-hoc network, where the network may be substantially more dynamic: nodes may move out of range, etc.

5.2.2 Database Exchange

OSPF database exchanges are intended to synchronise the link-state database between routers. In OSPF, database description packets are exchanged between two nodes through one node (the master) polling an other node (the slave). Both polls and responses have the form of database description packets containing a set of complete LSA headers, describing (a partial set of) the respective link-state databases of each of the two nodes. These database description packets are used by the nodes to compare their link-state databases. If any of the two nodes involved in the exchange detects it has out-of-date or missing information, it issues link-state request packets to request the pieces of information from the other node, which would update its link-state database.

In the context of wireless ad-hoc networks, wireless broadcast interfaces and a higher degree of node mobility are typically assumed. Therefore, inconsistencies between the link-state databases of the nodes in the network may occur more frequently, calling for more frequent database exchanges. Moreover, the broadcast nature of the network interfaces implies that the bandwidth in a region is shared among the nodes in that region and thus less bandwidth is available between any pair of nodes to conduct the database exchange.

5.3 Database Signature Exchange

In this section, we propose a mechanism for database exchange and reliable synchronisation, adapted to wireless ad-hoc networks. Specifically, we propose an extension to WOSPF.

The basic idea is to employ an exchange of compact "signatures" (hashing of the link state database) between neighbour nodes, in order to detect differences in the nodes' link state databases. When a discrepancy is detected, the bits of information required to synchronise the link state databases of the involved nodes are then identified and exchanged. The purpose of the exchange is to provide the nodes with a consistent view of the network topology – the task is doing so in an efficient way.

Our approach is somewhat inspired by IS-IS [3], in which packets which list the most recent sequence number of one or more LSAs (Sequence

Numbers packets) are used to ensure that neighbouring nodes agree on the most recent link state information. This means that, rather than transmitting complete LSA headers (as in OSPF), a more compact representation for database description messages is employed. Also, Sequence Numbers packets accomplish a function, similar to conventional acknowledgement packets.

The method proposed in this paper differs from the mechanism employed in IS-IS by the use of age. For example, it may be considered a waste of resources to check for databases consistency for LSAs issued from within a very dynamic part of a wireless ad-hoc network (e.g. RFID tags on products in a plant): LSAs from nodes within this domain should transmitted frequently and periodically, thus information describing these nodes is frequently updated and "with a small age". LSAs from a less mobile part of the wireless ad-hoc network (e.g. sensors on semi-permanent installations in the plant) might be updated less frequently. Thus consistency of the corresponding entries in the link-state databases should be ensured.

The following subsections outline how database signatures are generated, exchanged, interpreted and used for correcting discrepancies.

5.3.1 Definition of Link State Database Signatures

We define a signature message as a tuple of the following form:

$$\text{Signature Message} = (\text{Age Interval, Key, Prefix Signature}),$$

A signature features a set of prefix signatures:

$$\text{Prefix Signature} = (\text{Prefix, Sign(Prefix)}).$$

Each Sign(Prefix) results from hashing functions computed on the piece of the link state database matching the specified prefix, and represents this part of the database in the signature message.

More specifically, each Sign(Prefix) has the following structure:

$$\text{Sign(Prefix)} = (\text{Primary Partial Signature, Secondary Partial Signature,}$$
$$\text{Timed Partial Signature, \#LSA, Timed \#LSA}).$$

A primary partial signature (PPS) for a prefix is computed as a sum over all LSAs in a nodes link-state database, where the prefix matches the advertising router of the LSA:

$$PPS = \sum_{\text{prefixes}} (\text{Hash(LSA-identifier)}),$$

$\sum_{\text{prefixes}}$ denotes the sum over prefixes matching the advertising router of the LSA. The secondary partial signature (SPS) for a prefix is computed as a sum over all LSAs in a nodes link-state database, where the prefix matches the advertising router of the LSA:

$$SPS = \sum_{\text{prefixes}} (\text{Hash(LSA-identifier)}) \cdot \text{key},$$

$\sum_{\text{prefixes}}$ denotes the sum over prefixes matching the advertising router of the LSA. The timed partial signature (or TPS) for a prefix and an age interval is computed over LSAs in a nodes link-state database where:

- the prefix matches the advertising router of the LSA,
- the age falls within the age interval of the advertisement,

and has the following expression:

$$TPS = \sum_{\text{prefixes,time}} (\text{Hash(LSA-identifier)}),$$

with $\sum_{\text{prefixes,time}}$ denoting the sum over prefixes matching the advertising router of the LSA and where the age falls within the age interval of the advertisement. The LSA identifier is the string, obtained through concatenating the following LSA header fields:

- LS type,
- LS ID,
- Advertising router,
- LSA sequence number.

5.3.2 Signature Exchange

Signatures are exchanged between nodes in two forms: informational signatures, broadcast periodically to all neighbour nodes, and database exchange signatures, employed when a node requests a database exchange with one of its neighbours.

Informational Signature Exchange. Each node periodically broadcasts informational (info) signatures, as well as receives signatures from its neighbour nodes. This exchange allows nodes to detect any discrepancies between their respective link-state databases. Section 5.3.3 details how info signatures are generated; Section 5.3.4 details how signatures are employed to detect link-state database discrepancies.

Database Signature Exchange. Database exchange (dbx) signatures are directed towards a single neighbour only. The purpose of emitting a dbx signature is to initiate an exchange of database information with a specific neighbour node.

When a node detects a discrepancy between its own link-state database and the link-state database of one of its neighbours, a database exchange is desired. The node, detecting the discrepancy, generates a dbx signature, requesting a database exchange to take place. In OSPF terms, the node requesting the database exchange is the "master" while the node selected for receiving the dbx signature is the "slave" of that exchange. The dbx signature is transmitted with the destination address of one node among the discrepant neighbours. The node builds a dbx message signature, based on the information acquired from the info signature exchange.

5.3.3 Signature Message Generation

This section details how info and dbx signature messages are generated.

Info Signature Generation. An info signature message describes the complete link state database of the node that sends it. Absence of information in a signature indicates absence of information in the sending nodes link state database – in other words, if no information is given within an informational signature about a specific prefix, it is implicitly to be understood that the sending node has received no LSAs corresponding to that prefix.

The set of prefix signatures in an informative signature message can be generated with the following splitting algorithm, where the length L of the info signature (the number of prefix signatures in the message) can be chosen at will.

We define the weight of a given prefix as the function:

Weight(prefix) = # of LSAs whose originator matches the prefix.

And similarly, the timed weight as the function:

Timed Weight(prefix) = # of LSAs whose originator matches the prefix
and whose age falls inside the age interval.

Then, starting with the set of prefix signatures equal to $(0, \text{signature}(0))$, recursively do the following.

As long as:
$$|\text{set of prefix signatures}| < L$$

1. Find in the set of prefix signatures the prefix with largest timed weight, let it be called mprefix.
2. Replace the single (mprefix, signature(mprefix)) by the pair (mprefix0, signature(mprefix0)), (mprefix1, signature(mprefix1)).
3. If one of the expanded prefix of mprefix has weight equal to 0, then remove the corresponding tuple.

Dbx Signature Generation.   Dbx signatures serve to trigger an exchange of discrepant LSAs with one neighbour, known to have more up-to-date link-state information – the ideal is to pick the neighbour which has the "most complete" link-state database and which at the same time is going to remain a neighbour for a sufficient period of time. In WOSPF, database exchanges are to be conducted in preference with nodes selected as MPR.

The set of prefix signatures in a database exchange signature message can be generated with the following algorithm, where the length L of the dbx signature (the number of prefix signatures in the message) can be chosen at will.

1. Start with the same set of prefix signatures as one of the received info signature where the discrepancies were noticed.
2. Remove from that set all the prefix signatures such that signature(prefix) is not discrepant (with the LSA database). Use the same age interval and key used in the received info signature. Then use the recursive algorithm described above for info signatures, skipping step 3.

Indeed, contrary to info signature messages, the prefixes with zero weight are not removed here, since the signature is not complete, i.e. the signature might not describe the whole database. Therefore a prefix with empty weight may be an indication of missing LSAs.

5.3.4 Checking Signatures

Upon receiving a signature message from a neighbour, a node can check its local LSA database and determine if it differs with the neighbour's database. For this purpose, it computes its own prefix signatures locally using the same prefixes, time interval and key specified in the received signature message. A prefix signature differs with the local prefix signature when any of the following conditions occurs:

1. both the number of LSAs and the timed number of LSAs differ;
2. both the timed partial signatures and the (primary partial signature, secondary partial signature) tuples differ.

The use of a secondary signature based on a random key is a way to cope with the infrequent, but still possible, situations when the primary signatures agree although the databases differ. In this case, it can be assumed that using a random key renders the probability that both primary and secondary signatures agree while databases are different, to be very small.

5.3.5 Database Exchange

When a node receives a dbx signature with its own ID in the destination field, the node has been identified as the slave for a database exchange. The task is, then, to ensure that information is exchanged to remove the discrepancies between the link-state databases of the master and the slave.

Thus, the slave must identify which LSA messages it must retransmit, in order to bring the information in the master up-to-date. The slave must then proceed to rebroadcast those LSA messages.

More precisely, the slave rebroadcasts the LSA messages which match the following criteria:

- the age belongs to the age interval indicated in the dbx signature, AND
- the prefix corresponds to a signed prefix in the dbx signature, where the signature generated by the master differs from the signature as calculated within the slave for the same segment of the link-state database.

When a node is triggered to perform a database exchange it generates a new LSF with TTL equal to 1 (one hop only) and fills it with the update LSAs. These LSAs must indicate the age featured at the moment in the database, from which they are taken.

Optionally, the host can use a new type of LSF (denoted an LSF-D) which, contrary to the one hop LSF described above, is retransmitted as a normal LSF making use of MPRs. An LSF-D is transmitted with TTL equal to infinity. Upon receiving of such a packet, successive nodes remove from the LSF-D the LSAs already present in their database before retransmitting the LSF-D. If the LSF-D is empty after such a processing, a node will simply not retransmit the LSF-D. The use of LSF-D packets is more efficient for fast wide-area database updates in case of merging of two independent wireless networks.

5.4 Performance Evaluation

In this section we compare the performance of database signature exchange protocol with the full database exchange of OSPF. In this first analysis we consider the "cost" of the protocol when two databases differ on a single record. While this is a special case, it gives a good idea of what kind of performance gains we can obtain.

We denote by n the number of records in the database (typically n can range from a few tens to a thousand) and by Q the number of aggregated signatures contained in a signature message (typically $Q = 10$). Quantity b denotes the maximal size of the portion of database a node will transmit as a whole (i.e. without signature exchanges). To simplify we assume that a signature and a record exchange yields the same cost. Let this cost be the unit.

5.4.1 Retrieving a Single Mismatch

Let D_n being the average cost of database retrieval when the mismatch occurs on a random record among n. Let S_n be the average retrieval costs summed on all possible location of the mismatch in the database. Typically $S_n = nD_n$.

Theorem 1. *The average recovery cost of a single mismatch is:*

$$D_n = \frac{1}{\log Q}\left(Q + 1 - \frac{1}{Q}\right)\log n$$

$$+ \left(Q + 1 - \frac{1}{Q}H_{b-1} + \frac{Q-1}{Q^2}b\right)$$

$$+ P(\log n) + O\left(\frac{1}{n}\right),$$

where $H_k = \sum i = 1^k (1/i)$ *denotes the harmonic sum, and* $P(x)$ *is a periodic function of period* $\log Q$ *with very small amplitude.*

Proof. When $n \leq b$, then $D_n = n$ and $S_n = n^2$, since the database is exchanged as a whole in this case.

When $n > b$, elementary algebra on random partitions yields:

$$S_n = nQ + \sum_{n_1 + \cdots + n_Q = n} Q^{-n} \binom{n}{n_1 \cdots n_Q} (S_{n_1} + \cdots + S_{n_Q}).$$

Denoting

$$S(z) = \sum_n S_n \frac{z^n}{n!} e^{-z},$$

the so-called Poisson generating function of S_n, we get the functional equation:

$$S(z) = QS\left(\frac{z}{Q}\right) + Qz - \left(\left(Q + 1 - \frac{1}{Q}\right)ze_b(z) + \frac{(Q-1)}{Q^2} z^2 e_{b-1}(z)\right)e^{-z}$$

with the convention that

$$e_k(z) = \sum_{i \leq k} \frac{z^i}{i!}.$$

Let $D(z) = S(z)/z$. We therefore have the functional equation:

$$D(z) = D\left(\frac{z}{Q}\right) + Q - \left(\left(Q + 1 - \frac{1}{Q}\right)e_b(z) + \frac{(Q-1)}{Q^2} z e_{b-1}(z)\right)e^{-z}.$$

Using the Mellin transformation

$$D^*(s) = \int_0^\infty D(x)x^{s-1}dx,$$

defined for $\Re(s) \in \,]-1, 0[$, and using the fact that the Mellin transformation of $D(z/Q)$ is $Q^s D^*(s)$, we get to the closed form solution:

$$D^*(s) = \frac{\Gamma_b(s)(Q + 1 - 1/Q) + \Gamma_{b-1}(s+1)(Q-1)/Q}{1 - Q^s}$$

with the convention that $\Gamma_k(s)$ is the Mellin transformation of $e_b(z)e^{-z}$. We note, by the way, that

$$\Gamma_k(s) = \sum_{i \leq k} \frac{\Gamma(s+i}{i!}.$$

The reverse Mellin transformation then yields:

$$D(x) = \frac{1}{2i\pi} \int_{c-i\infty}^{c+i\infty} D^*(s) \exp(s \log x) ds$$

for any c such that $\Re(c) \in]-1, 0[$. When the integration path moves to the right, it encounters a succession of singularities on the vertical axis $\Re(s) = 0$. There is a double pole on $s = 0$ and there are single poles on

$$s_k = \frac{2ik\pi}{\log Q}$$

for k being integer. Therefore, by virtue of singularity analysis, we have for any m:

$$D(x) = -\mu_0 \log x - \lambda_0$$
$$- \sum_k \lambda_k \exp\left(-2ik\pi \frac{\log x}{\log Q}\right)$$
$$+ O\left(\frac{1}{x^m}\right),$$

where λ_0 and μ_0 are, respectively, the first and second order residus of $D^*(s)$ at $s = 0$, where and λ_k is the first order residus at $s = s_k$. Identifying residus is a trivial matter. Notice that

$$P(x) = -\sum_k \lambda_k \exp\left(-2ik\pi \frac{\log x}{\log Q}\right),$$

which is periodic of period $\log Q$. The estimate is true for every m, since there are no more singularities in the right half plan.

We use the depoissonization theorem to assess that $S_n = nD_n = nD(n) + O(1)$ when $n \to \infty$, which terminates the proof of the theorem. □

Figure 5.1 shows the asymptotic behaviour of retrieval cost with $Q = b = 16$ for n varying from 100 to 1,000. It is compared with full database retrieval cost.

5.5 Applicability of the Database Signature Exchange Mechanism

This section outlines the applicability of the specified mechanisms in a set of common scenarios. One application has been discussed previously: ensuring

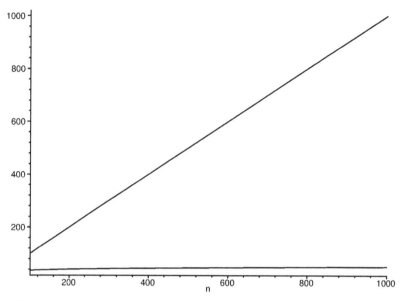

Figure 5.1 Signature retrieval cost (bottom) compared with full database retrieval cost (top) with a single record mismatch versus database size: $Q = 16, b = 16$.

that information from LSA messages, originating from attached wired networks with potentially long intervals between LSA message generation, is maintained in all nodes in the wireless ad-hoc network. The scenarios outlined in this section go beyond that situation, and consider how the database signature exchange may apply even in pure wireless scenarios.

5.5.1 Emerging Node

When a new node emerges in an existing network, the initialisation time for that node is the time until it has acquired link-state information, allowing it to participate fully in the network. Ordinarily, this time is determined solely by the frequency of control traffic transmissions. In order to reduce the initialisation time, the database exchange mechanisms can be employed as soon as the node has established a relationship with one neighbour node already initialised. This emerging node will select a neighbour as slave and transmit a dbx signature of the form ([age min, age max],(*,signature(*)), "*" implying an empty prefix. The slave will respond by, effectively, offering its entire link-state database to the master. In particular in situations where the some LSAs are not transmitted frequently (outside LSAs would be an example of such), this mechanism may drastically reduce the initialisation time of new nodes in the network.

5.5.2 Merging Wireless Clouds

Two disjoint sets of nodes, employing [1] as their routing protocol, may at some point merge or join – i.e. that a direct (radio) link is established. Prior to the merger, the respective clouds are "stable", periodically transmitting consistent info signatures within their respective networks. At the point of merger, at least two nodes, one from each network, will be able to establish a direct link and exchange control traffic. The combined network is now in an unstable state, with great discrepancies between the link-state databases of the nodes in the formerly two networks. Employing signature and database exchanges through the LSF-D mechanism, the convergence time until a new stable state is achieved can be kept at a minimum.

5.5.3 Reliable Flooding

If a node wants a specific LSA to be reliably transmitted to its neighbour, the db signature mechanism can be employed outside of general periodic signature consistency check. The node transmitting the LSA message broadcasts an info signature, containing the full LSA-originator ID as signed prefix and a very narrow age interval, cantered on the age of the LSA which is to be reliably transmitted. A neighbour which does not have the LSA in its database will therefore automatically trigger a database exchange concerning this LSA and send a dbx signature containing the LSA-originator ID signed with an empty signature. The receiving of such a dbx signature will trigger the first node to retransmit the LSA right away with a new LSF to ensure that the LSA does get through.

5.6 Conclusion

In this paper, we have introduced the notion of database exchange and reliable synchronisation in the context of wireless ad-hoc networks. Inspired by the mechanisms from the routing protocol OSPF, we have argued that in the form, present in OSPF, these mechanisms are not suitable for the wireless ad-hoc domain. While OSPF is designed for relatively static networks, the potentially very dynamic nature of wireless ad-hoc networks imply that database inconsistencies may arise more frequently with less available network capacity for alleviating the inconsistencies – and that acknowledgement-based reliability is unsuitable since the correct interpretation of an acknowledgement depends on being received in a specific context.

Consequently, we have deviced an mechanism for database exchange, adapted for the the specific environment of wireless ad-hoc networks. The mechanism is proposed as an extension to the OSPF interface type WOSPF [1]. The mechanism allows an efficient way of detecting and alleviating database inconsistencies, and can furthermore be employed as a way of providing "context-independent selective acknowledgements" for reliable synchronisation and link-state diffusion.

We have, analytically, compared the performance of our proposed mechanism to the performance of the mechanisms for database exchange in OSPF, and found it to be superior in terms overhead. We have furthermore outlined a couple of scenarios, where application of the mechanisms deviced in this paper may be advantageous for a wireless ad-hoc network.

Ongoing and future work on this topic involves extending the analytical performance evaluation of this mechanism, as well as conducting exhaustive simulations and experimental testing, comparing the performance of WOSPF with or without database exchange and reliable synchronisation mechanisms.

Appendix

A Packet Formats

Info and dbx signatures share the same packet format, detailed in this section.

A.1 Signature Packet Format

Version #, Packet length, Router ID, Area ID, Checksum, AuType and Authentication fields are the OSPF control packet header as described in [2].

AgeMin, AgeMax

AgeMin and AgeMax defines the age interval [AgeMin,AgeMax], used for computing the timed partial signatures in the prefix signatures as described in Section 5.3.3.

Type

Specifies if the signature is an info or a dbx signature, according to the following:

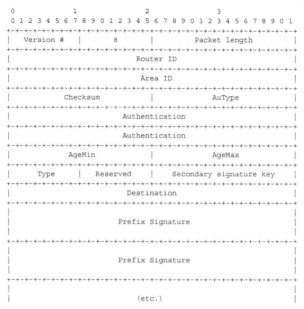

Figure 5.2 Signature packet format.

Value	Type
1	info (informative)
2	dbx (database exchange)

Reserved

Must be set to "00000000" for compliance with this specification.

Secondary signature key

The key of the secondary signature is a random number of 32 bits. Used for computing the secondary partial signature as described in Section 5.3.1.

Destination

– *If the signature is of type* = 2, then this field contains the address of the slave, with which a database exchange is requested.
– *If the signature is of type* = 1, then this field must be zeroed.

Prefix signature

The set of prefixes signatures contains the sub-signatures for different parts of the link-state database. The layout of the prefix signatures is detailed in Section 5.3.3.

A.2 Prefix Signature Format

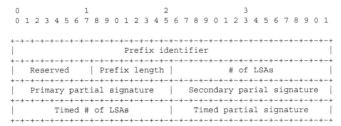

Figure 5.3 Prefix signature format.

Prefix identifier and Prefix length
Indicates the length of the prefix for the part of the link-state database, as well as the exact prefix.

of LSAs
The number of LSAs in the emitting nodes link-state database, matching by the prefix identifier and prefix length.

Primary partial signature
The arithmetic sum of the hashing of each string made of the concatenation of sequence number and LSA-originator ID fields of the tuples (LSA-originator ID, LSAsequence-number, LSA-age) from the emitting nodes link-state database such that the LSA-originator ID and prefix ID has same prefix of length prefix-length.

Secondary partial signature
The arithmetic sum of the XOR between the secondary signature key and each of the hashing of each string made of the concatenation of sequence number and LSA-originator ID fields of the tuples (LSA-originator ID, LSAsequence-number, LSA-age) from the emitting nodes link-state database such that the LSA-originator ID and prefix ID has same prefix of length prefix-length.

Timed # of LSAs

The number of LSAs in the emitting nodes link-state database, matching by the prefix identifier and prefix length and satisfying the condition that the LSA age is between AgeMin and AgeMax.

Timed partial signature

The arithmetic sum of the hashing of each string made of the concatenation of sequence number and LSA-originator ID fields of the tuples (LSA-originator-ID,LSA sequence-number, LSA-age) from the emitting nodes link-state database such that:

- Prefix ID and LSA-originator ID has same prefix of length prefix-length
- LSA-age is between AgeMin and AgeMax.

References

[1] J. Ahrenholz, T. Henderson, P. Spagnolo, P. Jacquet, E. Baccelli and T. Clausen, OSPFv2 wireless interface type, draft-spagnolo-manet-ospf-wireless-interface-00.txt, Internet Engineering Task Force, November 2003.

[2] J. Moy, OSPF version 2, RFC 2328, http://ietf.org/rfc/rfc2328.txt, 1998.

[3] D. Oran, OSI IS-IS intra-domain routing protocol, RFC 1142, http://ietf.org/rfc/rfc1142.txt, 1990.

[4] T. Clausen and P. Jacquet, Optimized link state routing protocol, RFC 3626, http://ietf.org/rfc/rfc3626.txt, 2003.

[5] S. Corson and J. Macker, Mobile Ad hoc Networking (MANET): Routing protocol performance issues and evaluation considerations, RFC 2501, http://ietf.org/rfc/rfc2501.txt, 1999.

[6] F. Baker, M. Chandra, R. White, J.Macker, T. Henderson and E. Baccelli, MANET OSPF problem statement, Internet draft: draft-baker-manet-ospf-problem-statement-00.txt, October 2003.

[7] C. Adjih, E. Baccelli and P. Jacquet, Link state routing in wireless ad hoc networks, in *Proceedings of MILCOM 2003 – IEEE Military Communications Conference*, Boston, USA, October, vol. 22, no. 1, pp. 1274–1279, 2003.

[8] A. Qayyum, L. Viennot and A. Laouiti, Multipoint relaying: An efficient technique for flooding in mobile wireless networks, INRIA Research Report RR-3898, March 2000.

6

Flow Transfer Times in Wireless Multihop Ad Hoc Networks

Tom Coenen[1,*], Hans van den Berg[2,3] and Richard J. Boucherie[1]

[1]*Department of Applied Mathematics, University of Twente, P.O. Box 217, 7500 AE Enschede, The Netherlands; e-mail: t.j.m.coenen@utwente.nl*
[2]*Department of Computer Science, University of Twente, P.O. Box 217, 7500 AE Enschede, The Netherlands*
[3]*TNO Information and Communication Technology, P.O. Box 217, 7500 AE Enschede, The Netherlands*

Abstract

A flow level model for multihop wireless ad hoc networks is presented in this paper. Using a flow level view, we show the main properties and modeling challenges for ad hoc networks. Considering different scenarios, a multihop WLAN and a serial network with a TCP-like flow control protocol, we investigate how capacity is allocated between the users of a network. This leads to two different Processor Sharing models, BPS and DPS, which are discussed and compared. Simulation is used to validate the proposed models. The flow level view leads to new insights into the capacity of ad hoc networks, opening new opportunities for analyzing this type of network.

Keywords: Ad hoc network, IEEE 802.11, multi-hop, capacity-allocation.

6.1 Introduction

Ad hoc networks are characterized by their multihop wireless connections and lack of infrastructure. This multihop property of the network presents

---
* Author for correspondence.

D. D. Kouvatsos (ed.), Performance Modelling and Analysis of Heterogeneous Networks, 113–132.

new challenges regarding the effectiveness of this type of network. For instance, whenever a transmission is started between two nodes in the network, other nodes in the neighbourhood are not allowed to transmit to avoid collisions and the loss of packets. This problem of interference at the MAC layer does not occur in wired networks and raises questions about the impact on the performance of ad hoc networks. A commonly used MAC protocol is IEEE 802.11, which uses Carrier Sense Multiple Access with Collision Avoidance (CSMA/CA). This protocol is set up to try and prevent collisions from occurring by letting all users sense if the network is available before allowing them to transmit data, which has a big impact on the use of the available network capacity and the division of it over all the users in the network.

The specific characteristics of multihop ad hoc networks call for new models to analyze their performance. In this paper important performance measures, such as the system throughput and the transfer time of flows, are investigated. The focus is put on the allocation of the capacity over the different network nodes, as this plays an important role on these performance measures.

We extend the successful approach for analyzing flow transfer times in (single hop) WLANs as presented in [1] to multihop networks. Two different network scenarios are considered. In the first scenario flows may have different path lengths (in terms of number of hops), but follow disjunct routes. The other scenario deals with a serial network in which multiple flows may travel through a particular node. Considering the capacity allocation in both scenarios, which turns out to be quite similar, we propose two processor-sharing (PS) models for describing the behaviour of the network at flow level. The first model, Batch Processor Sharing (BPS), deals with a queueing system where batches of jobs arrive. All jobs in the BPS model are served at the same time and are given an equal share of the capacity of the server. In the second model, Discriminatory Processor Sharing (DPS), jobs arrive one by one and all jobs in the queue are served at the same time, but some jobs get a bigger share of the capacity of the server than others. The batch sizes in the BPS model and the capacity shares in the DPS model reflect the different path lengths of the flows in the ad hoc network scenarios. The modeling results are compared with results obtained by simulation.

The rest of this paper is constructed as follows. In the remaining part of this section, we will give a review of related literature on the subject. Next, in Section 6.2 the IEEE 802.11 protocol is described. Section 6.3 presents the two main ad hoc network scenarios under consideration, and investigates the distribution of the capacity over the users in the network. The resulting processor-sharing models for analyzing flow transfer times are shown in Section 6.4. These models are validated by simulation in Section 6.5. Finally, Section 6.6 summarizes and concludes the paper.

6.1.1 Literature Review

Many papers have been devoted to the capacity and throughput of wireless (multihop) networks. Most of them use results from simulation to describe the characteristics of ad hoc networks, whereas analytical studies are scarce. The impact of MAC layer interference on the capacity of ad hoc networks as addressed in this paper has been studied in several settings. For instance [2] uses simulation to show that capacity can be very low in ad hoc networks. Scaling appears only to be possible if the distance between the source and destination remains small as the network grows. An analytical approach for determining the capacity is presented in [3], where it is shown how the throughput depends on the number of users in the network (when this number becomes large). A paper that focuses more on the multihop property of ad hoc networks is [4], which gives asymptotic results for a wireless network under a relay traffic pattern, whereas [5] considers mesh networks, slightly different from ad hoc networks. Both focus on the throughput of a chain of users processing flows in one direction over multiple hops. A bottleneck is found which determines the throughput that the network can achieve.

In the work of Litjens et al. [1], an integrated packet/flow level approach is used to analyze flow transfer times in a single hop WLAN scenario. Considering the system throughput at the packet level, and taking the system dynamics at flow level into account, leads to a processor-sharing (PS) type of queueing model. This PS model captures the equal allocation of transmission capacity among the active flows. Using known results for this PS model an approximation for the mean flow transfer time is proposed. Simulation results show that the approximation is very accurate.

Modeling bandwidth sharing in fixed communication networks by PS systems has been done by amongst others Bonald and Nunez Queija. In the papers by Bonald [6, 7], the main notion is that modeling the network with processor-sharing can lead to balanced fairness, which means that each user in the network receives an equal share of the available network resources. This type of PS network is analyzed by considering the bottleneck node and distributing the capacity there first. All nodes servicing the same flow adjust their capacity allocation accordingly, avoiding congestion. The capacity allocated to each flow is determined analytically. In his dissertation, Nunez

Queija discusses many different PS models for integrated services networks [8].

Batch arrival processor-sharing models have been investigated extensively in the past. It was Kleinrock who started with this approach. In his paper with Muntz and Rodemich [9] a start was made in giving a complete analytical approach to determine the throughput of a PS network. A discriminatory processor-sharing model has also been used for modeling networks. Kleinrock [10] already started with this in 1967 which created much interest in this type of network. In 1980, Fayolle et al. [11] built on the work of Kleinrock. New results have been obtained in [12, 13] where the queue length distribution and sojourn times for PS models are determined. The results presented hold for general service requirements and are used to analyze WLANs with Quality of Service support in [14].

6.2 IEEE 802.11 MAC Layer Protocol

Ad hoc networks are characterized by their multihop wireless connections and lack of infrastructure. An important aspect influencing the performance of such networks is interference. Signals transmitted by a user will not only be heard by the intended receiver, but also by all other nodes in the vicinity of the sender. If multiple signals reach a node at the same time, a collision occurs and the signals cannot be received correctly and are lost. To reduce the number of transmissions that collide and the impact this has on the throughput of the network, IEEE 802.11 uses Carrier Sense Multiple Access with Collision Avoidance (CSMA/CA). IEEE 802.11 can function in infrastructure or ad hoc mode, depending if an access point is being used, which however has no implications on the MAC layer.

When a node wants to initiate a transmission, it will first sense the network to find out if other nodes are already transmitting: carrier sensing. If other nodes are transmitting, the node will refrain from transmitting, while it keeps sensing the network. When the network becomes available, the node waits for a certain time (DIFS) to make sure the network really is available and if the network is then still free, a timer is started to avoid collisions. This timer is paused as soon as the node senses a transmission from another node. When the network becomes free again and stays free for a DIFS, the timer continues. When the timer ends, transmission starts. This approach does not make it completely certain that collisions will not occur, as timers can expire at the same moment. Therefore, instead of sending the packets of the data immediately, a node first transmits a request-to-send (RTS). The

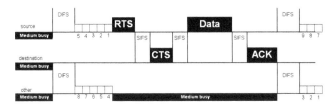

Figure 6.1 CSMA/CA with RTS-CTS.

receiver replies to this RTS by sending a clear-to-send message (CTS). The time between these transmissions (SIFS) is smaller than DIFS, so that the silence is not mistaken for the network being available. After receiving the CTS, transmission of the data starts. This approach is used so that in case of a collision, only the RTS is lost, and not a much bigger packet containing data. This way the impact of a collision on the throughput of the network is reduced. When a collision occurs, the node starts a timer again but it is set to a time taken from a window that is twice as big as before. When the transmission was successful, the procedure repeats as long as there still are packets that need to be transmitted. The operation of CSMA/CA is shown in Figure 6.1.

Under CSMA/CA, all nodes that want to transmit compete for network resources. In the multihop wireless ad hoc network we are considering, packets from a multihop flow are present at multiple nodes. Such a flow is competing for the network resources through multiple nodes at the same time. Hence even if there is only one multihop flow in the network, there is interference between the different users that are involved in the transmission. All flows in the network will have to share the MAC layer capacity, and the capacity allocation over these flows will influence the throughput of the network and the transfer times of the flows.

6.3 Scenarios

In this section, first, the analysis of the single hop WLAN considered in [1] is shortly reviewed, after which two multihop scenarios are described. For the analysis of these scenarios we use a similar approach as used for the WLAN in [1], with the extension of considering the multihop aspects involved in these scenarios.

6.3.1 Single Hop WLAN Scenario

In [1] a single hop WLAN is considered, in which new flow transmissions are initiated according to a Poisson process. Flow sizes are random variables with general distributions. The network operates under the IEEE 802.11 DCF MAC protocol as described in Section 6.2. First an analysis is made on the packet level of the aggregate system throughput, the total amount of data sent in a time interval in the complete network, that can be reached in a WLAN with a fixed number of persistent flows, flows that continuously try to transmit packets. Using Markov-chain analysis, the probability that a node is transmitting is computed, as well as the probability that a transmission fails. From this, the aggregate system throughput is derived, including the influence of the headers and control packets used by the protocol. Simulation validates the result that the average system throughput is about 87% of the total capacity on the MAC layer, and slightly dependent on the number of present flows. Next, using the results from the first step, the transfer time is analyzed, taking the flow level dynamics into account. The assumption is made that the service rate per flow is found by giving each flow an equal share of the aggregate system throughput computed for the persistent number of flows in the network. This leads to a processor-sharing (PS) model with state dependent service rates which is analytically tractable. Simulation shows that the results obtained through this approach are very accurate.

6.3.2 Multihop Ad Hoc Scenario

The model presented in [1] only considers single hop flows. This model is expanded by also allowing multihop flows. Where at first a flow was completed if the packets were sent from a node to the access point or vice versa, now there is the possibility that all packets of the flow are forwarded to another node before the transfer is completed.

The situation in Figure 6.2 is a WLAN cell in ad hoc mode with connections of only one or two hops. A transmission of any of the nodes in the cell, by which we mean the area under consideration, can be heard by all other nodes in the cell. This means that no two transmissions can take place at the same time, since the data will be lost due to a collision. A second hop is used to connect to a user outside the cell, for example to connect to the internet or a wired network, as in mesh networks. Whenever a node is relaying a flow, this node will only compete for the network resources if there are packets available that need to be sent. If at one point there are no packets available, the node will remain idle until new packets arrive that need to be forwarded.

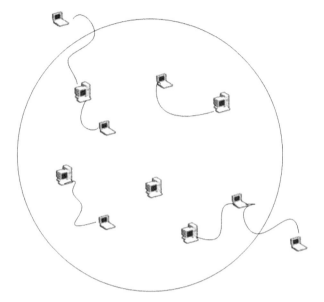

Figure 6.2 Multihop scenario.

Following the approach presented in [1], we first determine the aggregate system throughput of the network. The number of persistent users is taken to be the number of nodes active in sending the flows. This means that a flow over two hops, which needs two nodes, is counted as two users. However, there might be moments that the relaying nodes are not competing for the network since there are no packets available. This differs from the single hop WLAN situation described above where each node always has packets that need to be transmitted. Through simulation the aggregate system throughput is determined.

To take the flow level dynamics into account, the capacity allocation needs to be known. As in the WLAN model, we assume that every node receives an equal share of the aggregate system throughput. In general this is the case since all nodes behave according to the IEEE 802.11 protocol. For the flow level dynamics, we assume that flows are big enough to have packets available at all nodes they are going through, for most of the time. Then these nodes are continuously competing for the network and all nodes get an equal share of the capacity. A flow over two hops will get a share of the capacity for both the nodes it uses. Hence the capacity allocated to a flow over two hops will be double the amount of the capacity allocated to a flow over one hop,

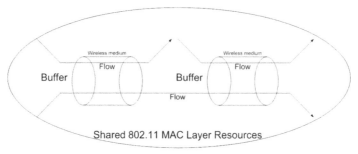

Figure 6.3 Serial model.

but note that each packet has to be sent twice, once for each hop. Just as for the WLAN model, this approach leads to a processor-sharing type of model which will be discussed in Section 6.4.

6.3.3 Multihop Serial Network Scenario

A different scenario is where nodes can serve more than one flow at the same time. Assuming that a flow will at most need two hops to reach its destination, such a network can be modeled as a network consisting of three nodes, with two connecting links. This network is represented by the model shown in Figure 6.3.

The flows through the wireless medium (the links), represented by the arrows through the tubes, will compete for the channel. There are three types of flows. Flows of type 0 will go over both links (as depicted by the lower arrow going through both tubes); flows of type 1 (2) will only use link 1 (2) (the upper left (right) arrow going through the left (right) tube). All flows consist of packets which first arrive in a buffer before being sent over the wireless medium. Because of interference, the links have to share the MAC layer resources.

Assume that the arrival process of the flows is according to a Poisson distribution and the flows consist of many packets, so that flows over multiple hops will again usually have packets in the queue of every node it passes. The packets of all flows join the same queue at a node. With all flows arriving simultaneously, the packets will be in the queue in a mixed order and are serviced according to a FCFS discipline. Hence all flows are serviced, packet by packet, in a more or less random order, as shown in Figure 6.4.

This way of servicing is approximated by a processor-sharing (PS) service discipline for the flows, where each link is assumed to have the same

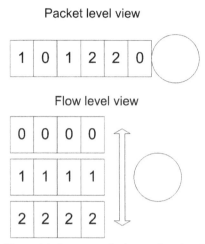

Figure 6.4 Processor sharing on flow level.

capacity. Just as in the previous scenario, the aggregate throughput needs to be determined, which is done through simulation.

An important aspect regarding the throughput of the different type of flows in the network is how the capacity is allocated among them. This depends on the flow control protocol used in the network. A commonly used transport protocol in (wired) networks is TCP. This protocol tries to avoid congestion by fairly sharing the available resources among active flows. A similar protocol can be used in wireless networks. Suppose that in the reference model there are n_0, n_1 and n_2 flows of type 0, 1 and 2 respectively. At first, the available capacity will be shared in a fair way over both nodes, each node receives half of the total capacity. Both nodes will then use this capacity to process the flows that are in their queues. We assume egalitarian processor-sharing, so each flow gets an equal share. Now the situation can occur that flows of type 0 get a different amount of the capacity over the first link than over the second link, as will be discussed later. If the capacity at the first link is higher, the queue at the second server will build up, since it cannot serve the flow as fast as it arrives. If the capacity at the first link is lower, then the queue at node two will often not contain any packets of the flow of type 0, since these are processed faster than the rate at which they arrive. These are unwanted situations, and so the flow control protocol will notify the sources to transmit at different rates for the flows that are causing the congestion. We assume a TCP-like flow control protocol to be active, which alters the use of capacity in case of queues building up or being empty most of the time. The flow control protocol resembles TCP, but assumes a perfect version with instantaneous rate adaptation, so it will be e.g. independent of the round trip times (RTT) of the flows.

In our model there can be a loss of packets due to a build up in the queue of node 2. This happens if the flows of type 0 get more capacity in node 1 than in node 2 for a long time. This is for example the case when there are flows of type 2, while there are no flows of type 1 in the system and both links get half of the total capacity. If the flows of type 0 keep being processed at this rate, packets will be lost. However, the flow control protocol will make sure that the rate at which packets are transmitted over link 1 is lowered. The rate to lower it to is the rate at which the flow is processed at node 2. The capacity used by node 1 hence drops due to the lowering of the transmission rate. This capacity can then be used by node 2.

Theorem 1. *The capacity a flow receives at a link is equal for any type of flow at any link, being $1/2n_0 + n_i$, when there are only flows of type 0 and one other type (i).*

Proof. Consider the network where only flows of type 0 and 1 are present. Let there be n_0 (n_1) flows of type 0 (1). If both links get half of the total capacity, which we set to be one, then the capacity allocated to a flow at node 1 will be equal to $\frac{1}{2}\frac{1}{n_0+n_1}$. At node 2, the flows of type 0 will receive a capacity of $\frac{1}{2}\frac{1}{n_0}$. If this situation persists, the queue at node 2 will often be empty, since there the flows of type 0 are processed faster than the rate at which they arrive. The flow control protocol therefore lowers the rate at node 2 and sets the capacity of a flow to $\frac{1}{2}\frac{1}{n_0+n_1}$. Node 1 can now use the residual capacity and a flow will get a capacity of $\frac{1}{n_0+n_1}(1 - \frac{1}{2}\frac{n_0}{n_0+n_1})$. This rate however is higher than the rate at node 2 and so the buffer will fill. The rate at node 2 is adjusted by the flow control protocol to $\frac{1}{n_0+n_1}(1 - \frac{1}{2}\frac{n_0}{n_0+n_1})$, which leaves a capacity for a flow at node 1 of $\frac{1}{n_0+n_1}(1 - \frac{n_0}{n_0+n_1}(1 - \frac{1}{2}\frac{n_0}{n_0+n_1}))$. This process continues (where each step is instantaneous by assumption) and it can easily be seen that this converges to a capacity allocation of $\frac{1}{2n_0+n_1}$ for each flow on any link. In total, a flow of type 0 will receive $\frac{2}{2n_0+n_1}$ of the capacity, whereas a flow of type 1 will get $\frac{1}{2n_0+n_1}$. The proof for the situation with only type 0 and type 2 flows follows in the same manner. □

If however there are flows of both type 1 and type 2 in the system, this situation will not occur. Supposing that there are more flows of type 2 than of type 1, a flow of type 0 will get a rate of $\frac{1}{n_0+n_2}$ at node 2, but receives a higher rate of $\frac{1}{n_0+n_1}$ at node 1. The protocol will hence lower the rate for all type 0 flows to $\frac{1}{n_0+n_2}$ at node 1. The capacity that now becomes available is not given to node 2, but the flows of type 1 will claim this extra capacity since the node has the right to use half of the total capacity. This is a different and interesting situation for further research, but we will not consider it any further in the analysis in this paper.

We see that in the first situation the flow control protocol makes sure that all flows get the same share of capacity per link. Of the total capacity, a flow over two links will get twice as much capacity as a flow over one link. This is equivalent to what we found for the two-hop WLAN model and hence can use a similar analysis, which as noted before leads to a processor-sharing model, which will be discussed in the next section.

The analogy between the scenario as shown in Figure 6.2 and the scenario of Figure 6.3 can be seen as follows. A flow over two hops coincides with a flow of type 0 and flows over one hop coincide with a flow of type 1 or 2 (but not both), which is arbitrary. In the serial network, just as in the two-hop WLAN scenario when a node relays, a flow taking the second hop will only compete for the network resources if it has packets available to be sent. If at one point no such packets are available in the buffer, no capacity is allocated to this flow at that node.

6.4 Flow Level Models

This section presents models that approximate the flow level dynamics of the multihop scenarios presented in the previous section. For the multihop ad hoc scenario, the equal share given to each of the nodes determines the capacity allocation. For the serial scenario, the transmission control protocol determines the capacity allocation at the MAC layer. This section presents two analytical models that capture the flow level dynamics of both scenarios. The two models take the capacity allocation into account by either varying the amount of jobs in an egalitarian processor-sharing queue or the priority and size of jobs in a discriminatory processor-sharing queue.

6.4.1 Batch Arrival Processor Sharing model

We can consider the network as a server with one queue, where all flows enter the queue, independent of the link(s) they have to be transmitted over. Since all flows are processed at the same time, we can consider the flows to be processed according to a processor-sharing discipline. As a flow over two hops gets the double amount of capacity, we can consider a flow over two hops as asking for capacity twice. Hence we can see the arrival of a flow over two hops as the arrival of two flows at the same time. A flow over two hops will be at both servers at the same time, so it will ask for capacity as if it were two different flows, which is captured in this abstract view. We thus arrive at a batch arrival processor-sharing model (BPS) with egalitarian processor-sharing, since the capacity for a flow is equal at every hop. A flow over a single hop is then equivalent to a single arrival, whereas an arriving flow over two hops is equivalent to two jobs arriving as a batch. It is important to note here is that we do need to consider that all jobs in a batch should not only have the same arrival time but also the same departure time, i.e. jobs in a single batch have the same service demand, since they represent only one flow.

Consider the $M^X/G/1$ PS queue where λ is the batch arrival rate, a is the average batch size, b is the average number of jobs that arrive in addition to a tagged job and $\overline{F}(x)$ is the complementary distribution function of the job size. The conditional response time of a job with service requirement x, $T(x)$, has to satisfy the system of differential equations [9]:

$$T'(x) = \lambda a \int_0^\infty T'(y)\overline{F}(x+y)dy + \lambda a \int_0^x T'(y)\overline{F}(x-y)dy + b\overline{F}(x) + 1.$$

The load in the system is given by

$$\rho = \lambda a E[X].$$

When flows have an exponential service requirement, solutions can be found [15]. For the $M^X/M/1$ PS queue, this leads to

$$T(x) = \frac{x}{1-\rho} + \frac{b(2-\rho)E[X]}{2(1-\rho)^2}\left[1 - \exp\left(\frac{-(1-\rho)}{E[X]}x\right)\right]$$

and bounds are given by

$$\frac{x}{1-\rho} \leq T(x) \leq \min\left(\frac{b+1}{1-\rho}x, \frac{x}{1-\rho} + \frac{b(2-\rho)E[X]}{2(1-\rho)^2}\right),$$

where the bounds coincide when

$$x^* = \frac{(2-\rho)E[X]}{2(1-\rho)}.$$

In this model, the departure moments of jobs inside a batch will not be the same. Only when the service times are deterministic, is this model applicable. Therefore, we also propose a more appropriate model.

6.4.2 Discriminatory Processor Sharing model

A flow over two hops receives more capacity than a flow over one hop, hence we can instead consider the processor-sharing not to be egalitarian. The jobs are then processed at the same time, but not all jobs get an equal share. As a flow over two hops takes twice the amount of capacity, it can be seen as being serviced twice as fast as a single hop flow. A flow over two hops however has an expected service requirement that is twice the expected service requirement of a single hop flow. We thus arrive at a discriminatory processor-sharing model (DPS). In this type of model, all jobs get processed at the same time, but not all jobs get the same amount of service. Customers are given a certain weight which shows how much more service they receive in comparison to other users. In our case, a job that represents a two-hop flow will get a weight twice as high as a job representing a single-hop flow.

Consider the $M/G/1$ DPS queue where λ_j denotes the arrival intensity of class j jobs, g_j denotes the 'weight' of class j customers and $F_j(x)$ the distribution of the required service with mean $1/\mu_j$ and a total of M classes. The conditional response time of a job of class k, given its size t, $W_k(t)$, satisfies the system of differential equations [10]:

$$W_k'(t) = 1 + \sum_{j=1}^{M} \int_0^\infty \frac{\lambda_j g_j W_j'(u)}{g_k} (1 - F_j(u + \frac{g_j t}{g_k})) du$$

$$+ \int_0^t W_k'(u) \sum_{j=1}^{M} \frac{\lambda_j g_j (1 - F_j(g_j(t-u)/g_k))}{g_k} du, \quad k = 1 \dots M.$$

For the $M/M/1$ DPS queue, we know that (for the derivation, see [11]):

$$W_k(t) = \frac{t}{1-\rho} + \sum_{j=1}^{m} \frac{g_k c_j \alpha_j + d_j}{\alpha_j^2} (1 - e^{-\alpha_j t/g_k}),$$

where the α_j's are the distinct roots of

$$1 - \Psi^*(s) = 1 - \sum_{j=1}^{M} \frac{\lambda_j g_j}{\mu_j g_j + s} = 0$$

and c_j and d_j are given by

$$c_j = (s + \alpha_j)a^*(s)|_{s=-\alpha_j} = \frac{\prod_{k=1}^{m}(g_k\mu_k - \alpha_j)}{-\alpha_j \prod_{k \neq j}^{m}(\alpha_k - \alpha_j)},$$

$$d_j = (s^2 + \alpha_j^2)\theta(s)|_{s^2=-\alpha_j^2}$$

$$= \frac{\left[\sum_{k=1}^{M} \lambda_k g_k^2/(\mu_k^2 g_k^2 - \alpha_j^2)\right]\left[\prod_{k=1}^{m}(g_k^2\mu_k^2 - \alpha_j^2)\right]}{\prod_{k \neq j}^{m}(\alpha_k^2 - \alpha_j^2)}.$$

Since each flow is represented by only one job in the system, the departure of a job is equivalent to the departure of a flow from the network. Therefore, this approach gives a better approximation of the situation that we want to model. When considering deterministic service requirements, we see that both models give equivalent results.

6.5 Numerical Results

To verify that the proposed model is an accurate approximation of the network under consideration, a simulation model has been constructed to obtain data on the sojourn time of a flow in the network. The simulation model mimics the transmissions as they occur in the scenario depicted in Figure 6.2. The simulation model uses the following standard settings for the parameters:

parameter	value	parameter	value	parameter	value
PHY	192 bit	Payload size	12 kbit	SIFS	10 μs
MAC	272 bit	r_{net}	1 Mbit/s	DIFS	SIFS + 2τ
RTS	PHY + 160 bit	n_{max}	100	$cw_{min / max}$	31/1023
CTS	PHY + 112 bit	δ	1 μs	r^*	5
ACK	PHY + 112 bit	τ	20 μs	r_{max}	6

Here r_{net} is the rate at which the network can transmit data, n_{max} is the maximum number of users, PHY, MAC, RTS, CTS and ACK give the sizes of the headers and interframe spaces, δ is the propagation delay, τ is the slot duration, cw are the values for the contention windows, r^* is the maximum number of times the contention window may be doubled and r_{max} is the maximum number of retransmissions for one packet. The payload size is set at

12 kbit, the maximum amount of data that can be sent within a packet. The probability that a flow is over two hops is given as input to the simulator. All flows are either single or double hop flows. Flows arrive at the system according to a Poisson process, where users try to transmit the packets according to the IEEE 802.11 protocol discussed earlier. When the first packet of a double hop flow has been sent over the first hop, a free user is assigned as the relaying node and this user will also start transmitting the packets that it receives from the first user. When the second user has no packets in its queue to relay, it will go into waiting, meaning that he will not compete for the channel. Note that we are assuming that no node is both a source and relay node, so that packets of flows do not mix at one node. The DCF function of IEEE 802.11 is incorporated in the simulation, where collisions are considered to be fatal, meaning that all packets involved in the collision are lost and retransmitted after backing of, unless the maximum number of retransmissions has been reached, in which case the packet is dropped.

The results of the simulation and the M^X/M/1 BPS and M/M/1 DPS model are compared, we hence are considering Poisson arrivals and exponentially distributed flow sizes. The model considers the network in a situation that the full capacity of the network can be used for the flows, which is not the case for the simulation program. Here headers are added to all the packets and the number of active users influence the aggregate system throughput as described in [1]. Following their approach, the aggregate throughput for persistent flows is determined. For the determination of this aggregate throughput the amount of double hop flows can be of influence. The interference that the flow causes for itself deteriorates the throughput of the network. However, the second node in a double hop flow may not always have packets to transmit, at which point it will not cause interference. Simulation is used to compare the results of single and double hop flows. In the following figure, the aggregate system throughput is computed for single hop flows, and compared with the aggregate system throughput of double hop flows, where the number of persistent flows is taken to be twice the amount of double hop flows.

Figure 6.5 clearly shows that the aggregate throughput is hardly influenced by double or single hop flows. For persistent flows over two hops, it is likely that most of the time there are packets to be transmitted over both hops, so that the flow is claiming capacity twice. For small sized flows, this may no longer be the case. The results presented in Figure 6.5 however shows that for determining the system throughput, we can use the results for single hops, but counting the double hop flows as if there are two users in the system.

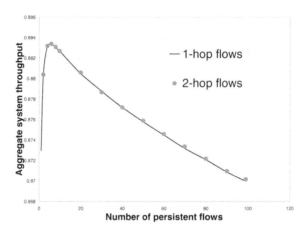

Figure 6.5 Aggregate system throughput.

Under the RTS/CTS mode, the aggregate throughput is roughly constant as was also found for the WLAN situation in [1]. As can be read from the figure only about 88% of the capacity can really be used. This is taken into account in the calculation of the average transfer time in the models. First we compare the average transfer time of a job in the system with the results from the BPS model. A drawback of this model is that it is not possible to make a distinction between the single and double hop flows in this model. Results are shown in Figure 6.6a for the situation where 30% and 70% of all the flows are double hop flows and the file sizes are exponentially distributed with a mean of 150 kbytes.

The figure shows that for a lower amount of double hop flows, the approximation is better. When 70% of all the flows are double hop flows, the difference becomes bigger as the error made by assuming that double hop flows are two separate flows with different sizes becomes bigger. In both cases, the model underestimates the average transfer time. The capacity that is allocated to single hop flows depends on the capacity taken by the double hop flows, which also depends on the size of these flows. Consider for example the situation where there is one single hop flow and one double hop flow. The single hop flow has a size of 100 kbit, and the double hop flow also has a size of 100 kbit. If this situation both flows get a third of the capacity and finish at the same time. Assuming the effeicive capacity of the network is 3 Mbit/sec, they all would have been sent after 0.1 sec. The model however would not assume that the arrival of the double hop flow are two identical flows. So let us assume they are modelled as the arrival of a 50 kbit and a 150 kbit flow. In

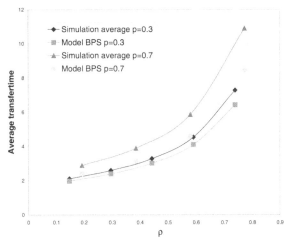

Figure 6.6a BPS versus simulation.

this case, at first again they all get a third of the capacity. After 0.05 seconds, the smallest job of 50 kbit would be done. From this point on, the other two flows will receive half of the total capacity. The single hop flow hence would be finished after 0.083 seconds and the remaining flow would finish after 0.1 seconds. So the assumption that the two flows are not of the same size causes the model to underestimate the flow transfer time of the single hop flow, as a smaller flow modelling part of a double hop flow would leave the system early, thus creating capacity for the single hop flow before the double hop flow actually is completed.

Next, the comparison is made between simulation and the DPS model, where we can distinguish between the different type of flows, which is shown in Figures 6.6b and 6.6c for 30% and 70% of the flows being double hop flows.

The approximations are accurate, independent of the amount of double hop flows in the system. It can be seen that for a higher load the approximation is slightly worse. Interesting is to see that the model overestimates the transfer time for single hop flows, whereas it underestimates the transfer time of double hop flows. This is due to the fact that the assumption that a double hop flow receives twice the capacity of a single hop flow is not exact, since packets are not always available at all the different nodes on a multihop path. Therefore the capacity allocated to double hop flows is slightly less than the double of single hop flows, resulting in the differences seen in the figures.

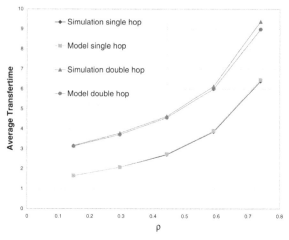

Figure 6.6b DPS versus simulation for $p = 0.30$.

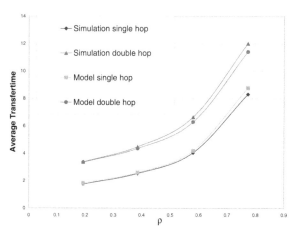

Figure 6.6c DPS versus simulation for $p = 0.70$.

With the load increasing, but the average file size remaining constant, there will be more active nodes in the network. The frequency at which packets arrive to a relaying node hence will decrease, as the capacity allocated to each node is smaller. The duration that a node remains empty in case this occurs will hence be longer, explaining that the effect of the approximation will be larger for higher loaded systems.

Another interesting aspect than can be viewed from the figures is the effect the different flows have on each other. Consider for example a single hop flow of 100 kbit being transfered over a 1 Mbit/sec network. This would take 0.1 seconds, whereas a double hop flow of the same size would take 0.2 seconds. With both flows present and starting at the same time, they all would need 0.3 seconds to be completed, simply the sum of the two. However, the increase in time needed due to the presence of the other flow is much higher for the single hop flow than for the double hop flow. This trend is also seen in the figures, as the distance between the lines of the types of flows is decreasing. Whereas for the lower loaded system the double hop flows have almost twice the transfer time of a single hop flow, this certainly is no longer the case for the higher loaded system.

6.6 Conclusion

This paper proposes considering ad hoc wireless networks from a flow level point of view. All flows in the network try to make use of the resources of the network at the same time. The total capacity of the network hence has to be shared between the flows. The amount of capacity allocated to a flow depends on the number of flows that are active, as well as on the number of hops taken by the flow. To model this allocation of capacity and determine the transfer times of flows, we propose two types of models, being a BPS and DPS model. The models use the $M^X/G/1$ PS and $M/G/1$ DPS queues, as these models can take the allocation of the capacity over the different flows in the network into account. This is done by adjusting the size of the batch arriving at the $M^X/G/1$ PS queue, or by the 'weight' given to a flow in the $M/G/1$ DPS queue. Several results for these types of queues have already been published, and some analytical results have been presented in this paper. When considering the network as a BPS queue, the problem arises that all flows inside a batch should have the same service requirement, which is not the case for the $M^X/G/1$ PS queue, making it impossible to distinguish between the different classes of flows. Therefore, it is more accurate to use the $M/G/1$ DPS queue for modeling the ad hoc network.

The flow level view as presented in the paper opens new opportunities for modeling ad hoc networks, taking into account interference at the MAC layer, especially self-interference within a flow. Through simulation the model has been validated, showing that the results obtained using the $M/M/1$ DPS queue gives a good approximation of the transfer time in an ad hoc network using IEEE 802.11 in RTS/CTS mode, independent of the percentage of double hop flows.

References

[1] R. Litjens, F. Roijers, J. L. van den Berg, R. J. Boucherie and M. Fleuren, Performance analysis of wireless LANs: An integrated packet/flow level approach, in *Proceedings of the 18th International Teletraffic Congress-Itc-18*, Berlin, Germany, pp. 931–940, September 2003.

[2] J. Li et al., Capacity of ad hoc wireless networks, in *Proceedings of the 7th ACM International Conference on Mobile Computing and Networking*, Rome, Italy, pp. 61–69, 2001.

[3] P. Gupta and P. R. Kumar, The capacity of wireless networks, *IEEE Transactions on Information Theory*, vol. IT-46, no. 2, pp. 388–404, March 2000.

[4] M. Gastpar and M. Vetterli, On the capacity of wireless networks: The relay case, in *Proceedings of IEEE INFOCOM '02*, New York, p. 1577–1586, 2002.

[5] J. Jun and M. L. Sichitiu, The nominal capacity of wireless mesh networks, *IEEE Wireless Communications*, vol. 10, no. 5, pp. 8–14, 2003.

[6] T. Bonald and A. Proutière, Insensitivity in processor-sharing networks, *Performance Evaluation*, vol. 49, nos. 1–4, pp. 193–209, 2002.

[7] T. Bonald and A. Proutière, Insensitive bandwidth sharing in data networks, *Queueing Systems*, vol. 44, pp. 69–100, 2003.

[8] R. Núñez Queija, Processor-sharing models for integrated-services networks, PhD Thesis, Eindhoven University of Technology, 2000.

[9] L. Kleinrock, M. M. Muntz and E. Rodemich, The processor sharing queueing model for time shared systems with bulk arrivals, *Networks: An International journal*, vol. 1, pp. 1–13, 1971.

[10] L. Kleinrock, Time-shared systems: A theoretical treatment, *Journal of the Association for Computing Machinery*, vol. 14, no. 2, pp. 242–261, 1967.

[11] G. Fayolle, I. Mitrani and R. Iasnogorodski, Sharing a processor among many job classes, *Journal of the Association for Computing Machinery*, vol. 27, no. 3, pp. 519-0532, 1980.

[12] S. K. Cheung, J. L. van den Berg and R. Boucherie, Decomposing the queue length distribution of processor-sharing models into queue lengths of permanent customer queues, *Performance Evaluation*, vol. 62, nos. 1–4, pp. 100–116, 2005.

[13] S. K. Cheung, J. L. van den Berg and R. Boucherie, Insensitive bounds for the moments of the sojourn time distribution in the M/G/1 processor-sharing queue, Research Memorandum No. 1766, Department of Applied Mathematics, University of Twente, 2005.

[14] S. K. Cheung et al., An analytical packet/flow-level modelling approach for wireless LANs with Quality-of-Service support, in *Proceedings of the 19th International Teletraffic Congress*, Beijing, China, August 29–September 3, 2005.

[15] K. Avrachenkov, U. Ayesta and P. Brown, Batch arrival M/G/1 processor sharing with application to multilevel processor sharing scheduling, INRIA Technical Report 5043, INRIA Sophia Antipolis, 2003.

PART THREE
WIRELESS SENSOR NETWORKS

7

CAAC Mechanism: A Cluster Address Auto-Configuration Mechanism

Alexandre Delye de Clauzade de Mazieux, Vincent Gauthier,
Michel Marot and Monique Becker

*SAMOVAR CNRS Research Lab, GET/INT: Institut National des
Télécommunications, 9 rue C. Fourier, 91011 Evry Cedex, France;
e-mail: alexandre.delye@int-edu.eu, {vincent.gauthier, michel.marot,
monique.becker}@int-evry.fr*

Abstract

This paper focuses on cluster creation, addressing mechanism and auto-routing. In fact, we propose an address assignment mechanism using node clusters. These multi-hop clusters are trees with a given depth, allowing to automatically assign addresses. We designed a new multi-hop performance criterion because we want the clusterheads to have a large degree. This address assignment mechanism is designed for wireless fixed sensor networks, which is usually the case for sensor networks, where self-organisation is crucial. Because of its scalability, it is particularly suitable for very large sensor networks, where automatic address assignment cannot be avoided. It is a fully distributed node allocation address, with no possible duplication. Moreover, it has the key feature to allow auto-routing. Actually, like in Banyan switching networks, we use a systematic pattern to assign the addresses, which leads to auto-routing by deriving locally on a node the next node on the route without the help of any routing protocol.

Keywords: Wireless sensor networks, clustered topology, addressing, auto-routing, performance evaluation.

*D. D. Kouvatsos (ed.), Performance Modelling and Analysis of Heterogeneous
Networks,* 135–152.

7.1 Introduction

Sensor networks consist of wireless nodes with limited energy. Because of their large number of applications to very diversified domains, this type of ad-hoc networks is more and more studied. Several ways of researches in this area as routing, MAC layer, time synchronisation or position-based algorithms, are investigated in order to reduce battery consumption.

An increasing number of articles deal with hierarchical networks. Indeed, hierarchical organisations let the network protocols to be scalable and permit to develop data fusion. Aggregating nodes into clusters allows to reduce the complexity of the routing algorithms, to optimise the resource of the shared medium, by enabling a clusterhead to locally manage the medium access, to make data fusion easier, to simplify the management of the network, and particularly the address assignment, to optimise the energy, and at last to make the network more scalable. Since Mhatre and Rosenberg [16] showed that multi-hop clustering might save energy, it is often more efficient to use multi-hop clusters than single-hop ones. Moreover, in such networks, the question of the design of address assignment algorithms cannot have the same answer as in classical networks. Indeed, the size of the network and the cost spent to ensure the uniqueness of these addresses throughout the network is prohibitive when using a classical mechanism. The addressing mechanism must be distributed and low energy consuming. Because of the network size and its location, which could not be accessible to human beings, an addressing mechanism should be adapted to these constraints in order to let the sensor do a self-configuration.

In this paper, an automatic addressing mechanism based on a clustered architecture is proposed. The proposed algorithm is completely distributed and allows to avoid address duplications. The address allocation is based on a systematic pattern. As with Banyan networks, it can be used to perform auto-routing. Actually, the auto-routing mechanism lets every node know every route to another one, and this, without the use of any routing algorithm. Routes are constructed at the same time as the cluster and the addressing scheme. Section 7.2 gives a review of the related work. In Section 7.3 a presentation of the algorithms is given, while Section 7.4 deals with the evaluation of the algorithms. Finally, on-going works will be presented in Section 7.6.

7.2 Related Work

7.2.1 Cluster Mechanisms

Cluster formation algorithms can be classified into:

- implicit (nodes constitute themselves into groups, without clusterhead), or explicit (nodes are grouped into clusters with a clusterhead);
- Active (clusters are the result of a dedicated protocol) or passive (clusters are built from information obtained by overhearing MAC messages used to transmit the data traffic);
- Hierarchical (clusters of clusters) or non-hierarchical;
- Centralised or distributed, distributed algorithms being themselves possibly emerging if they allow to obtain a foreseeable global result in a deterministic or stochastic fashion.

The easiest way to build clusters is to choose the lowest address between nodes. This was proposed by Gerla in [2] where a single-hop clusters construction algorithm is described (each node in a cluster can reach another node in the cluster through the clusterhead). The author's goal was to spatially reuse the bandwidth, by building clusters, to control the bandwidth and to have a more stable topology. Creating single-hops clusters allows to reuse power control algorithm of cellular network (cf. [3]).

To reduce the overhead due to the control traffic used to build the clusters, Gerla et al. [4] suggest an algorithm using a minimal information transmitted in the MAC data frames. The overhearing by all the nodes in a same neighbourhood of all the traffic allows them to obtain this information. This algorithm is well suited when nodes are highly mobile. Generally, cluster formation algorithms are decomposed into a building phase and a maintenance one, particularly when nodes are moving. Generally, the first phase assumes that all nodes are motionless. To relax this hypothesis, Basagni [5] (see also [7, 8] for some considerations on the performance) proposed an algorithm (DMAC) which assigns a weight to each node so that the clusterheads are the nodes having the greater weights and can never be neighbours. The weight is variable, and can depend on the speed or the power level.

In [9], Kawadia and Kumar proposed an integrated routing, power control and implicit clusterisation algorithm, CLUSTERPOW and tunnelled CLUSTERPOW, for networks for which the distribution of the nodes is heterogeneous. It is a multi-hop routing algorithm for which each node has several power levels and chooses the smallest one to reach the destination. Each power level then defines a cluster: to reach a remote destination, the node must send the information with its higher power level, which comes down to send it to another cluster because the network is not uniformly distributed.

LEACH' authors [10] proposed an auto-configurable architecture based on clusters by minimising the energy of the nodes. The clusterheads transmit the data directly to a base station, and so spend more energy than the other nodes. To balance the energy consumption, the authors suggest an algorithm where each node becomes periodically clusterhead with a probability which increases with the time spent since the last time when it was a clusterhead, said probability being chosen so that the mean number of clusterheads is a parameter of the algorithm. Since this algorithm gives only mean guarantees for the number of clusters and their locations, a centralised version of this algorithm (LEACH-C) is proposed. It allows to derive the optimal configuration to minimize the spent energy from the exact location of the nodes with a simulated annealing (the problem being NP-hard).

To solve the problem of the absence of guarantee on the good formation of the clusters inherent to LEACH, Younis and Fahmy [11] proposed the HEED algorithm which allows the selection of a clusterhead as a function of its energy and a cost function, depending on the desired objectives, on the number of its neighbours, on their proximity or on the mean value of the minimal power necessary to be reached by its neighbours. Depending on the objectives, very dense clusters or balanced ones can be obtained. In [12] it is also allowed to obtain dense clusters in a completely distributed fashion.

Finally, Chan and Perrig [13] proposed an algorithm which allows to obtain perfectly homogeneous clusters by minimising overlaps, the complexity of which depends only on the node density. Periodically, each node declares itself as clusterhead. It then counts the number of "loyal followers", which is the number of nodes for which it would be the unique possible clusterhead if it would be clusterhead. If this number is higher than a threshold, it becomes clusterhead. By counting the number of "loyal followers", and not only the number of nodes which can belong to several clusters, the chosen candidate clusterhead is the one for which the obtained cluster produces a minimal overlap. This generates a repulsing effect between clusters, and so a better distribution of the clusters.

While the existing clustering algorithms allowed us to build only one-hop clusters, in [14] the authors proposed a d-hop clustering algorithm based on the address of each node. The idea is to select the highest address in the d-neighbouring of a node and to choose the node with the highest address as the clusterhead of the network. This algorithm is then highly scalable and the time required to form the entire network does not delay the other communications.

The impact of the density of clusters has been studied in [15] and a hierarchical clustering algorithm has been proposed. The authors showed how aggregation of data can be fruitfully used through hierarchical clustering to save energy and they investigate the optimal number of clusters. The proposed algorithm allows each node to become a level i clusterhead with a given probability pi; it elaborates a method to optimally choose the pi and the cluster depth. Data aggregation is also investigated in [16], with a comparison of the performance of a multi-hop routing network versus a single-hop routing network.

At last, clustered networks can be made of specific nodes playing the role of clusterheads, with higher capabilities and performance than other network nodes. In this case, the network is said heterogeneous. Otherwise, all the nodes are of the same type and any node can become clusterhead. In [16], the authors study the performance of a heterogeneous network versus a homogeneous one, in terms of energy. They also compare the multi-hop case with the single-hop one, and propose a variant of LEACH adapted to the multi-hop case called M-LEACH.

In this work, a clustering technique based on the Min-Max algorithm presented in [14] is developped. It aims to build clusters which are trees with a given depth allowing to automatically assign addresses. The cost function to become clusterhead (as in [11]) is the number of neighbours, in order to minimize the address space. The underlying idea of Bandyopadhyay and Coyle [15] is to use both multi-hop routing and hierarchical aggregation, in order to minimize the energy. With the algorithm proposed here, it will be possible to use the existence of trees, with a parametrisable depth, inside the clusters, in order to implement a multi-hop hierarchical data aggregation.

7.2.2 Addressing Mechanisms

There are two different types of approaches for addressing mechanisms [17]: (1) stateful protocols and (2) stateless protocols. Whereas stateless protocols rely on the probability of uniqueness of an address, stateful protocols ensure this uniqueness at the cost of a lot of exchange between nodes. Stateful protocols can again be subdivided in three categories, depending on whether the address table is centralised (CAC, i.e. Centralised Auto-Configuration), distributed (DAC, i.e. Distributed Auto-Configuration), or disjoint (DAT, i.e.

Disjoint Allocation Table). Three categories can again be separated from stateless protocols: passive DAD, weak DAD and query DAD.

7.2.2.1 Stateful Protocol

The most famous CAC protocol is DHCP (Dynamic Host Configuration Protocol) [18]. DHCP cannot be used with clustered sensor network because DHCP server needs to be at one hop to its client. In [19], a mechanism is proposed wherein a centralised management of the allocation table is used by periodically electing a node responsible for maintaining the allocation table of all the already assigned addresses, their correspondence with the MAC addresses, and their expiration times. This node periodically sends the allocation table in network verification packets.

MANETconf [20], Boleng's and ProphetAllocation [21] belong to this category of stateful protocols, and have a similar behaviour, except that the allocation table maintenance is distributed. Each node maintains a table of the allocated addresses and another one of the addresses being allocated. When a node needs an address, it chooses one of its neighbours and sends it a request. This neighbour assigns an address and informs the network. If this address is not already allocated, the allocation process is terminated and the allocating node sends an information message to the network announcing that this address is actually allocated; otherwise another possible free address must be tried.

The DAT (Disjoint Allocation Tables) protocol splits the address space into several sub-spaces which are distributed among all the nodes of the network. When a node needs an address, it sends a request to another node which can assign a unique address without any possible duplication. In order to avoid to loose address spaces when a node fails, synchronisation mechanisms are implemented, which supposes that nodes periodically broadcast their address space tables.

7.2.2.2 Stateless Protocol

Contrary to stateful protocols, stateless protocols do not use allocation tables. They are distributed protocols. Instead, they use a randomly generated address or the result of a given series. The uniqueness of the chosen address is verified with an address collision detection mechanism, called DAD (Duplicated Address Detection). The difference between the different protocols of this category is mainly due to the difference of the performance of the DADs.

With the QDAD (Query DAD), when a node needs an address, it assigns itself a random one, and asks the network whether this address is free or not.

If this address is already allocated, the corresponding node informs the new node.

With the WDAD (Weak DAD) protocol [22], each node generates an initialisation key, which is distributed in the same time as its address in the routing packets. These keys are stocked by all the nodes and redistributed in all routing packets. If a node receives a routing packet containing an address being also in its routing table, it compares both keys. If they are different, an address collision is detected and the entry is marked in the routing table. If two nodes choose a same key for a same address, the conflict cannot be detected by WDAD. The probability of the non-detection of collision decreases with the increase of the size of the value space of the key, but in this case the size of the routing packets increases: a trade off must be found.

With the PDAD (Passive DAD), the information about addressing is no more sent in an active way. The collision is detected only by observing the routing tables. Actually, some events can exist only when collisions occur. For instance, when exchanging routing information, if a node "1" receives an indication that a node "2" is between "1" and another node, it can be deduced if "1" has no neighbour numbered "2" that it exists another node "1" in the network.

A distributed address allocation mechanism, with no possible duplication, is presented below.

7.3 Presentation of the Algorithms

A wireless fixed sensors network is considered. Sensors are nodes with limited capacity and limited power. Network protocols must be optimised to reduce overheads in order to maximize their life duration. Cluster formation mechanism impacts on the life of nodes. Indeed, location of clusterheads must be optimised in order to save energy by relaying cluster nodes packets to the gateway. Thus, the starting mechanism of a clustered network, as the addressing mechanism too, is very important to save energy. A cluster creation mechanism is proposed which aims to put the clusterhead in the center of its cluster. The addressing mechanism impacts on the energy because each addressing mechanism loses less or more energy by sending less or more packets. The mean time required for obtaining an address is quite important too. An addressing mechanism which is totally distributed and requires only two packets exchanges for a node to get a unique address is proposed. Then, it is obvious that the routing mechanism is very impacting. Routing mechanism needs to discover routes, to keep routing table, etc. An auto-routing mechanism which is only based on the previous addressing mechanism is developed.

7.3.1 Cluster Mechanism

The objective is to develop a heuristic for clustering a network which is scalable, not time-delaying and which permits to choose clusterheads in a large and dense fixed sensor network.

A mechanism which sets up multi-hop clusters based on the degree (number of neighbour) of the nodes in the network is presented. The mechanism is based on the heuristic presented in [14]. This algorithm leads to a multi-hop cluster formation based on node's addresses. A new selection criterion of a node in order to be the clusterhead is designed, so as to choose the clusterhead with the largest number of neighbours and to expect it to be more or less in the center of its cluster. We choose the clusterheads according to their node degree rather than choosing them according to their address as in [14].

A node has to maintain three lists: WINNER, OWNER and SENDER. We define a Cluster Address Auto-Configuration (CAAC) packet by the information (degree of a node, identification of this node). The heuristic runs for $2d$ rounds, where d is the maximum number of hops allowed in the future cluster.

Flood-max

At the beginning of the first round, each node locally broadcasts a CAAC packet which contains its own degree (initial WINNER value) and identification (initial OWNER value). After all CAAC packets of neighbouring nodes have been heard for this round, the node chooses the largest value among its own WINNER value and the WINNER values received from its neighbour. This value is then its new WINNER value and its new OWNER value is the owner value associated with the new WINNER value. The new SENDER value is the identification of the neighbour which sent the new WINNER value. This process continues for d rounds.

Flood-min

Each node chooses the smallest WINNER value after all neighbouring nodes have been heard and updates the OWNER and the SENDER values. This process continues for d rounds.

7.3.1.1 Clusterhead Selection Rules

The following rules explain the logical steps of the heuristic that each node i runs on the logged entry. Rule 1 will tell when i node knows that it is a clusterhead. Rule 2 and 3 will tell when i node knows which other node is the clusterhead it is linked to.

Clusterhead Selection Rules

Rule 1: First, i looks for its address id in the 2nd d rounds of flooding. If it gets it, then it elects itself as a clusterhead and skips the rest of the heuristic.

Rule 2: Else, i looks for a node pair (a node id which appears at least once in the Flood-max and once in the Flood-min). It selects the minimum node pair to be the clusterhead. If no node pair is found, it proceeds to Rule 3, else it skips the rest of the heuristic.

Rule 3: If i found no node pair, the node with the highest degree is elected as the clusterhead.

Let us define a "gateway" node as a node that belongs to two different clusters. If all neighbours of a node have the same clusterhead selection, then this node is not a "gateway" node. If there are neighbouring nodes with clusterhead selection that are different, then these nodes are "gateway" nodes. Once a node has identified itself as a "gateway" node, it then begins to inform (convergecast) its clusterhead by sending a list formed with its node id, all neighbouring "gateway" nodes and their associated clusterhead to its father. A node uses its SENDER table to determine its father. The process continues with the father which adds its own id to the previous list and sends it to its own father. When the clusterhead has heard each of its neighbours, it knows all the links between it and nodes in its cluster. Moreover it knows all the link between its cluster and the other neighbouring clusters thanks to the data provided by the "gateway" nodes.

Rule 4: This heuristic can generate a clusterhead that is on the path between a node and its elected clusterhead. In this case, the first clusterhead to receive the convergecast will adopt the node as one of its children.

In [23], the authors present a generalised Max-Min heuristic to any criterion (instead of using the address) and prove the correctness of the generalised heuristic. The heuristic proposed in today's paper is a particular case of the

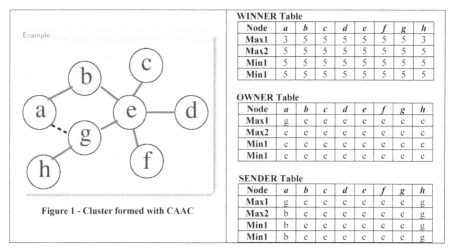

Figure 7.1 Cluster formed with CAAC.

generalised Max-Min heuristic. Moreover, the authors prove than the output of this generalised algorithm is a set of connex component and all these components are trees with a maximal depth of d. The roots of these trees are the clusterheads elected by the proposed heuristic.

Once the creation of clusters is finished, we are now able to assign a single address to each node. The clusterhead degrees of a cluster will permit us to choose the optimal parameter of the addressing algorithm which is presented below.

7.3.2 Addressing Mechanism

Let us propose an addressing mechanism which may be classified as a stateless mechanism, in which there is no need for an addressing table, but where uniqueness is ensured.

The originality of this mechanism is the possibility to address nodes with a unique address in the network, by simply exchanging two packets. Nodes are addressed thanks to the tree generated by the previous algorithm. Let us consider a tree and denote CH (ClusterHead) the root of this tree. Let us denote C as the maximum number of children in this tree. Let d, called "tree module", be an integer strictly greater than C: $d > C$.

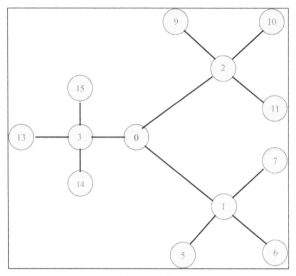

Figure 7.2

Mechanism

Let $@_i$ be the address of node i and e_i its number of children ($e_i \leq C$). Suppose that a node j, without address, is in the neighbourhood of node i and seeks to obtain an address from i.

Rule 1: if $e_i < C$, then $e_i = e_i + 1$ and $@_j = d * @_i + e_i$

Rule 2: if $e_i = C$, then i does not give address to j.

Example

Figure 7.2 shows a cluster with a maximum number of children equal to 3. The tree module is chosen to 4 ($d = 4$). At the beginning the address of the CH is 0. Then a node asks node 0 for an address. Node 0 does not have a child. The address it assigns to the new node is: $1 = 0 * 4 + 1$. When the future node 14 asks for an address to the node 3, node 3 already has a child which is node 13. The node 14 will get the address $14 = 3 * 4 + (1 + 1)$.

The proof of the uniqueness of the CAAC addresses is obvious (Uniqueness of the Euclidean division).

Suppose that the clusterhead has the maximum degree of its cluster. Therefore, the maximum number of children in such a cluster will be the clusterhead degree. In this case, the tree module d will be chosen equal to the degree of the clusterhead plus one.

As a consequence, *Rule 2* is not useful when the previous cluster creation mechanism is used because in this case, the clusterhead has the maximum degree of its cluster. Moreover, the clusterhead has not to determine C in its cluster since it knows that C is equal to its degree.

The uniqueness is ensured. The algorithm only needs two packets in order to assign a unique address. This addressing mechanism is the base of the auto-routing mechanism which is proposed below.

7.3.3 Auto-routing mechanism

The main feature is that this addressing mechanism lets every node know the route between every other node without using a routing mechanism. The father's address of a node is obtained by dividing the node address by the module tree. By using Euclidean division, the route between them may be derived from the following algorithm:

Lemma 1. *Let us denote i a node with address $@_i$. Let us denote j its father if it exists. $@_j = @_i/d$ where "/" stands for the Euclidian division.*

Intra-routing Algorithm

Let us denote i a node with address $@_i$ and j another node with address $@_j$. An auto-routing algorithm which leads to a route between i and j is presented below.

Node i sets the list of j and node j fathers. Then it sets a list including itself and its own fathers. The first node which appears in both lists is the linking node.

Example of the auto-routing algorithm

Suppose that node 5 wants to join node 13. The node 5 calculates the list of its fathers (5, 1, 0) and the one of node 13 fathers (13, 3, 0). The first father which appears in both lists is the linking node between 5 and 13. In this example, 0 is the linking node.

Route from 5 to 13: {5, 1, 0, 3, 13}

The auto-routing mechanism can be improved by keeping a table of the nodes already heard. The optimised rule becomes: when a first node wants to send a packet to a second one, it first looks for a father of the second one in its "listening" table and sends the packet to the father if it exists.

> *Example of the improved auto-routing algorithm*
> Suppose that node 5 has already heard node 3 and wants to send a packet to node 13. Node 5 first calculates the list of node 5 fathers and looks for a node it has heard. Then node 5 will send to the first father it knows, node 3. Then, node 3 will relay the packet to its child, node 13.
>
> Route from to 3:{5, 3, 13}

7.4 Performance Evaluation

7.4.1 Cluster creation with CAAC

Let us define $N_c(n, R, A)$ as the number of clusters formed by the CAAC algorithm in a network topology over an area A, with n nodes and a transmission radius equal to R. Let us evaluate the algorithm for dense network.

First let us generate n random nodes and apply the CAAC algorithm and compute the mean number of clusterheads for 20000 generated sets of nodes.

Figure 7.3 shows that the number of clusters formed by our algorithm converges towards a constant when the number of nodes increases. Indeed, for a given area and a given transmission range, when the density increases, the number of nodes a cluster can contain is not upper bounded. A consequence is that the number of clusters does not increase with the number of nodes.

The density of such a network may be defined by $\lambda = N/A$. $\lambda \pi R^2$ is the real entry of this algorithm. Indeed, $\lambda \pi R^2$ is expected to be the average number of neighbours in such a network which means that two networks are topologically equivalent if and only if this value is the same for the both of them. We will then study the evolution of the number of clusters formed by the CAAC algorithm as a function of $\lambda \pi R^2$. The results of both algorithms are presented in Figure 7.4. The difference between the two results is not very important, but let us note that the result of our algorithm does not depend on the node addresses. Indeed, the heuristic of Amis et al. [14] does not give, for the same topology, the same number of clusters since it depends on the addresses of nodes.

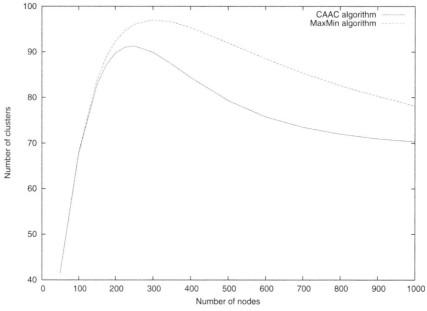

Figure 7.3 Number of clusters with $S = 100 \times 100$, $R = 5$, $d = 3$.

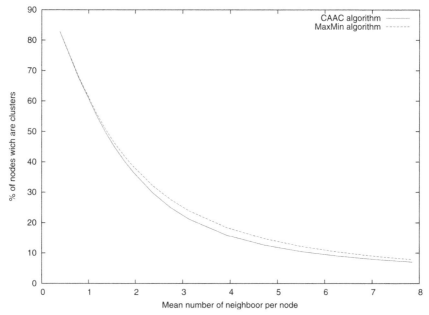

Figure 7.4 Percentage of nodes which are clusters with $S = 100 \times 100$, $R = 5$, $d = 3$.

Figure 7.4 shows that the performance of Amis' Max–Min algorithm and our algorithm are almost identical, when focusing on the number of cluster-heads. Amis' Max–Min algorithm is slighty better than our heuristic. This is explained by the fact that adding the constraint to have clusterheads with the highest possible degree necessarily leads to a larger number of clusters for given depth than without this constraint. However, this implies, referring to [14], that the performance of our algorithm is better than the LCA (Linked Cluster Algorithm) heuristic [24] (remember that the proposed clusterhead creation algorithm has a time complexity in $O(d)$ whereas LCA has a time complexity in $O(n)$ for large wireless networks).

LEACH leads to single-hop clusters, which is not very useful for aggregation, data fusion. The d parameter of CAAC mechanism is therefore an interesting point compared to LEACH. Moreover, LEACH leads to a larger number of clusters (because of single-hop), which means more expensive exchanges between clusterheads and the base station.

Though they claim that their algorithm could be extended to the multi-hop case, the actual algorithm which is presented by authors of HEED in their paper [11] is only designed for the single-hop case and the way to extend it to the multi-hop case does not seem easy. Moreover even if it is possible, the way to put the maximal number of hops as a parameter of their algorithm is not clear.

The network topology can be tuned by choosing adequate value of d (the tree depth): the number of elected clusterheads or the density of clusterheads in the network can be modified. This allows a control of the network topology

7.5 Addressing Mechanism

This addressing mechanism is distributed and scalable. Indeed, the time needed to configure a cluster is in $O(n)$ where n is the number of nodes in the cluster. In the worst case, the time needed to configure a network of N nodes is in $O(N)$. In average, the time configuration of this mechanism scales with $O(1/d_{CH})$, where d_{CH} is the clusterhead density: $d_{CH} =$ (number of clusters)/(number of nodes). This addressing mechanism leads to the auto-routing mechanism which is proposed below.

The number of exchanges for addressing a cluster of N nodes is $2N - 1$ (one request and one answer are necessary when inserting one node). A node address is constituted by its CAAC address and the cluster address. This identifier may be the MAC address of the clusterhead. In a cluster, all the routes from a node to its clusterhead (upload) are implicit: each node sends its packet to its father. However it should be useful to make possible for gateways or clusterheads to send packets from them to a node. This is required when a gateway wants to send a command to a terminal node (e.g. in actuator networks). This is very easy with our algorithm, due to the knowledge by every node of the father of every other node. Moreover, communications between nodes are possible. Gateways between clusters could be imagined. The same algorithm will permit these communications.

7.6 Conclusion

In this paper, an address assignment mechanism was proposed, using node clusters. This algorithm is based on the Max–Min algorithm to build the clusters, but it has been adapted to better choose the clusterheads in order to have clusterheads with a high number of neighbours. This address assignment mechanism is designed for wireless fixed sensor networks, which is usually the case for sensor networks. It is particularly suitable for very large sensor networks, where automatic address assignment cannot be avoided, because of its scalability. This mechanism has the key feature to lead directly autorouting. Actually, like in Banyan networks, the systematic pattern used to assign the addresses allows to derive locally on a node the next node on the route without the help of any routing protocol.

In future work, we plan to determine the amount of energy we could save by using power control inside clusters and between clusters and the gateway by using the CAAC algorithm. We will study the effect of a node arrival or departure on the stability of the CAAC mechanism. The algorithm could be run periodically at large time intervals, but we plan to design an optimized maintenance mechanism.

References

[1] V. Mhatre and C. Rosenberg, Design guidelines for wireless sensor networks: Communication, clustering and aggregation, *Ad Hoc Networks Journal*, vol. 2, pp. 45–63, 2004.

[2] T. J. Kwon and M. Gerla, Clustering with power control, in *Proceedings of IEEE MILCOM'99*, Atlantic City, NJ, 31 October–3 November, vol. 2, pp. 1424–1428, 1999.

[3] C. R. Lin and M. Gerla, Adaptive clustering for mobile wireless networks, *IEEE Journal on Selected Areas in Communications*, vol. 15, pp. 1265–1275, 1997.

[4] M. Gerla, T. J. Kwon and G. Pei, On demand routing in large ad hoc wireless networks with passive clustering, in *Proceedings of IEEE WCNC*, Chicago, IL, 23–28 September, vol. 1, pp. 100–105, 2000.

[5] S. Basagni, Distributed clustering for ad hoc networks, in *Proceedings of the IEEE International Symposium on Parallel Architectures, Algorithms, and Networks (I-SPAN)*, Perth/Fremantle, WA, Australia, 23–25 June, pp. 310–315, 1999.

[6] S. Basagni, Distributed and mobility-adaptive clustering for multimedia support in multi-hop wireless networks, in *Proceedings of the IEEE 50th International Vehicular Technology Conference, VTC 1999-Fall*, Amsterdam, the Netherlands, 19–22 September, vol. 2, pp. 889–893, 1999.

[7] S. Basagni, M. Mastrogiovanni and C. Petrioli, A performance comparison of protocols for clustering and backbone formation in large scale ad hoc network, in *Proceedings of the 1st IEEE International Conference on Mobile Ad Hoc and Sensor Systems, MASS*, Fort Lauderdale, FL, 25–27 October, pp. 70–79, 2004.

[8] C. Bettstetter, The cluster density of a distributed clustering algorithm in ad hoc networks, in *Proceedings of IEEE International Conference on Communications (ICC)*, Paris, France, 20–24 June, vol. 7, pp. 4226–4340, 2004.

[9] V. Kawadia and P. R. Kumar, Power control and clustering in ad hoc networks, in *Proceedings of IEEE INFOCOM*, San Francisco, CA, 30 March–3 April, vol. 1, pp. 459–469, 2003.

[10] W. B. Heinzelman, A. Chandrakasan and H. Balakrishnan, An application-specific protocol architecture for wireless microsensor networks, *IEEE Transactions on Wireless Communications*, vol. 1, pp. 660–667, 2002.

[11] O. Younis and S. Fahmy, Distributed clustering in ad-hoc sensor networks: A hybrid, energy-efficient approach, in *Proceedings of IEEE Infocom*, Hong Kong, China, 7–11 March, vol. 1, p. 640, 2004.

[12] C. Wen and W. A. Sethares, Automatic decentralized clustering for wireless sensor networks, *EURASIP J. Wireless Commun. Netw.*, vol. 5, no. 5, pp. 686–697, 2005.

[13] H. Chan and A. Perrig, ACE: An emergent algorithm for highly uniform cluster formation, in *Proceedings of European Workshop on Wireless Sensor Networks (EWSN 2004)*, Berlin, Germany, pp. 154–171, 2004.

[14] A. D. Amis, R. Prakash, T. H. P. Vuong and D. T. Huynh, Max-Min D-cluster formation in wireless ad hoc networks, in *Proceedings of INFOCOM*, Tel Aviv, Israel, 26–30 March, pp. 32–41, 2000.

[15] S. Bandyopadhyay and E. Coyle, An energy efficient hierarchical clustering algorithm for wireless sensor networks, in *Proceedings of the 22nd Annual Joint Conference of the IEEE Computer and Communications Societies (Infocom2003)*, San Francisco, CA, 30 March–3 April, vol. 3, pp. 1713–1723, 2003.

[16] V. Mhatre and C. Rosenberg, Homogeneous vs heterogeneous clustered sensor networks: A comparative study, in *Proceedings of IEEE International Conference on Communications*, Paris, France, 20–24 June, vol. 6, pp. 3646–3651, 2004.

[17] K. Weniger and M. Zitterbart, Address autoconfiguration in mobile ad hoc networks: Current approaches and future directions, *IEEE Network*, vol. 18, pp. 6–11, 2004.

[18] R. Droms, Dynamic host configuration protocol, RFC 2131, 1997.

[19] M. Gunes and J. Reibel, An IP address configuration algorithm for zeroconf mobile multihop ad hoc networks, in *Proceedings of the International Workshop on Broadband Wireless Ad Hoc Network and Services*, Sophia Antipolis, France, September 2002.

[20] S. Nesargi and R. Prakash, MANETconf: Configuration of hosts in a mobile ad hoc

network, in *Proceedings of IEEE INFOCOM*, New York, vol. 2, pp. 1059–1068, 2002.

[21] H. Zhou, L. M. Ni and M. W. Mutka, Prophet address allocation for large scale MANETs, in *Proceedings of 22nd Annual Joint Conference of the IEEE Computer and Communication Societies (Infocom2003)*, San Francisco, CA, 30 March–3 April, vol. 2, pp. 1034–1311, 2003.

[22] N. H. Vaidya, Weak duplicate address detection in mobile ad hoc networks, in *Proceedings of the 3rd ACM International Symposium on Mobile Ad Hoc Networking & Computing*, Lausanne, Switzerland, pp. 206–216, 2002.

[23] A. Delye de Clauzade de Mazieux, M. Marot and M. Becker, Correction, generalisation and validation of the max-min cluster formation heuristic, in *Proceedings of TC6/LNCS Networking 2007*, Atlanta, USA, 14–18 May, pp. 1149–1152, 2007.

[24] D. J. Backer and A. Ephremides, The architectural organization of a mobile radio network via a distribute algorithm, *IEEE Transactions on Communications*, vol. 29, no. 11, pp. 1694–1701, 1981.

8

Placide: Ad Hoc Wireless Sensor Network for Cold Chain Monitoring

Rahim Kacimi, Riadh Dhaou and André-Luc Beylot

IRIT-ENSEEIHT, University of Toulouse, 2 Rue Charles Camichel, 31071 Toulouse Cedex 7, France; e-mail: {rkacimi, dhaou, beylot}@enseeiht.fr

Abstract

In this paper, we present new protocols for the deployment of ad hoc wireless sensor networks in a cold supply chain monitoring. The following protocols are conceived to be implemented over a ZigBee/IEEE802.15.4 protocol stack and aim at saving energy in order to increase the sensor lifetime and as a result the global network longevity within the given context. Indeed, the limited availability of energy within network nodes generates critical issues in wireless sensor networks (WSN). Therefore making good use of energy is fundamental. The solution is original for two reasons. Firstly, no base station is needed. Secondly, the energy save is integrated to the conception of application level functionalities.

Keywords: Wireless sensor networks, ZigBee/IEEE 802.15.4, energy saving, cold supply chain.

8.1 Introduction

A cold chain is a temperature-controlled supply chain and unbroken cold chain is an uninterrupted series of storage and distribution activities which maintain a given temperature range. The cold chain monitoring is based on complementary devices. There are many logistic solutions within the cold

D. D. Kouvatsos (ed.), Performance Modelling and Analysis of Heterogeneous Networks, 153–167.

chain area, but to get a cheap and reusable solution, we propose to integrate directly the sensors into the pallets. The users' needs reveal that many existing solutions enable to supervise cold in warehouses with already operative alarm systems. However, the phase of transport is much more problematic because of the loading and unloading periods, but also because trucks are not as well equipped as warehouses.

In addition to application constraints, as in many WSN solutions, it is essential to conserve energy to increase sensors lifespan. Several propositions aim at optimizing network operations at various levels in order to reduce energy consumption. The first approach operates at MAC layer. Primary techniques have been considered in MAC layer designs. S-MAC [1], TMAC [2] and TRAMA [3] are for instance based on listening schedules. The other technique is low-power listening adopted by B-MAC and WiseMAC [4, 5]. Topology control is a second approach to conserve energy, and is specific to dense sensornets [6, 7]. With topology control, some nodes shut down for extended periods of time, but the network colludes to ensure that enough nodes remain active to guarantee coverage and full connectivity. Recently a third category of applications has emerged, that of "mostly-off applications" [8], like, for instance, equipment monitoring for extended periods [9] and seismic monitoring of underwater oil fields [10]. In these applications, nodes are active only during brief periods to collect data.

This paper proposes a self-organizing and energy-efficient network solution to ensure the cold chain monitoring during transport phase. The remainder of the paper is organized as follows. In Section 8.2, we pose the problematic in the cold chain context and introduce the studied network scenario. In Section 8.3, the protocols constituting the Placide solution are described and an evaluation is presented in Section 8.4. Section 8.5 introduces some performance metrics and shows the results obtained by simulations. Section 8.6 provides some conclusions and points out aspects that will be considered in future research.

8.2 Problematic

To equip trucks with a WSN solution, we must consider several constraints of exploitation, deployment and energy consumption. Indeed, the pallets can be leased, sold and especially moved during the transport phase. Therefore, a centralized network solution or incorporating a base station is not very suitable. Moreover, in the studied case, we do not have real time constraints.

So, to implement a solution that does not need any specific equipment inside the trucks, it is excluded to have a system with a base station which would manage the whole communications and which could be interfaced with external communication systems. Then, the basic idea is the following. Sensors must be self-organized to exchange information during the trip. The absence of communication with external world makes useless real time alarm triggering; the sensors functions are then limited to measure the temperature and to communicate in delayed time their possible alarms.

Let us consider the following scenario. Each pallet is equipped with a sensor. Pallets are loaded in a truck and sensors are activated simultaneously by an external device. Sensors will then exchange information to have a complete knowledge of the load composition (presence of sensors). During the transport, pallets could be removed or added. The network must be able to adapt to these modifications. Some sensors suffering from low battery level will warn of their imminent deactivation.

At the end of the trip, an external operator comes to gather information. He should be able to get information from any sensor which must have a complete knowledge of all what occurred during transport: presence of sensors, alarms, breakdowns of sensors. To save the batteries, each sensor must stop its radio as long as possible.

8.3 Proposed Solution

The above described scenario leads us to the following conclusions:

- A direct visibility can be considered between all the sensors inside the truck. Consequently, only broadcast communications may be used.
- The non real time processing of information, even when the temperature increases strongly, allows the use of periodic alarm clocks with exchanges of messages between the nodes aiming at preserving a global vision of the network by communicating possible alarms. Moreover, it will ensure a higher lifespan of the batteries; the sensors will, most of the time, turn their radio off.

Our protocols are based on the basic Zigbee/IEEE 802.15.4 stack [11, 12]. The IEEE 802.15.4 MAC layer is parameterized according to the specified protocols. The network layer will be minimal. All the nodes will be FFDs (Full Function Devices) i.e. they will have maximum functionalities. The absence of base station (sink) prevents any other solution. To carry out the fixed objectives, it is thus necessary to specify application protocols and their associated PDU (Protocol Data Unit).

So, we can distinguish various phases of the exchanges which lead to the following protocols:

- Initialization (P1): after loading pallets into the truck, the sensors will start a phase of mutual recognition to set up an exhaustive vision of the network state.
- Addition of a sensor (P2): during the transport phase, sensors could be added to the loading and thus will have to participate to communications.
- Planned sensor removal (P3): when a sensor has a low battery level, it has to inform other nodes about its extinction.
- Not planned sensors removal (P4): during the trip, some pallets can be unloaded, the system must remain operational.
- Steady state (P5): the nodes will awake regularly to announce their presence, to gather alarms.
- Data collection (P6): send all the collected data during transport to the operator.

8.3.1 Solution

The proposed solution is based on effective algorithms and protocols which reduce the energy consumption. The objective is to design a powerful scheduling which organizes the cycle of sleep and activity of network nodes since the activity duration of a node must be much lower than inactivity. Moreover, the radio antenna consumes more energy than any other sensor component. In this work, we distinguish between two phases (Figure 8.1): the initialization phase (P1) and the phase of maintenance of information (steady operation). In this second phase, we try to schedule sensors communications in order to minimize collisions and retransmissions.

These choices have primarily an impact on the application layer protocols implemented and possibly on the parameter setting of the MAC layer. Here, each sensor chooses its activity periods according to its predecessor in the chain. Each sensor thus calculates its waking date in order to disseminate its information during the following cycle, towards its predecessor in the current cycle.

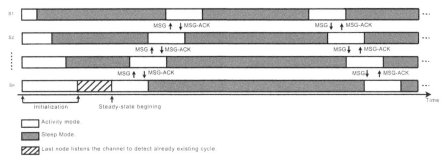

Figure 8.1 Placide cycle.

8.3.1.1 Initialization and Configuration of the Chain

This is probably the most complicated phase. An electromagnetic signal starts all the sensors. Each node picks a random duration (T_r) during which it senses the carrier. At the application layer, it corresponds to an idle period. The sensor which has the smallest (T_r) sends a synchronization message (SYNC) by providing its MAC address, it allots the number 1 in the cycle and indicates the date of its next waking.

In the best case, no other node sends a SYNC at the same time and there is no error of transmission. So, all the active nodes receive this message. They will pick a new random duration. At the end of this duration, one of them will send an acknowledgment (SYNC-ACK) by allotting number 2, providing its MAC address and calculating its date of waking.

The first node of the cycle may then sleep again and the construction of the cycle can continue. Once the last node transmits its message, none answers. It can fall asleep again. The cases of losses, collisions and errors are detailed in [13]. This phase is illustrated in Figure 8.2.

8.3.1.2 Steady State

After the network initialization part, the cycle is constructed and the last sensor added to the cycle builds a first message MSG with its date of waking and sends it to the following sensor. The first cycle starts in the opposite direction of its formation because of possible radio range problems. The last sensor is the only one to have all the information, particularly the number of sensors in the network.

We can distinguish three behaviors in the steady state, each one corresponding to the sensor position in the chain: at the beginning, in the middle or

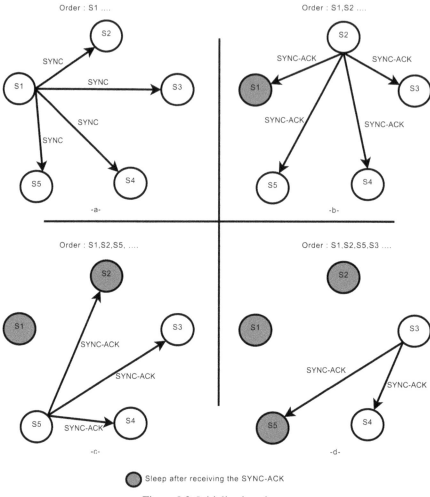

Figure 8.2 Initialization phase.

at the end. The main differences come from the fact that the first node (resp. the last node) of a cycle has no preceding node (resp. following node).

Steady State of a Sensor in the Chain   Two distinct parts are to be considered. Indeed, a sensor being located in the middle of the chain initially receives a message from its predecessor, handles it, adds its own information, and then sends it to its successor in the chain.

In the best case, sensor S_i receives an MSG, answers by an MSG-ACK, sends the MSG to the following and returns to sleep mode after receiving an MSG-ACK from its successor. Otherwise, S_i can handle two cases of losses: loss of S_{i-1} for lack of MSG reception; loss of S_{i+1} for lack of MSG-ACK from its successor.

8.3.1.3 Addition of One or Several New Nodes

When one or more nodes during an activity period, they sense the channel during their initialization phase. Thus, this period must be sufficiently long to receive messages from nodes of an already established cycle (in order to avoid the creation of a second cycle and the unavoidable addition of collisions and inconsistency in the protocol). When they receive a message indicating the presence of another cycle, they wait until the end of the current cycle. The PDU should contain the necessary information allowing the nodes to prepare their fusion with the existing cycle.

8.3.1.4 Unforeseen Loss of a Sensor

The unforeseen loss of a sensor results from the lack of answer that it should provide to its predecessor in the cycle.

If node S_i is lost, close nodes must react (S_{i-1} and S_{i+1}). As S_{i+1} is still sleeping, node S_{i-1} will try to replace S_i. Firstly, S_{i-1} tries to send messages to S_i, it discovers whereas the loss, so it extends its activity period for this cycle.

We note that for reliability reasons (transmission error in the messages and/or the acknowledgment), it is necessary to retransmit the message several times. These treatments could be considered either at MAC layer, or at the application layer. In our approach, we inhibit any acknowledgment at the MAC layer. Consequently, it is the application layer which will carry out this treatments. After some attempts, S_{i-1} decrees that S_i disappeared and communicates directly with S_{i+1}. The problem is settled, node S_{i+1} receives the messages and the chain continues its functioning. During the following cycles, the other nodes will be informed of the disappearance and will update their database.

8.3.1.5 Announced Loss of a Sensor

Another functionality which we considered is scheduled disappearance. When the remaining battery level reaches a given threshold, the node will switch off its radio. However, to keep a good functioning of the network, it should inform the others in the cycle preceding its stop. That can be done as follows:

- When it receives the message from its predecessor, it introduces into its acknowledgment (MSG-ACK) its disappearance during the following cycle.
- When it sends its message (MSG) to its successor, it informs it of its disappearance at the following cycle.

During the following cycle, the predecessor node will directly send its message to the successor node. In the following cycle, all the downstream nodes from the stopped node will be informed and thus their waking dates could be shifted for the next cycle.

The Placide solution has several advantages: it avoids collisions because the number of nodes communicating at the same time is very reduced; it allows to save energy because the activity duration of a node is much lower than its sleep period; and it is based on a topology where all the nodes have an identical role. Indeed, there is no particular node like a base station which represents sometimes an handicap, if this station breaks down all the network architecture collapses.

8.4 Evaluation

We implement and evaluate our Placide Protocols by using QNAP2 (Queuing Networks Analysis Package) [14]. This package allows the performance analysis of queuing networks by analytical and simulation methods. Since Placide is based on a simplified IEEE802.15.4 MAC protocol (mainly in contention mode) we initially implemented the CSMA protocol without acknowledgments because we have acknowledgments at application level (MSG-ACK). Thereafter, we implemented above, the protocols of initialization and steady-state.

8.4.1 Initialization Time

In order to reduce the number of collisions, after their activation, the nodes initially sense the channel during a random duration T_r uniformly distributed in the interval $[0, T_{max}]$. It is thus difficult to determine an optimal value of the timer parameter (which depends on the number of sensors). Indeed, the larger the network, the higher the probability that collisions occur. Consequently,

Table 8.1 Simulation parameters.

Message duration (ms)	
T_{SYNC}	1.44
$T_{SYNC-ACK}$	1.69
T_{MSG}	4.25
$T_{MSG-ACK}$	1.69
Current (mA)	
I_{TX}	19.5
I_{RX}	21.8
I_L	21.8
N: [10–100], $\theta = 3$, $T_A = 0$ ms	
T_{max}: [15–120] ms	

our objective is to determine the duration of the timer which leads to good results for various numbers of sensors. So, our simulations aim at dimensioning this parameter.

8.4.2 Energy Consumption

To exchange information with its neighbors, we point out that a node is subject to three modes: listening, reception and sending. To estimate the mean energy consumption we need to add the various consumptions corresponding to each mode. Thus the energy spent by each sensor is:

$$E = \sum E_{TX} + \sum E_{RX} + \sum E_L. \qquad (8.1)$$

These sums of energy depend on the current intencities according to the three modes: reception (I_{RX}), sending (I_{TX}) and listening (I_L) (cf. Table 8.1). We use the Telos sensor parameters [15].

8.5 Simulation Results

The simulation parameters of the messages duration and the intensities of the electrical current are listed in Table 8.1.

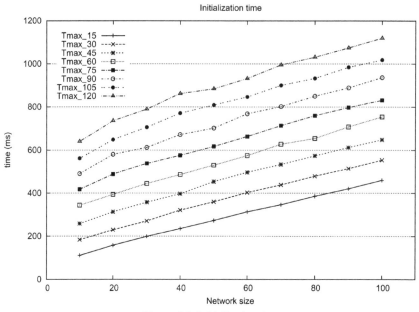

Figure 8.3 Initialization time.

Figure 8.3 shows the mean initialization time of a sensor network while varying the number of nodes N: [10–100] and T_{max} value: [15–120] ms. In this figure, we point out that the initialization time of the network increases linearly with its size (N). Moreover, for a given network size, the initialization time increases also linearly when T_{max} value grows.

In the same way, we represent the energy consumption (of the last node) during the initialization phase. Figure 8.4 shows this consumption expressed in mAh according to the size of the network and T_{max} value.

Let us consider a network of 33 nodes (truck of 33 pallets). Therefore, for this network size, we estimate the mean initialization times of the nodes and their energy consumption during this phase. These results are expressed respectively in ms and mAh according to the node position in the cycle. That is illustrated in Figures 8.5 and 8.6.

It is shown that initialization times and energy consumption linearly grow according to the node position, then in an exponential way for the last nodes. The reason is the growth of the temporization T_r caused by the reduction in the remaining number nodes.

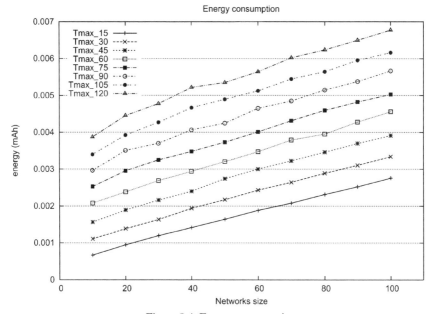

Figure 8.4 Energy consumption.

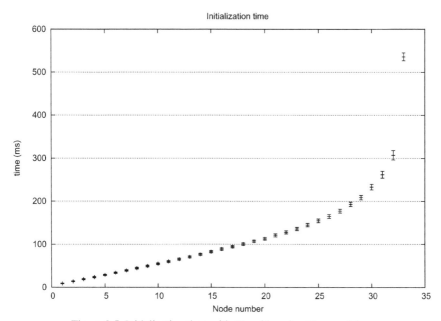

Figure 8.5 Initialization time with $N = 33$ nodes, $T_{\max} = 75$ ms.

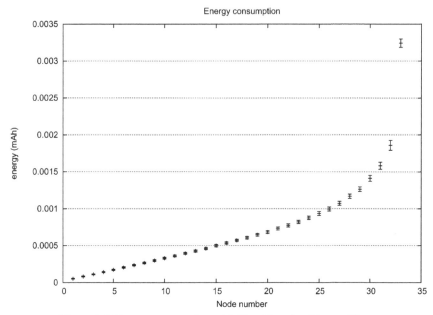

Figure 8.6 Energy consumption with $N = 33$ nodes, $T_{\max} = 75$ ms.

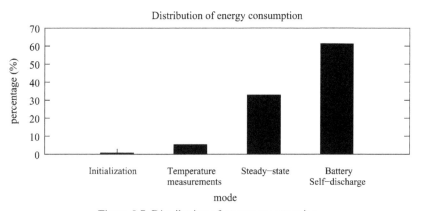

Figure 8.7 Distribution of energy consumption.

From our simulation results, let us consider the distribution of the energy consumption according to these modes: Initialization, Steady-state, Measurement of the temperature and battery self-discharge phenomenon. We consider a typical day of functioning of the system (24 hours): one initialization, a steady-state and temperature measurements every 10 minutes.

Figure 8.7 illustrates the distribution of the energy consumption of a sensor according to the various modes enumerated above. Initially, we notice that the proposed solution does not consume much energy. Indeed, the initialization and steady-state phases represent only $0.63\% + 32.84\%$ of the daily energy consumption of a sensor. Then, the measurements of the temperature represent 5.34% and the phenomenon of battery self-discharge remains the largest consumer of energy (61.19%).

8.6 Conclusion

Cold chain control is increasingly important because of the growing number of drugs and food labeled to be held under refrigerated and more tightly controlled conditions. Indeed, the inadequate control and/or monitoring of storage and transportation of temperature-sensitive products costs the industry enormously and causes food poisoning in the population. Temperature control cannot be guaranteed at every point of the cold chain, in particular during the transport phase.

In this work we proposed a wireless sensor network solution for refrigerated trucks. Sensors are integrated directly into the pallets. They exchange their data using broadcast transmissions. To save energy, sensors alternate between two operational modes: sleep and active mode. Simulation results show the accuracy of the proposed solution. We also note that Placide solution is very efficient to save energy and increase the sensors lifetime. It organizes the sensors, without a base station, in a cycle where the sleep periods are longer than the activity periods and the number of sensors communicating at the same time is very reduced. So, this solution is easily useable because after deployment it is self-organizing and does not require external intervention. It is energy-efficient and with the obtained results, our protocols guarantee several months for the sensors lifespan. Placide is based on IEEE802.15.4 standard which facilitate cooperation with other networks solutions where necessary.

As a perspective, we think that an analytical model can endorse the simulation results, then we will try to find other application areas to the Placide. In fact, we think that this solution will be very suitable in many monitoring fields which use wireless sensor networks.

Acknowledgement

This research is supported by the "Capteurs" grant, a National Telecommunication Research Network project. It is the French National Research and Innovation Programme for Telecommunication. Currently, this project is in deployment phase.

References

[1] W. Ye, J. Heidemann and D. Estrin, An energy-efficient MAC protocol for wireless sensor networks, in *Proceedings INFOCOM 2002, Twenty-First Annual Joint Conference of the IEEE Computer and Communications Societies*, vol. 3, pp. 1567–1576, 2002.

[2] T. Van Dam and K. Langendoen, An adaptive energy-efficient MAC protocol for wireless sensor networks, in *SenSys'03: Proceedings of the 1st International Conference on Embedded Networked Sensor Systems*, I. F. Akyildiz, D. Estrin, D. E. Culler and M. B. Srivastava (Eds.), ACM Press, New York, pp. 171–180, 2003.

[3] V. Rajendran, K. Obraczka and J. J. Garcia-Luna-Aceves, Energy-efficient collision-free medium access control for wireless sensor networks, in *SenSys '03: Proceedings of the 1st International Conference on Embedded Networked Sensor Systems*, I. F. Akyildiz, D. Estrin, D. E. Culler and M. B. Srivastava (Eds.), ACM Press, New York, pp. 181–192, 2003.

[4] A. El-Hoiydi, J.-D. Decotignie, C. C. Enz and E. Le Roux, Wisemac: an ultra low power mac protocol for the wisenet wireless sensor network, in *SenSys'03: Proceedings of the 1st International Conference on Embedded Networked Sensor Systems*, I. F. Akyildiz, D. Estrin, D. E. Culler and M. B. Srivastava (Eds.), ACM Press, New York, pp. 302–303, 2003.

[5] A. El-Hoiydi, J. Decotignie and J. Hernandez, Low power MAC protocols for infrastructure wireless sensor networks, in *Proceedings of the Fifth European Wireless Conference*, Barcelona, Spain, pp. 563–569, 2004.

[6] Y. Xu, J. Heidemann and D. Estrin, Geography-informed energy conservation for ad hoc routing, in *MobiCom '01: Proceedings of the 7th Annual International Conference on Mobile Computing and Networking*, ACM Press, New York, pp. 70–84, 2001.

[7] B. Chen, K. Jamieson, H. Balakrishnan and R. Morris, Span: An energy-efficient coordination algorithm for topology maintenance in ad hoc wireless networks, *Wireless Networks*, vol. 8, no. 5, pp. 481–494, 2002.

[8] L. Yuan, W. Ye and J. Heidemann, Energy efficient network reconfiguration for mostly-off sensor networks, in *Proceedings of Sensor and Ad Hoc Communications and Networks, SECON '06*, IEEE, vol. 2, pp. 527–535, 2006.

[9] N. Ramanathan, M. Yarvis, J. Chhabra, N. Kushalnagar, L. Krishnamurthy and D. Estrin, A stream-oriented power management protocol for low duty cycle sensor network applications, in *Proceedings of the Second IEEE Workshop on Embedded Networked Sensors*, pp. 53–62, 2005.

[10] J. Heidemann, W. Ye, J. Wills, A. Syed and Y. Li, Research challenges and applications for underwater sensor networking, in *Proceedings of the IEEE Wireless Communications and Networking Conference*, Las Vegas, pp. 228–235, 2006.

[11] ZigBee Alliance, *Zigbee Specifications*, April 2005.

[12] Wireless LAN medium access control (MAC) and physical layer (PHY) specification for low rate wireless personal area networks (LR-WPANs), 2003.

[13] R. Kacimi, R. Dhaou and A.-L. Beylot, *Placide: Specifications*, March 2007.

[14] SIMULOG, *Qnap2 User's Manual*, 1992.

[15] J. Polastre, R. Szewczyk and D. Culler, Telos: Enabling ultra-low power wireless research. in *IPSN: Information Processing in Sensor Networks*, pp. 364–369, 2005.

9

Energy Balancing by Combinatorial Optimization for Wireless Sensor Networks

J. Levendovszky, G. Kiss and L. Tran-Thanh

*Department of Telecommunications, Budapest University of Technology
and Economics, Magyar tud. krt. 2, H-1117 Budapest, Hungary;
e-mail: {levendov, kissg}@hit.bme.hu, ttl@cs.bme.hu*

Abstract

The paper is concerned with developing new energy balancing protocols for wireless sensor networks (WSNs) to maximize the life-span of the system. Optimal packet forwarding mechanisms from the nodes to the base station (BS) are derived which minimize the energy consumption of WSN and the statistical traffic characteristics of the sensed quantities are also taken into account. The tail distribution of the energy consumption is estimated by the tools of large deviation theory and the concept of generalized statistical bandwidth has been introduced to estimate the energy need of the network. Based on this estimated energy need sensor nodes do not always send packets to their nearest neighbours (having a "chain" type forwarding mechanism to the BS) but randomly decide to send either to the nearest neighbour or directly to the BS. In this way, the nodes located far from the BS may consume more energy (by being engaged with long-distance packet transmissions directly to BS), but the nodes being close to the BS are relieved from forwarding each packet handed down on the chain. As result, energy consumption can get smoothly distributed along the nodes which may increase the lifetime of the network. The optimal probability distribution of deciding to forward to packet to the nearest neighbour or the BS has been found by using a modified form of the Chernoff bound and some combinatorial optimization tools. Both

D. D. Kouvatsos (ed.), Performance Modelling and Analysis of Heterogeneous Networks, 169–182.

the theoretical results and the simulations demonstrate that the lifespan of WSN can be greatly increased by the new protocols.

Keywords: Sensor networks, energy balancing.

9.1 Introduction

Due to the recent advances electronics and wireless communication, the development of low-cost, low power, multifunctional sensors have received increasing attention [1]. These sensors are compact in size and besides sensing they also have limited signal processing and communication capabilities. However, the limitation in size and energy makes WSNs different from other wireless and ad-hoc networks [2]. As a result, new protocols must be developed with special focus on energy balancing in order to increase the lifetime of the network which is crucial in case applications (e.g. military field observations, living habitat monitoring, etc.; for more details see [4]), where recharging of the nodes is out of reach.

The paper, addresses energy balancing in WSN and develops novel packet forwarding mechanisms to increase the lifetime of the system. A random class of protocols will be investigated, where the sensor nodes randomly select other nodes for packet forwarding, subject to a probability distribution. For example, node i can choose to forward the neighbouring node closer to the base station (labeled as $i - 1$) with probability $(1 - a_i)$, or directly send the packet to the BS with probability a_i.

In the paper, the optimal p.d.f. a_i, $i = 1, 2, \ldots, N$ is found which maximizes the tail of life-time distribution, based on large deviation theory by using two approaches: (i) large deviation theory and (ii) iterative calculations. In this way, the new protocol can ensure longer WSN lifespan than the traditional packet forwarding mechanisms.

9.2 The Model

After the routing protocol (e.g. PEDAP [6, 7]) has found the path to the base station the subsequent nodes participating in the packet transfer can be regarded as a one dimensional chain labeled by $i = 1, \ldots, N$ and depicted by Figure 9.1.

The model is characterized as follows:

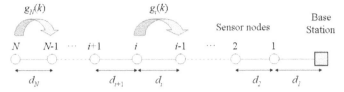

Figure 9.1 One dimensional chain topology of WSN packet forwarding.

- the topology is uniquely defined by a distance vector $\mathbf{d} = (d_1, \ldots, d_N)$, where d_i ($i = 1, \ldots, N$) denotes the distance between node i and $i - 1$, respectively;
- the energy needed to transmit packet over distance d is given as

$$g = \frac{d^\alpha \Theta \sigma_Z^2}{- \ln p_r} + g_{\text{Elec}}$$

dictated by the Rayleigh model, where d is the distance, α depends on the propagation type, p_r is the reliability of correct reception, Θ is the modulation coefficient, σ_Z^2 is the noise energy, while g_{Elec} represents the consumption of the electronics during transmitting and receiving;
- the initial battery power on each node is the same and denoted by C;
- the traffic state of the network is described by a binary vector $\mathbf{y} \in \{0, 1\}^N$ where $y_i = 1$ if sensor i is sending a packet to the BS;
- we assume that each sensor generates packets subject to an On/Off model, i.e. packet generation occurs with probability $P(y_i = 1) = p_i$, whereas the node does not generate packet with probability $P(y_i = 0) = 1 - p_i$;
- the traffic state of the network is represented by an N dimensional binary vector $\mathbf{y} \in \{0, 1\}^N$ and the corresponding the probability of a traffic state is given as $p(\mathbf{y}) = \prod_{i=1}^N p_i^{y_i} (1 - p_i)^{1 - y_i}$ assuming independence among the sensed quantities;
- the nodes are assumed to have finite buffers and the buffer length associated with node i is denoted by l_i.

We investigate two possible operations of a sensorial node: (i) the node does not use any buffer but transmits the packet to a destination upon receiving one (bufferless case); (ii) packets can queue up in the finite-length buffer waiting for transmission (buffered case).

The nodes operate in a time synchronous manner where the discrete time (clock signal) is denoted by $k = 0, 1, 2, \ldots$. As a result, a WSN is fully characterized by vectors **g**, **p**, **c**, and **l** respectively.

When analyzing the lifespan of the network, the following packet forwarding mechanisms are taken into account:

Chain protocol: Each node transmits packet to its neighbour lying closer to the BS. In this way, each node consumes minimal energy being engaged with short range energy transmission. However, each packet sent the BS consumes energy on each node along its path to the BS.

Random shortcut protocol: Node i can choose to forward the packet to its neighbouring node closer to the base station (labeled as $i - 1$) with probability $1 - a_i$, or directly send the packet to the BS with probability a_i.

Single-hop protocol: Each node sends its packet directly to the BS.

The paper is concerned with evaluating the lifetime of chain forwarding (case 1) and of random shortcut (case 2) protocols. Furthermore, our aim is to optimize probability vector $\mathbf{a} = (a_1, \ldots, a_N)$ in order to minimize energy consumption and thus maximizing the lifespan for WSNs operating with the random shortcut protocol.

9.3 Lifespan Estimation by Large Deviation Theory for the Bufferless Case

Let us first assume that the chain protocol is in effect. The energy consumed by sending a packet generated on node i to the BS is given as

$$G_i := \sum_{j=1}^{i} g_j \qquad (9.1)$$

and the average energy consumption up to time instant K is given as

$$\sum_{k=1}^{K} \frac{1}{N} \sum_{i=1}^{N} y_i(k) G_i \qquad (9.2)$$

The lifespan of node denoted by $\tilde{K}$ is defined as

$$\tilde{K} : P\left(\sum_{k=1}^{K} \frac{1}{N} \sum_{i=1}^{N} y_i(k) G_i < C \right) = e^{-\alpha} \qquad (9.3)$$

where $e^{-\alpha}$ is close to one and α is a reliability parameter.

By using the complementary probability

$$P\left(\sum_{k=1}^{K} \frac{1}{N} \sum_{i=1}^{N} y_i(k) G_i > C\right) = 1 - e^{-\alpha} \tag{9.4}$$

the life time evaluation is cast as a tail estimation problem, where bounds like the Chernoff inequality can be used to upperbound the tail

$$P\left(\sum_{k=1}^{K} \frac{1}{N} \sum_{i=1}^{N} y_i(k) G_i > C\right) \le \exp\left(\sum_{i=1}^{N} \mu_i(\hat{s}, G_i) - \frac{\hat{s}NC}{K}\right) \tag{9.5}$$

Here

$$\mu_i(s, G_i) := \log\left(E\left[e^{sy_iG_i}\right]\right) = \log\left(1 - p_i + p_i e^{sG_i}\right)$$

and

$$\hat{s} : \min_{s} K \sum_{i=1}^{N} \mu_i(s, G_i) - \frac{sNC}{K}.$$

By using the estimation above, one obtains

$$\exp\left(\sum_{i=1}^{N} \mu_i(\hat{s}, G_i) - \frac{\hat{s}NC}{K}\right) = 1 - e^{-\alpha} \tag{9.6}$$

and the lifespan of the simple chain protocol can finally be estimated by the following formula:

$$\tilde{K} = \frac{\hat{s}NC}{\sum\limits_{i=1}^{N} \mu_i(\hat{s}, G_i) + \log(1 - e^{-\alpha})} \tag{9.7}$$

If the random shortcut protocol is in effect then the packet generated by node i will travel in the chain till the first shortcut to BS. Let the node in which the shortcut takes place be denoted by λ_i. The distribution of λ_i is given as

$$P(\lambda_i = l_i) = a_{i-l_i} \prod_{j=i-l_i+1}^{i} (1 - a_j) \tag{9.8}$$

In this case the packet consumes

$$V_i := \sum_{j=i-l_i+1}^{i} g_j + \gamma_{i-l_i}$$

energy, where γ_{i-l_i} is the shortcut energy from node $i - l_i$ (i.e. the energy required to transmit the packet from node $i - l_i$ directly the BS). As a result, the average energy consumption is given as

$$\sum_{k=1}^{K} \frac{1}{N} \sum_{i=1}^{N} y_i \left(\sum_{j=i-\lambda_i+1}^{i} g_j + \gamma_{i-\lambda_i} \right) \tag{9.9}$$

Thus the lifespan is defined as follows:

$$\tilde{K} : P\left(\sum_{k=1}^{K} \frac{1}{N} \sum_{i=1}^{N} y_i \left(\sum_{j=i-\lambda_i+1}^{i} g_j + \gamma_{i-\lambda_i} \right) > C \right) = 1 - e^{-\alpha} \tag{9.10}$$

The probability in equation (9.10) can be rewritten as

$$P\left(\sum_{k=1}^{K} \frac{1}{N} \sum_{i=1}^{N} y_i \left(\sum_{j=i-\lambda_i+1}^{i} g_j + \gamma_{i-\lambda_i} \right) > C \right)$$

$$= \sum_{l_1} \cdots \sum_{l_N} P\left(\sum_{k=1}^{K} \frac{1}{N} \sum_{i=1}^{N} y_i \left(\sum_{j=i-\lambda_i+1}^{i} g_j + \gamma_{i-\lambda_i} \right) \right.$$

$$> C \,|\, \lambda_1 = l_1, \ldots, \lambda_N = l_N) \, P\left(\lambda_1 = l_1, \ldots, \lambda_N = l_N \right)$$

$$= \sum_{l_1} \cdots \sum_{l_N} P\left(\sum_{k=1}^{K} \frac{1}{N} \sum_{i=1}^{N} y_i \left(\sum_{j=i-l_i+1}^{i} g_j + \gamma_{i-l_i} \right) > C \right) \prod_{i=1}^{N} P\left(\lambda_i = l_i \right)$$

$$= \sum_{l_1} \cdots \sum_{l_N} \exp\left[\sum_{i=1}^{N} \mu_i(s, V_i) - \frac{sNC}{K} \right] \prod_{i=1}^{N} \left(a_{i-l_i} \prod_{j=i-l_i+1}^{i} (1 - a_j) \right)$$

$$= e^{-sNC/K} \sum_{l_1} \cdots \sum_{l_N} \prod_{i=1}^{N} e^{\mu_i(s, V_i)} \left(a_{i-l_i} \prod_{j=i-l_i+1}^{i} (1 - a_j) \right)$$

$$= \sum_{l_1} \cdots \sum_{l_N} \exp\left[\sum_{i=1}^{N} \mu_i(s, V_i) - \frac{sNC}{K} \right] \prod_{i=1}^{N} \left(a_{i-l_i} \prod_{j=i-l_i+1}^{i} (1 - a_j) \right)$$

$$= e^{-sNC/K} \prod_{i=1}^{N} \sum_{l_i} \left(e^{\mu_i(s, V_i)} \left(a_{i-l_i} \prod_{j=i-l_i+1}^{i} (1 - a_j) \right) \right) \tag{9.11}$$

where
$$\mu_i(s, V_i) := \log\left(E[e^{sy_i V_i}]\right) = \log(1 - p_i + p_i e^{sV_i}).$$

Introducing the extended logarithmic moment generation function as

$$\beta_i(s, V_i) := \log\left(\sum_{l_i}\left(e^{\mu_i(s,V_i)}a_{i-l_i}\prod_{j=i-l_i+1}^{i}(1-a_j)\right)\right) \tag{9.12}$$

one can write

$$P\left(\sum_{k=1}^{K}\frac{1}{N}\sum_{i=1}^{N}y_i\left(\sum_{j=i-\lambda_i+1}^{i}g_j+\gamma_{i-\lambda_i}\right)>C\right)$$

$$\leq e^{-sNC/K}\prod_{i=1}^{N}e^{\beta_i(s,V_i)} = \exp\left[\sum_{i=1}^{N}\beta_i(s,V_i)-\frac{sNC}{K}\right] \tag{9.13}$$

Comparing the bound with $1 - e^{-\alpha}$, we obtain

$$\exp\left[\sum_{i=1}^{N}\beta_i(\hat{s}, V_i)-\frac{\hat{s}NC}{K}\right]=1-e^{-\alpha} \tag{9.14}$$

where
$$\hat{s} : \min_s \sum_{i=1}^{N}\beta_i(s, V_i)-\frac{sNC}{K}.$$

The lifespan is the solution of the following equation:

$$\tilde{K} : \sum_{i=1}^{N}\beta_i(\hat{s}, V_i) = \frac{\hat{s}NC}{K}+\log(1-e^{-\alpha}) \tag{9.15}$$

As one can see, the above equation determines the lifespan as a function of vector **a**, the components of which represent the probabilities of shortcut on a given node. This relationship is denoted by $\tilde{K} = \Psi(\mathbf{a})$. Using equations and to evaluate $\Psi(\mathbf{a})$ for a given **a** vectors, protocol optimization can take place by searching in the space of **a**-vectors to find the optimal shortcut probabilities. This can be done by a gradient descent type of optimization given as follows:

$$a_i(n+1) = a_i(n)-\Delta\text{sgn}\left\{\frac{\Psi(\mathbf{a}(n))-\Psi(\mathbf{a}(n-1))}{a_i(n)-a_i(n-1)}\right\}, \quad i=1,\ldots,N \tag{9.16}$$

As a result, protocol optimization has been carried out in the following steps:

Given:	N - number of nodes, **p** - packet generate probability vector, **g** - energy vector, **c** - initial battery power vector;
Step 1.	select an initial **a**(0) shortcut probability vector;
Step 2.	evaluate the value of $\tilde{K} = \Psi(\mathbf{a})$ by solving the equation $\tilde{K} : \sum_{i=1}^{N} \beta_i(\hat{s}, V_i) = \frac{\hat{s}NC}{K} + \log(1 - e^{-\alpha});$
Step 3.	Perform the gradient search $a_i(n+1) = a_i(n) - \Delta\operatorname{sgn}\left\{\frac{\Psi(\mathbf{a}(n)) - \Psi(\mathbf{a}(n-1))}{a_i(n) - a_i(n-1)}\right\},$ where $i = 1, \dots, N$;
Step 4.	Check the stopping criterion (i.e. $\|\mathbf{a}(n+1) - \mathbf{a}(n)\| \le \varepsilon$) and go back to Step 2. if it is not met.

In the case of single-hop protocol we have $a_i = 1$, $i = 1, \dots, N$. Here, we obtain

$$\tilde{K} : P\left(\sum_{k=1}^{K} \frac{1}{N} \sum_{i=1}^{N} y_i(k)\gamma_i < C\right) = e^{-\alpha} \tag{9.17}$$

which leads to the following life span:

$$\tilde{K} = \frac{\hat{s}NC}{\sum_{i=1}^{N} \mu_i(\hat{s}, \gamma_i) + \log(1 - e^{-\alpha})} \tag{9.18}$$

where

$$\mu_i(s, \gamma_i) := \log\left(\mathrm{E}[e^{s y_i \gamma_i}]\right) = \log(1 - p_i + p_i e^{s\gamma_i})$$

is the energy required by the shortcut.

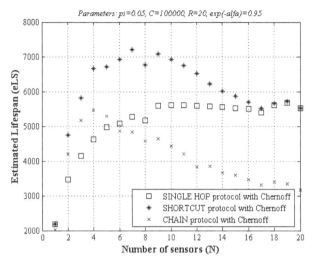

Figure 9.2 Estimated lifespan in the function of the number of sensors.

9.4 Performance Analysis and Numerical Results for the Bufferless Case

In this section a detailed performance analysis is given using the chain, the shortcut and the single-hop protocols. The aim is to evaluate the lifespan of a sensor network containing N number of sensors placed in an equidistant manner.

Figure 9.2 shows how the lifespan changes as the function of the number of nodes (N) in the case of the three methods described above. The distance between the base station and the farthest node was 20 meters.

One can see, that there is a maximum lifespan in the cases of chain and random shortcut protocols with the optimal number of nodes $N_{\text{Chain}} = 4$ and $N_{\text{Shortcut}} = 7$, respectively. Figure 9.2 shows that when the network is sparsely installed, both methods result in almost the same lifespan, while departing form the optimal number of nodes (either decreasing or increasing the number of sensors), the shortcut model definitely gives much higher (it is more than 37% in the case of $N = 7$) relative lifespan.

Figure 9.3 demonstrates the accuracy of lifespan estimation at the different protocols. One can see, that the Chernoff yields a relatively sharp estimation.

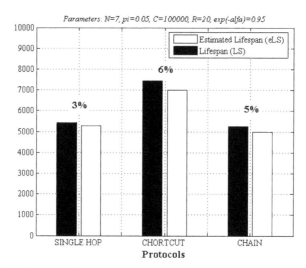

Figure 9.3 Lifespan and estimated lifespan values achieved by different protocols.

9.5 Lifespan Evaluation by Iterating the Node Energy Consumption Distribution for the Buffered Case Helpful

In this section, the lifespan will be calculated by iterating the probability distributions of the energy consumptions of the nodes. We use a more general model, where we consider a sensor node with capacity C and queue length L. A node can send packets to a set of nodes D, with a set of cost g_j, $j \in D$, and with a set of probability s_j, $j \in D$. To evaluate the distribution of the energy consumption of a single node, two iteration methods will be discussed.

9.5.1 Approximation No. 1

Let $F_i^{(k)}(X)$ denote the probability of that the energy consumption of node i from the initial state to the time instant k (but before sending a packet in time instant k) is less than X. This probability can be iterated as follows:

$$F_i^{(k+1)}(X) = P_0(k) \cdot F_i^{(k)}(X)$$
$$+ \sum_{j \in D} (1 - P_0(k)) \cdot s_j \cdot F_i^{(k)}(X - g_j) \qquad (9.19)$$

with the initial values $F_i^{(0)}(X) = 0$, if $X \leq 0$, and $F_i^{(0)}(X) = 1$, if $X > 0$.

Our objective is to compute $F_i^{(k)}(C)$ recursively. To achieve this end, let us define a set of control points with a set of value a_r, $r \in [1, M]$, $0 \leq$

$a_r \leq C$. We compute formula (9.15) in these control points and we use linear interpolation for approximating (9.19) on the other points.

9.5.2 Approximation No. 2

We assume that the probability distribution of packets sent by a node is known, and it is stored in an array called SPD (Sent Packets' Distribution). Let l_{max} denote the maximum number of packets which a node can send. We define an array SPC (Sent Packets' Coefficients), having $l_{max} + 1$ elements, the jth element of which stores the probability that after sending j packets, the energy consumption of the node is less than C. This probability (i.e. the probability that the energy consumption of the node is less than C) can be expressed as

$$\sum_{k=0}^{l_{max}} SPC[k]\,SPD[k] \qquad (9.20)$$

We determine the value of $SPC[k]$ as follows:

Let us divide the interval $[0, C]$ into e energy consumption levels. We define $I_k(m)$, $0 \leq m < e$ as the probability that after the kth time instant, the value of the energy consumption visits the mth level. Let $S(m, n)$ denote the set of numbers defined as follows: $j \in D$ is an element of $S(m, n)$ if the value of the energy consumption jumps from level n to level m by sending a with the cost g_j. To calculate $I_k(m)$, we use the formula below:

$$I_{k+1}(m) = \sum_{n=0}^{m} \sum_{j \in S(m,n)} s_j \cdot I_k(n) \qquad (9.21)$$

The initial values are: $I_0(0) = 1$, and for each $1 \leq m < e$ $I_0(m) = 0$. Let

$$SPC[k] = \sum_{n=0}^{e-1} I_k(n) \qquad (9.22)$$

Based on these approximations the lifetime of the network with reliability parameter ε is given as the first time instance $\tilde{K}$ when

$$\min_{i} \{P_i(\text{consumption})\} < \varepsilon \qquad (9.23)$$

where $P_i(\text{consumption})$ is the probability that the energy consumption of node i is less than C capacity, and is calculated by one of the approximations discussed above.

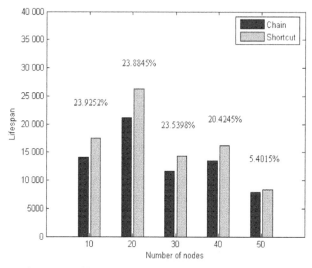

Figure 9.4 Lifespan in the function of equidistant sensors.

9.6 Performance Analysis and Numerical Results for the Buffered Case

9.6.1 Equidistant Nodes

In this section a detailed performance analysis is given by using the chain, the shortcut and the generalized random protocols. For determining the energy consumption of transmitting a packet, we assume that a packet is of 512 bytes length and it is transmitted at 10 dBm. Throughout the simulations we used $\alpha = 2$ as distance-exponent, assuming the second approximation model for evaluating the lifespan of the network.

The following numerical results indicate the lifespan in the case of chain and random shortcut protocols. When applying the random shortcut protocol the optimal **a** vector is found by the Matthias algorithm.

In this case, we assume that the sensors are placed in an equidistant manner. Figure 9.4 shows how the lifespan changes in the function of the number of nodes (N) in the case of the two protocol methods described in Section 9.2.

One can see, in most of the observed networks, the improvement in the value of network lifespan in almost 25%.

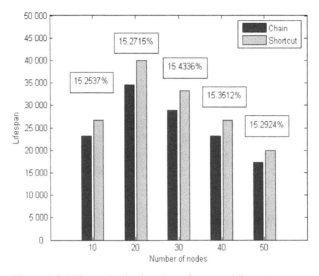

Figure 9.5 Lifespan in the function of non-equidistant sensors.

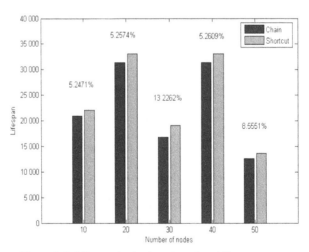

Figure 9.6 Lifespan in the function of equidistant sensors.

9.6.2 Non-Equidistant Nodes

Here, we observed networks that have nodes with non-equidistant arrangement. Due to the modified topology, the improvement differs from the previous case. The results are shown in Figure 9.5. If we assume that there is no receiver energy cost, then the results have changed as we can see in Figure 9.6.

9.7 Conclusion

In this paper, new packet forwarding mechanisms have been developed for energy balancing of WSN. A novel "random shortcut" protocol has been introduced and the optimal probability distribution for selecting destination for packet forwarding has been found by using large deviation theory. By this new protocol, the lifespan of WSN can be considerably increased.

Acknowledgements

The research reported here was partly supported by the Katholieke Universiteit Leuven under the grant GOA/98/06 of Research Fund K.U. Leuven, by INTAS Project 00-738 and by the Mobile Innovation Centre (MIK) in Hungary.

References

[1] C. Y. Chong and S. P. Kumar, Sensor networks: Evolution, opportunities, and challenges, in *IEEE Proceedings*, pp. 1247–1254, August 2003.
[2] A. Goldsmith and S. Wicker, Design challenges for energy-constrained ad hoc wireless networks, *IEEE Wireless Communications Magazine*, vol. 9, pp. 8–27, August 2002.
[3] Self-healing Mines, http://www.darpa.mil/ato/programs/SHM/
[4] A. Mainwaring, J. Polastre, R. Szewczyk, D. Culler and J. Anderson, Wireless sensor networks for habitat monitoring, in *Proceedings of the First ACM Workshop on Wireless Sensor Networks and Applications*, Atlanta, GA, September 2002, pp. 88–97, 2002.
[5] D. Puccinelli and M. Haenggi, Wireless sensor networks – Applications and challenges of ubiquitous sensing, *IEEE Circuits and Systems Magazine*, vol. 5, pp. 19–29, August 2005.
[6] W. Heinzelman, A. Chandrakasan and H. Balakrishnan, Energy-efficient communication protocols for wireless microsensor networks, in *Proceedings of the Hawaiian International Conference on Systems Science*, January 2000, pp. 1–10, 2000.
[7] H. O. Tan and I. Korpeoglu, Power efficient data gathering and aggregation in wireless sensor networks, *ACM SIGMOD Record*, vol. 32, no. 4, pp. 66–71, December 2003.

10

Estimation of Packet Transfer Time at Sensor Networks: A Diffusion Approximation Approach

Tadeusz Czachórski[1], Krzysztof Grochla[1] and Ferhan Pekergin[2]

[1]*IITiS PAN, ul. Bałtycka 5, 44-100 Gliwice, Poland;*
e-mail: {tadek, kil}@iitis.gliwice.pl
[2]*LIPN, Université Paris-Nord, 93430 Villetaneuse, France;*
e-mail: pekergin@lipn.univ-paris13.fr

Abstract

We propose a model based on diffusion approximation to estimate the probability density function of the distribution of a packet travel time in a multihop wireless sensor network. In its general form, the model assumes that the propagation medium and the distribution of relay nodes may be heterogeneous in space and that the system characteristics may change over time. It considers also the retransmission in case of a packet loss.

Keywords: Diffusion approximation, transient analysis, wireless networks, sensor networks.

10.1 Introduction

Prediction of a packet travel time in wireless sensor networks is still an open issue. The sensor networks, see e.g. [1] consist of a large number of simple nodes scattered randomly over a certain area, having ability to route packets to their neighbours and finally to the sink which collects the data sent to it via multihop transmission.

The topology of such networks is in most of cases uncertain and it changes in time (due to nodes movement or failures), hence special routing

D. D. Kouvatsos (ed.), Performance Modelling and Analysis of Heterogeneous Networks, 183–195.

algorithms were proposed, e.g. [5, 6] to face this situation. As it is also hard to introduce global addressing, the routing decision must be made without the complete information about the network. We consider the same network model as in [4] – a packet wireless network in which nodes are distributed over an area, but where we do not know about the presence, exact location, or reliability of nodes. The packets are forwarded to a node witch is most probably nearer to the destination, but it is also possible that a transmission may actually move the packet further away from the sink or send it to a node which is in the same distance to the destination (see for example [8]). It may also happen that a packet cannot be forwarded any further, that the inter-mediate node has a failure, or that the packet is lost through noise or some other transient effect. In that case, the packet may be retransmitted after some time-out period has elapsed, either by the source or from some intermediate storage location on the path which it traversed before it was lost.

10.2 Model Formulation

Recently Gelenbe [4] proposed a model based on diffusion approximation to estimate the mean transmission time from a source to destination in a random multihop medium. In this model a value of the diffusion process represents the distance defined as the number of hops between the transmitted packet and its destination (sink). Due to complex topology and transmission constraints, it is not sure that each one-hop transmission makes this distance shorter and the changes of the distance may be considered as random process. This jus-tifies the use of diffusion process to characterise it. Diffusion approximation is a classical mcthod used in queueing theory to represent a queue length or queueing time e.g. [3], in case of general independent distributions of interrarival and service times. Diffusion process is a continuous stochastic process but it is used to approximate some discrete processes, see [2], like – as mentioned above – the number of customers in a queue; here it represents the number of hops remaining to packet to the destination.

If $N(t)$ denotes the number of hops remaining to destination at time t, we construct a diffusion process $X(t)$ such that its density function $f(x, t; x_0)$ approximates probability distribution $p(n, t; n_0)$ of the process $N(t)$, $N(0) = n_0$: $f(n, t; n_0) \approx p(n, t; n_0)$. The density function $f(x, t; x_0)$

$$f(x, t; x_0)dx = P[x \leq X(t) < x + dx \mid X(0) = x_0]$$

is defined by the diffusion equation

$$\frac{\partial f(x, t; x_0)}{\partial t} = \frac{\alpha}{2} \frac{\partial^2 f(x, t; x_0)}{\partial x^2} - \beta \frac{\partial f(x, t; x_0)}{\partial x}, \tag{10.1}$$

where the parameters β and α define respectively the mean and variance of infinitesimal changes of the diffusion process. To maintain them similar to the considered process $N(t)$, they should be chosen as

$$\beta = \lim_{\Delta t \to 0} \frac{E[N(t + \Delta t) - N(t)]}{\Delta t},$$

$$\alpha = \lim_{\Delta t \to 0} \frac{E[(N(t + \Delta t) - N(t))^2] - (E[N(t + \Delta t) - N(t)])^2}{\Delta t}.$$

In general, the parameters may depend on time and on the current value of the process, $\beta = \beta(x, t)$ and $\alpha = \alpha(x, t)$, as the propagation medium and distribution of relay nodes may be heterogeneous in space and the system characteristics may change over time. We include this case in the proposed approach.

Gelenbe in [4] constructs an ergodic process going repetitively from starting point to zero and considers its steady-state properties. Here, to obtain the distribution (and not only the mean transmission time as given in [4]), we use transient solution of diffusion equation and we consider only a single process. Let us repeat that the process starts at $x_0 = N$ and ends when it successfully comes to the absorbing barrier at $x = 0$; the position x of the process corresponds to the current distance between the packet and its destination, counted in hops.

10.3 Model without Deadline and without Losses

In this simplest case we consider diffusion equation (10.1) with constant coefficients, supplemented with absorbing barrier at $x = 0$. This barrier is expressed by the boundary condition $\lim_{x \to 0} f(x, t; x_0) = 0$. The process starts at x_0: $X(0) = x_0$ and ends when it comes to the barrier. The diffusion process is defined at the interval $(0, \infty)$. Let us denote the solution of the diffusion equation in this case by $\phi(x, t; x_0)$; it is obtained using mirror method, see e.g. [2]

$$\phi(x, t; x_0) = \frac{e^{\frac{\beta}{\alpha}(x - x_0) - \frac{\beta^2}{2\alpha}t}}{\sqrt{2\Pi\alpha t}} \left[e^{-\frac{(x - x_0)^2}{2\alpha t}} - e^{-\frac{(x + x_0)^2}{2\alpha t}} \right].$$

f(x,t;x0) ────

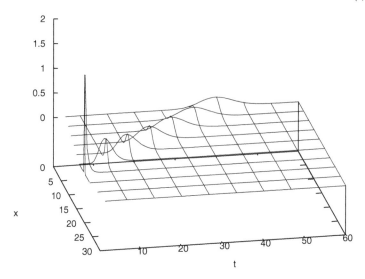

Figure 10.1 Density function $\phi(x, t; x_0)$ of the diffusion process with absorbing barrier, $x_0 = 20$, $\alpha = 0.1$, $\beta = -0.5$.

Figure 10.1 presents a plot of the function $\phi(x, t; x_0)$. The function allows us to determine the first passage time from $x = x_0$ to $x = 0$ and to estimate this way the density of a packet transmission time through x_0 hops from a node to the sink:

$$\gamma_{x_0,0}(t) = \lim_{x \to 0} \left[\frac{\alpha}{2} \frac{\partial}{\partial x} \phi(x, t; x_0) - \beta \phi(x, t; x_0) \right]$$

$$= \frac{x_0}{\sqrt{2\Pi \alpha t^3}} e^{-\frac{(x_0 - |\beta|t)^2}{2\alpha t}}.$$

Some exemplary curves of $\gamma_{x_0,0}(t)$ are presented in Figure 10.2.

10.4 Introduction of the Deadline

Denote by T the time after which a packet is considered lost and is retransmitted by the source. Knowing the density $\gamma_{x_0,0}(t)$ of the travel time from x_0 to 0, we can determine the probability $p_T = \int_T^\infty \gamma_{x_0,0}(t)dt$ that a packet at the moment T is still on its way – see Figure 10.3.

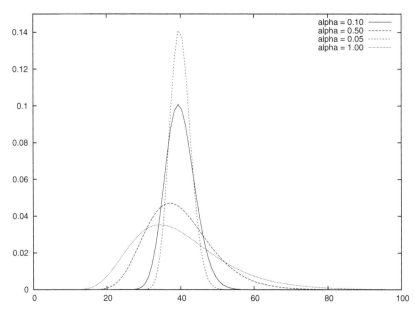

Figure 10.2 Distribution of first passage time from x_0 to 0, $\gamma_{x_0,0}(t)$, $x_0 = 20$, $\beta = -0.5$, $\alpha = 0.05, 0.1, 0.5, 1.0$.

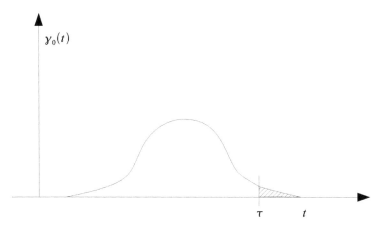

Figure 10.3 Probability $p_T = \int_T^\infty \gamma_{x_0,0}(t)dt$ that a packet at the moment T is still on its way.

In the model, at $t = T$ we shift this probability mass p_T to x_0 and we restart the diffusion process. We may of course introduce an additional delay before the restart.

10.5　Modelling Heterogeneous Medium and Losses

To reflect the fact that the transmission conditions may be different for each hop, the diffusion interval is divided into unitary intervals corresponding to single hops. The subintervals are separated by fictive barriers allowing us to balance the probability density flows between them. We limit the whole interval to a value corresponding to the size of the network $x \in [0, D]$, the starting point x_0 is somewhere inside this interval. As in general $\beta < 0$ (i.e. a packet has a tendency of going towards the sink), the probability of reaching the right barrier by the diffusion process is small. If however the process reaches the right barrier, it is immediately sent to the point $x = D - \varepsilon$ and the process is continued.

An interval i, $x \in [i - 1, i]$ represents the packet transmission when it is i hops distant from the sink. We assume that parameters β_i, α_i are proper to this interval and we assume also the loss probability l_i within this interval.

When the process approaches one of these barriers, for example the barrier i, it acts as an absorbing one, but then immediately the process reappears at the other side of the barrier with probability $(1 - l_i)$ (probability of successful transmission) or with probability l_i it comes to the node that it visited previously, i.e. to the barrier at $x = i + 1$ or at $x = i - 1$.

Let $\gamma_i^L(t)$ represent the flow coming to the barrier placed at $x = i$ from its left side and $\gamma_i^R(t)$ be the flow coming to this barrier from its right side. The flows start diffusion processes at both sides of the barrier, respectively $\gamma_i^R(t)$ reappears at $x = i - \varepsilon$ and $\gamma_i^L(t)$ at $x = i + \varepsilon$ but the intensities of the trespassing flows are reduced by flows corresponding to the loss of packet during the previous hop transmission. Thus, inside the interval i the process starts with intensities

$$g_{i-1+\varepsilon}(t) = (1 - l_{i-1})\gamma_{i-1}^L(t) + l_i\gamma_i^L(t) * l(t),$$

$$g_{i-\varepsilon}(t) = (1 - l_{i+1})\gamma_i^R(t) + l_i\gamma_{i-1}^R(t) * l(t),$$

where $g_{i-1+\varepsilon}(t)$ and $g_{i-\varepsilon}(t)$ are the probability densities that the diffusion process starts at time t at the point $x = i - 1 + \varepsilon$ and $x = i - \varepsilon$.

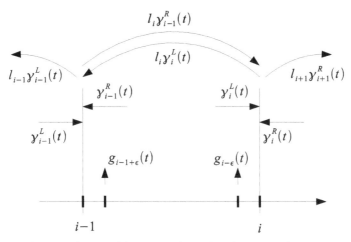

Figure 10.4 Diagram of probability mass circulation due to nonhomegonous diffusion parameters and due to losses with probability l_i at the i-th hop, $i = 2, \ldots, D - 1$.

If we assume that the loss may be repaired by sending the lost packet from the neighbouring node, the flow $\gamma_i^L(t)l_i$ is sent to $x = i - 1 + \varepsilon$ and the flow $\gamma_i^R(t)l_{i+1}$ is sent to $x = i + 1 - \varepsilon$.

The circulation of probability mass for the i-th interval, representing the i-th hop, is presented in Figure 10.4.

If the lost packets are retransmitted with a certain delay, e.g. after a random time distributed with density function $l(t)$ [if this time is constant and equal to r, then $l(t) = \delta(t - r)$], we rewrite the above equations as

$$g_{i-1+\varepsilon}(t) = (1 - l_{i-1})\gamma_{i-1}^L(t) + l_i\gamma_i^L(t) * l(t)$$

$$g_{i-\varepsilon}(t) = (1 - l_{i+1})\gamma_i^R(t) + l_i\gamma_{i-1}^R(t) * l(t),$$

where $*$ denotes the operation of convolution.

Within each subinterval we have diffusion process with two absorbing barriers, e.g. for i-th interval at $x = i - 1$ and $x = i$ and with two points when the process is started, at $i - 1 + \varepsilon$ with intensity $g_{i-1+\varepsilon}(t)$ and at $i - \varepsilon$ with intensity $g_{i-\varepsilon}(t)$.

The density of the diffusion process started at x_0 within an interval $(0, N)$ having the absorbing barriers at $x = 0$ and $x = N$ has the form (see [2])

$$\phi(x, t; x_0) = \begin{cases} \delta(x - x_0), t = 0 \\[2ex] \dfrac{1}{\sqrt{2\Pi\alpha t}} \displaystyle\sum_{n=-\infty}^{\infty} \left\{ \exp\left[\dfrac{\beta x_n'}{\alpha} - \dfrac{(x - x_0 - x_n' - \beta t)^2}{2\alpha t} \right] \right. \\[3ex] \left. - \exp\left[\dfrac{\beta x_n''}{\alpha} - \dfrac{(x - x_0 - x_n'' - \beta t)^2}{2\alpha t} \right] \right\}, t > 0 , \end{cases}$$

where $x_n' = 2nN$, $x_n'' = -2x_0 - x_n'$.

The density $f_i(x, t; \psi)$ may be expressed as a superposition of functions $\phi_i(x, t; x_0)$ at the interval $(i - 1, i)$

$$f_i(x, t; \psi_i) = \phi(x, t; \psi_i) + \int_0^t g_{i-1+\varepsilon}(\tau)\phi(x, t - \tau; i - 1 + \varepsilon)d\tau$$

$$+ \int_0^t g_{i-\varepsilon}(\tau)\phi(x, t - \tau; i - \varepsilon)d\tau .$$

where the function ψ_i represents the initial conditions.

The flows $\gamma_{i-1}^L(t)$ and $\gamma_i^R(t)$ for the i-th interval are obtained as

$$\gamma_{i-1}^R(t) = \lim_{x \to (i-1)} \left[\frac{\alpha_i}{2} \frac{\partial f_i(x, t; \psi_i)}{\partial x} - \beta_i f_i(x, t; \psi_i) \right]$$

$$\gamma_i^L(t) = -\lim_{x \to (i)} \left[\frac{\alpha_i}{2} \frac{\partial f_i(x, t; \psi_i)}{\partial x} - \beta_i f_i(x, t; \psi_i) \right].$$

It is much easier to solve the system of all the above equations when they are inverted with the use of Laplace transform: all convolutions become in this case products of transforms, the Laplace transform of the function $\phi(x, t; x_0)$ is

$$\bar{\phi}(x, s; x_0) = \frac{\exp[\frac{\beta(x-x_0)}{\alpha}]}{A(s)} \sum_{n=-\infty}^{\infty} \left\{ \exp\left[-\frac{|x - x_0 - x_n'|}{\alpha} A(s) \right] \right. $$

$$\left. - \exp\left[-\frac{|x - x_0 - x_n''|}{\alpha} A(s) \right] \right\} ,$$

where $A(s) = \sqrt{\beta^2 + 2\alpha s}$.

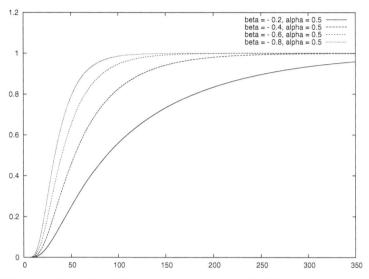

Figure 10.5 Probability $p(0, t)$ if the starting point is $x_0 = 10$, diffusion interval $x \in [0, 20]$, $\alpha = 0.5$, and β is variable ($\beta = -0.2, -0.4, -0.6, -0.8$).

The final solution $f_i(x, t; \psi_i)$ is obtained by numerical inversion of its Laplace transform $\bar{f}_i(x, s; \psi_i)$. In examples below we used Stehfest algorithm [7] where a function $f(t)$ is obtained from its transform $\bar{f}(s)$ for any fixed argument t as

$$f(t) = \frac{\ln 2}{2} \sum_{i=1}^{H} V_i \, \bar{f}\left(\frac{\ln 2}{t} i\right),$$

where

$$V_i = (-1)^{H/2+i} \times \sum_{k=\left\lfloor \frac{i+1}{2} \right\rfloor}^{\min(i, H/2)} \frac{k^{H/2+1}(2k)!}{(H/2 - k)!k!(k - 1)!(i - k)!(2k - i)!}.$$

H is an even integer and depends on a computer precision; we used $H = 14$.

Figure 10.5 displays the probability $p(0, t)$ that the transmission process ends before time t if it is started at the point $x_0 = 10$, the diffusion interval islimited to $x \in [0, 20]$, $\alpha = 0.5$, and $\beta = -0.2, -0.4, -0.6, -0.8$. Figure 10.6 gives the same probability for the fixed value of $\beta = -0.4$ and variable values of α ($\alpha = 0.1, 0.3, 0.5, 0.7, 1, 2, 5$).

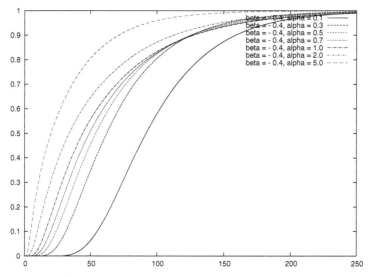

Figure 10.6 Probability $p(0, t)$ if the starting point is $x_0 = 10$, diffusion interval $x \in [0, 20]$, $\beta = -0.4$, and α is variable ($\alpha = 0.1, 0.3, 0.5, 0.7, 1, 2, 5$).

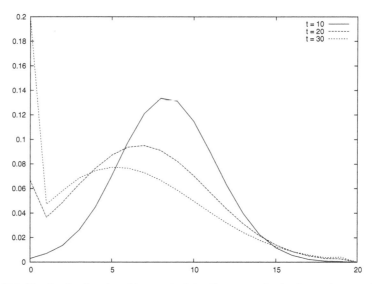

Figure 10.7 The density function $f(x, t; x_0)$ of the distance to destination at time $t = 10, 20$ and 30; $\beta = -0.4$ and $\alpha = 0.54$; $x_0 = 10$.

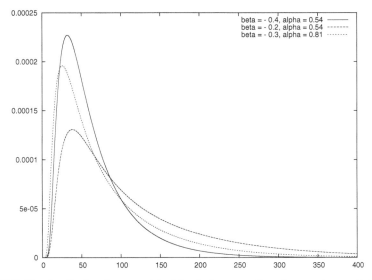

Figure 10.8 The density function $\gamma_{x_0,0}(t)$ of the first passage time from x_0 to $x = 0$, i.e. the approximation of the transmission time density.

Figure 10.7 shows the density $f(x, t; x_0)$ of the remaining distance to complete the transfer calculated for time moments $t = 10, 20$ and 30 if $\beta = -0.4$ and $\alpha = 0.54$. The packet transmission started at the distance $x_0 = 10$ from its destination.

Figure 10.8 displays the density $\gamma_{x_0,0}(t)$ of the first passage time from x_0 to $x = 0$, hence the approximation of the transmission time density.

As mentioned previously, parameters of the diffusion equation are determined by the mean and variance of the approximated discrete process. If

π_{-1} is the probability to advance (to go to a node nearer to the sink by one hop),

π_0 is the probability to stay at the same distance from the sink,

π_{+1} is the probability to go to a node more distant by one hop from the sink,

then

$$\beta = \frac{\pi_{-1} \times (-1) + \pi_0 \times (0) + \pi_{+1} \times (+1)}{1 \text{ time unit}}$$

and

$$\alpha = \frac{\pi_{-1} \times (-1)^2 + \pi_0 \times 0^2 + \pi_{+1}(+1)^2}{1 \text{ time unit}} - \beta^2.$$

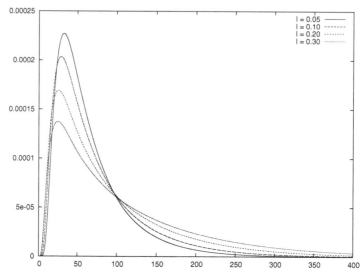

Figure 10.9 The impact of loss ratio l ($l = 0.05, 0.1, 0.2, 0.3$) on the density function of transfer time at a network with $\beta = -0.4$ and $\alpha = 0.54$.

For example, in displayed numerical examples, the correspondance between these sets of parameters is as follows

$$\beta = -0.4, \quad \alpha = 0.54 \quad \longrightarrow \quad (\pi_{-1}, \pi_0, \pi_{+1}) = (0.55, 0.30, 0.15),$$
$$\beta = -0.2, \quad \alpha = 0.54 \quad \longrightarrow \quad (\pi_{-1}, \pi_0, \pi_{+1}) = (0.40, 0.40, 0, 20),$$
$$\beta = -0.3, \quad \alpha = 0.81 \quad \longrightarrow \quad (\pi_{-1}, \pi_0, \pi_{+1}) = (0.60, 0.10, 0.30).$$

Figure 10.9 presents the influence of the loss rate l ($l = 0.05, 0.1, 0.2, 0.3$) on the density of transmission time for parameters $\beta = -0.4$ and $\alpha = 0.54$.

10.6 Conclusions

Owing to the introduction of the transient state analysis, the presented model seems to capture more parameters (time-dependent and heterogeneous transmission, different loss rate for each hop, etc.) of a sensor network transmission time than the existing models, also based on diffusion approximation. It gives also more detailed results: the density function of a packet travel time instead of its mean value. Numerical results prove that the model is operational.

References

[1] I. F. Akyildiz, Y. S. Sand and E. Çayirci, A survey on sensor networks, *IEEE Commun. Mag.*, vol. 40, no. 8, pp. 102–114, 2002.

[2] R. P. Cox and H. D. Miller, *The Theory of Stochastic Processes*, Chapman and Hall, London, 1965.

[3] E. Gelenbe, On approximate computer systems models, *J. ACM*, vol. 22, no. 2, pp. 261–269, 1975.

[4] E. Gelenbe, A diffusion model for packet travel time in a random multihop medium, *ACM Trans. on Sensor Networks*, vol. 3, no. 2, Article 10, June 2007.

[5] W. R. Heinzelman, A. Candrakasan and H. Balakrishnan, Energy-efficient communication protocols for wireless microsensor networks, in *Proceedings of the Hawaiian International Conference on Systems Science*, January 2000.

[6] E. Royer and C. Toh, A review of current routing protocols for ad-hoc mobile wireless networks, *IEEE Personal Commun.*, vol. 6, no. 2, pp. 46–55, 1999.

[7] H. Stehfest, Algorithm 368: Numeric inversion of Laplace transform, *Comm. of ACM*, vol. 13, no. 1, pp. 47–49, 1970.

[8] M. Zorzi and R. Rao, Geographic random forwarding (GERAF) for ad hoc and sensor networks: Multihop performance, *IEEE Trans. Mobile Comput.*, vol. 2, no. 4, pp. 337–348, 2003.

PART FOUR
WIRELESS CELLULAR NETWORKS

11

Efficient Computation of Optimal Capacity in Multiservice Mobile Wireless Networks

Vicent Pla, Jorge Martinez-Bauset and Vicente Casares-Giner

Department of Communications, Universidad Politécnica de Valencia (UPV),
ETSIT Camí de Vera s/n, 46022 Valencia, Spain;
e-mail: {vpla,jmartinez,vcasares}@dcom.upv.es

Abstract

In this paper we propose a new algorithm for computing the optimal parameters setting of the Multiple Fractional Guard Channel (MFGC) admission
policy in multiservice mobile wireless networks. The optimal parameters
setting maximizes the offered traffic that the system can handle while meeting certain QoS requirements. The proposed algorithm is shown to be more
efficient than previous algorithms appeared in the literature.

Keywords: Mobile wireless network, multiservice, admission control, capacity, algorithms.

11.1 Introduction

The enormous growth of mobile telecommunication services, together with
the scarcity of radio spectrum has led to reducing the cell size in cellular
systems. Smaller cell size entails a higher handoff rate having an important
impact on QoS and radio resource management. During the last two decades
a considerable number of papers have addressed this topic (see, for instance
[1–3]). Moreover, forthcoming 3G networks will establish a new paradigm

*D. D. Kouvatsos (ed.), Performance Modelling and Analysis of Heterogeneous
Networks,* 199–214.

with a variety of services having different QoS needs and traffic characteristics.

Admission control in the presence of mobility or multiple services is quite well studied. However, this new paradigm where multiservice and mobility meet has not received attention from researchers until very recently.

In [4] Li et al. propose an extension of the well-known *Guard Channel* (GC) [1] mechanism where multiple service types are considered. Bartollini and Chlamtac [5] considered a more general policy than that of [4]. More recently, Heredia et al. [6–8] had proposed an extension of the *Fractional Guard Channel* (FGC) [9] scheme. The structure of optimal admission policies in single service cellular networks under different criteria is studied in [9, 10]. In [5] the authors show that the optimal admission policy (with respect to a certain cost function) in a multiservice cellular network does not belong to any of the types mentioned above; instead it belongs to the wider family of stationary policies [11]. In [12] several types of admission policies for cellular multiservice networks (including MFGC and randomized stationary) have been compared.

In this paper we propose a new algorithm for computing the optimal parameters setting of the of Multiple Fractional Guard Channel (MFGC) admission policy in multiservice mobile wireless networks. As it will be explained later, in MFGC the policy parameters control the amount of system resources that each call type can access. The optimal parameters setting maximizes the offered traffic that the system can handle while meeting certain QoS requirements. To the best of our knowledge only one algorithm for this purpose has been proposed [8] in the literature and its computational performance is substantially improved by the one proposed here. Besides, we observe that a further enhancement of both algorithms is possible by eliminating the iterative procedure of computing the handoff arrival rates.

The remaining of the paper is structured as follows. In Section 11.2 the system model is described and its mathematical analysis is outlined in Section 11.3. Section 11.4 describes in detail the new proposed algorithm. Computational complexity of the algorithm is comparatively evaluated in Section 11.5. Finally, Section 11.6 concludes the paper.

11.2 Model Description

The system has a total of C resource units. The physical meaning of a unit of resources will depend on the specific technological implementation of the radio interface.

The system offers N different classes of services. For each type of service new and handoff call arrivals are distinguished so that there are N types of services and $2N$ types of arrivals. Arrivals are numbered in such manner that for service i new call arrivals are referred to as arrival type i, whereas handoff arrivals are referred to as arrival type $N + i$.

For the sake of mathematical tractability we make the common assumptions of Poisson arrival processes and exponentially distributed random variables for cell residence time and call duration.

The arrival rate for new (handoff) calls of service i is λ_i^n (λ_i^h). A request of service i consumes b_i resource units, $b_i \in \mathbb{N}$.

The call duration of service i is exponentially distributed with rate μ_i^c. The cell residence time of a service i customer is exponentially distributed with rate μ_i^r. Hence, the resource holding time in a cell for service i is exponentially distributed with rate $\mu_i = \mu_i^c + \mu_i^r$.

Recent papers present more accurate modeling of the cell residence time [13], channel holding time [14, 15], arrival processes [16–18] and time within the handoff area [19, 20]. Logically, these models add an extra complexity to the analysis, making it highly intricate or simply infeasible. Some analytical results for the single service case are reported in [21–23]. Notwithstanding, the exponential assumption represents a good performance approximation. Essentially, only the average cell dwell time matters. When the average cell dwell time is small compared to the call duration, there is no expected difference between the exponential assumption and the gamma one. When cell dwell times are large, the difference becomes more noticeable, but the exponential assumption indicates general performance trends [24]. The exponential assumption can also be considered a good approximation for the time in the handoff area [25] and for the interarrival time of handoff requests [26].

Anyhow, the main contribution of this paper is an algorithm to determine the optimal capacity of the system, which relies on a method to compute the system blocking probabilities. Our proposal, however, does not depend on any specific method to find the blocking probabilities and hence it could be substituted – for instance if different assumptions are made for the underlying model – without affecting the proposed algorithm.

If we denote by $\boldsymbol{p} = (P_1, \ldots, P_{2N})$ the blocking probabilities for each of the $2N$ arrival streams, the new call blocking probabilities is $P_i^n = P_i$, the handoff failure probability is $P_i^h = P_{N+i}$ and the the forced termination probability of accepted calls under the assumption of homogeneous cell [1] is

$$P_i^{ft} = \frac{P_i^h}{\mu_i^c/\mu_i^r + P_i^h}$$

The system state is described by an N-tuple $x = (x_1, \ldots, x_N)$, where x_i represents the number of type i calls in the system, that were initiated either as new or handoff calls. Let $b(x)$ represent the amount of occupied resources at state x, $b(x) = \sum_{i=1}^{N} x_i b_i$.

11.2.1 Admission Policy (MFGC)

The MFGC policy operates in a manner that the maximum number of resource unit that stream i can dispose of is, on average, t_i. In order to decide on the acceptance of a request of type i, upon its arrival the system compares the amount of resources that will be occupied if it is accepted with the corresponding threshold t_i. The following decisions can be taken

$$b(x) + b_i \begin{cases} \leq \lfloor t_i \rfloor & \text{accept} \\ = \lfloor t_i \rfloor + 1 & \text{accept with probability} \quad t_i - \lfloor t_i \rfloor \\ > \lfloor t_i \rfloor + 1 & \text{reject} \end{cases}$$

11.3 Mathematical Analysis

The model of the system is a multidimensional birth-and-death process. The set of feasible states for the process is

$$S := \left\{ x : x_i \in \mathbb{N}; \sum_{i=1}^{N} x_i b_i \leq C; x_i b_i \leq \lceil t_i \rceil, 1 \leq i \leq N \right\}$$

Let r_{xy} be the transition rate from x to y and let e_i denote a vector whose entries are all 0 except the i-th one, which is 1.

$$r_{xy} = \begin{cases} a_i^n(x)\lambda_i^n + a_i^h(x)\lambda_i^h & \text{if } y = x + e_i \\ x_i \mu_i & \text{if } y = x - e_i \\ 0 & \text{otherwise} \end{cases}$$

The coefficients $a_i^n(x)$ and $a_i^h(x)$ denote the probabilities of accepting a new and handoff call of service i respectively. Given a policy setting $(t_1, \ldots, t_{2N})$

these coefficients can be determined as follows

$$a_i^n(x) = \begin{cases} 1 & \text{if } b(x) + b_i \le \lfloor t_i \rfloor \\ t_i - \lfloor t_i \rfloor & \text{if } b(x) + b_i = \lfloor t_i \rfloor + 1 \\ 0 & \text{if } b(x) + b_i > \lfloor t_i \rfloor + 1 \end{cases}$$

and

$$a_i^h(x) = \begin{cases} 1 & \text{if } b(x) + b_i \le \lfloor t_i \rfloor \\ t_{N+i} - \lfloor t_{N+i} \rfloor & \text{if } b(x) + b_i = \lfloor t_{N+i} \rfloor + 1 \\ 0 & \text{if } b(x) + b_i > \lfloor t_{N+i} \rfloor + 1 \end{cases}$$

From the above, the global balance equations can be written as

$$p(x) \sum_{y \in S} r_{xy} = \sum_{y \in S} r_{yx} p(y) \qquad \forall x \in S \tag{11.1}$$

where $p(x)$ is the state x stationary probability. The values of $p(x)$ are obtained from (11.1) and the normalization equation. To obtain the stationary state distribution we used the Gauss–Seidel method. From the values of $p(x)$. the blocking probabilities are obtained as

$$P_i = P_i^n = \sum_{x \in S} \left(1 - a_i^n(x)\right) p(x) \qquad P_{N+i} = P_i^h = \sum_{x \in S} \left(1 - a_i^h(x)\right) p(x)$$

If the system is in statistical equilibrium the handoff arrival rates are related to the new call arrival rates and the blocking probabilities (P_i) through the expression [2]

$$\lambda_i^h = \lambda_i^n \frac{1 - P_i^n}{\mu_i^c / \mu_i^r + P_i^h} \tag{11.2}$$

The blocking probabilities do in turn depend on the handoff arrival rates yielding a system of non-linear equations which can be solved using a fixed point iteration method as described in [1,2].

11.4 Optimal Capacity: Algorithm

We pursue the goal of computing the system capacity, i.e. the maximum offered traffic that the network can handle while meeting certain QoS requirements. These QoS requirements are given in terms of upper-bounds for the new call blocking probabilities (B_i^n) and the forced termination probabilities (B_i^{ft}). Let $\lambda^T = \sum_{1 \le i \le N} \lambda_i^n$ be the aggregated call arrival rate and let f_i

$(0 \leq f_i < 1, \sum_{1 \leq i \leq N} f_i = 1)$ represent the fraction of λ^T that correspond to service i, i.e. $\lambda_i^n = f_i \lambda^T$, the capacity optimization problem can be formally stated as follows:

Given: $C, b_i, f_i, \mu_i^c, \mu_i^r, B_i^n, B_i^{ft} ; i = 1, \ldots, N$

Maximize: λ^T
    by finding the appropriate MFGC parameters $t_i; i = 1, \ldots, 2N$

Subject to: $P_i^n \leq B_i^n, P_i^{ft} \leq B_i^{ft} ; i = 1, \ldots, N$

We propose an algorithm to work out this capacity optimization problem. Our algorithm has a main part (Algorithm 1 capacity) from which the procedure solveMFGC (see Algorithm 2) is called. The procedure solveMFGC does, in turn, call another procedure (MFGC) that calculates the blocking probabilities. For the sake of notation simplicity we introduce the 2N-tuple $\boldsymbol{p}_{\text{max}} = (B_1^n, \ldots, B_N^n, B_1^h, \ldots, B_N^h)$ as the upper-bounds vector for the blocking probabilities, where the value of B_i^h is given by

$$B_i^h = \frac{\mu_i^c}{\mu_i^r} \frac{B_i^{ft}}{1 - B_i^{ft}} \tag{11.3}$$

Following the common convention bold-faced fonts were used to represent array variables in the pseudo-code of the algorithms.

The algorithm capacity is essentially a binary search of λ_{max}^T that calls solveMFGC at each iteration to find out whether, for the tested value of the aggregated new call arrival rate λ^T, there exists a policy configuration (t) that fulfills the QoS constraints $(\boldsymbol{p}_{\text{max}})$. If it exists (solveMFGC returns possible=TRUE) the lower limit of the interval that encloses λ_{max}^T is increased as $L := \lambda$; and otherwise (solveMFGC returns possible=FALSE) the upper limit of the interval is decreased as $U := \lambda$.

In order to find a policy configuration that fulfills the QoS constraints, or decide that such configuration does not exist, the algorithm solveMFGC proceeds as follows. All t_i's are initialized with a small value (δ).[1] Then the algorithm cyclically checks for each stream $i = 1, \ldots, 2N$ whether its QoS constraint $(p(i) \leq \boldsymbol{p}_{\text{max}}(i))$ is met for the current policy setting, and if not $(p(i) > \boldsymbol{p}_{\text{max}}(i))$ the value of t_i is increased so that $(1 - \varepsilon)\boldsymbol{p}_{\text{max}}(i) \leq p(i) \leq \boldsymbol{p}_{\text{max}}(i)$. This process continues until either the QoS goal is achieved $(p(i) \leq$

---

[1] According to the philosophy of the algorithm an initial value of zero should have been used. However, a non-zero value was used due to implementation reasons.

Algorithm 1 $(\lambda_{max}^T, t_{opt}) = \texttt{capacity}(\boldsymbol{p_{max}}, \boldsymbol{f}, \boldsymbol{\mu_c}, \boldsymbol{\mu_r}, \boldsymbol{b}, C)$

$\varepsilon_1 := <$ desired precision $>$
$L := 0$
$U := <$ high value $>$
$(possible, t) := \texttt{solve\_MFGC}(\boldsymbol{p_{max}}, U\boldsymbol{f}, \boldsymbol{\mu_c}, \boldsymbol{\mu_r}, \boldsymbol{b}, C)$
atLeastOnce:=FALSE;

while possible **do**
  $L := U$
  $t_L := t$
  atLeastOnce:=TRUE
  $U := 2U$
  $(possible, t) := \texttt{solve\_MFGC}(\boldsymbol{p_{max}}, U\boldsymbol{f}, \boldsymbol{\mu_c}, \boldsymbol{\mu_r}, \boldsymbol{b}, C)$
end while{it makes sure that $U > \lambda_{max}^T$}

repeat
  $\lambda := (L + U)/2$
  $(possible, t) := \texttt{solve\_MFGC}(\boldsymbol{p_{max}}, \lambda\boldsymbol{f}, \boldsymbol{\mu_c}, \boldsymbol{\mu_r}, \boldsymbol{b}, C)$
  if possible **then**
    $L := \lambda$
    $t_L := t$
    atLeastOnce:=TRUE;
  else
    $U := \lambda$
  end if
until $(U - L)/L \le \varepsilon_1$ AND atLeastOnce
$\lambda_{max}^T := L$
$t := t_L$

$\boldsymbol{p_{max}}(i), \forall i$ and the algorithm returns possible=TRUE); or the algorithm gives up as it realizes that the QoS goal is unattainable (then the algorithm returns possible=FALSE). The algorithm decides that the QoS goal is achieved if after a complete cycle ($i = 1, \ldots, 2N$) the QoS constraint was met for all streams without requiring to increment the corresponding thresholds t_i. The algorithm decides that the QoS goal is unattainable if it happens that for a stream the QoS constraint can not be met even if the corresponding threshold is set to its maximum value $t_i = C$.

Algorithm 2 (possible, t) =solveMFGC($p_{\max}, \lambda_n, \mu_c, \mu_r, b, C$)
(calculates MFGC parameters)

INPUTS: $p_{\max}, \lambda_n, \mu_c, \mu_r, b, C$
OUTPUTS: possible, t

```
1:
2:    ε₂ :=< desired precision >
3:    δ :=< small value >
4:    t := (δ, δ, ..., δ)
5:    p := MFGC(t, λₙ, μc, μr, b, C)
6:
7:    repeat
8:      canConverge:=TRUE;
9:      i := 1;
10:
11:     repeat
12:       if p(i) > pmax(i) then
13:         t' := t; t'(i) := C
14:         p' := MFGC(t', λₙ, μc, μr, b, C)
15:
16:         if p'(i) > pmax(i) then
17:           canConvege:=FALSE;
18:         else
19:           L := t(i); U := C
20:           repeat
21:             t(i) := (L + U)/2
22:             p := MFGC(t, λₙ, μc, μr, b, C)
23:             if p(i) > pmax(i) then
24:               L := t(i)
25:             else
26:               U := t(i)
27:             end if
28:           until (1 − ε₂)pmax(i) ≤ p(i) ≤ pmax(i)
29:         end if
30:
31:       end if
32:       i := i + 1
33:     until (i > 2N) OR ( NOT(canConverge))
34:
35:     if canConverge then
36:       if p(i) ≤ pmax(i)   ∀i then
37:         possible:=TRUE; exit:=TRUE;
38:       else
39:         exit:=FALSE;
40:       end if
41:     else
42:       possible:=FALSE; exit:=TRUE;
43:     end if
44:
45:   until exit
```

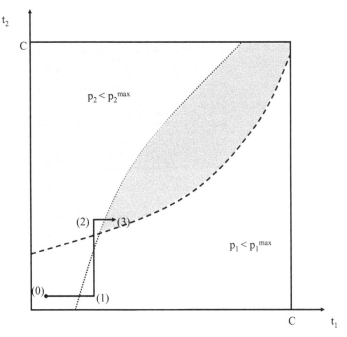

Figure 11.1 Graphical traces of a `solveMFGC` run, $\lambda_1^T \leq \lambda_{\max}^T$.

Figures 11.1 and 11.2 show an example illustrating the basic behavior of our algorithm. We used a rather simple case with only one type of service (two types of arrivals) in order to represent it graphically. Each figure represents one execution of the algorithm `solveMFGC` with a fixed value of λ^T. Figure 11.1 shows an example where the value λ^T of was relatively low and then the possible solutions of t was rather wide. Figure 11.2 shows another run of `solveMFGC` using a higher value of λ^T; again a policy setting that fulfills the QoS constraints can be found. Note, however, that increasing λ^T had the effect of shrinking the solution region. Finally, Figure 11.3) shows an example where $\lambda^T > \lambda_{\max}^T$ and then no feasible solution for t exists.

11.4.1 On the Procedure `MFGC`

The procedure `MFGC`, which is invoked in the inner-most loop of our algorithm, is used to obtain the blocking probabilities ($p := \text{MFGC}(t, \lambda_n, \mu_c, \mu_r, b, C)$). For this computation an iterative procedure is required in order to obtain the value of the handoff request rates

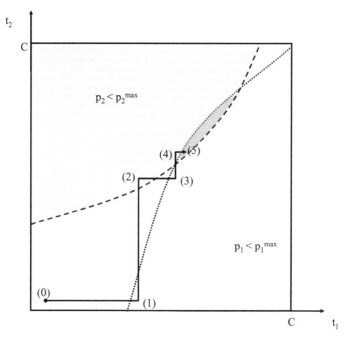

Figure 11.2 Graphical traces of a `solveMFGC` run, $\lambda_1^T \leq \lambda_2^T \leq \lambda_{max}^T$.

(see the end of Section 11.3). At each iteration a multidimensional birth-and-death process is solved. Solving this process, that in general will have a large number of states, constitutes the most computationally expensive part of the algorithm.

The following observation can be used to speed up the algorithm since it permits to eliminate the above mentioned iterative procedure. Each run of `solveMFGC` tries to find t so that $p = p_{max}$ (within tolerance limit). Thus, instead of using (11.2) to compute λ_i^h we use the expression

$$\lambda_i^h = \lambda_i^n \frac{1 - B_i^n}{\mu_i^c / \mu_i^r + B_i^h} \tag{11.4}$$

Although (11.2) and (11.4) look very similar there is a substantial difference between the two. In Eq. (11.4), λ_i^h is explicitly defined whereas in (11.2) it is not as P_i^n and P_i^h depend on λ_i^h.

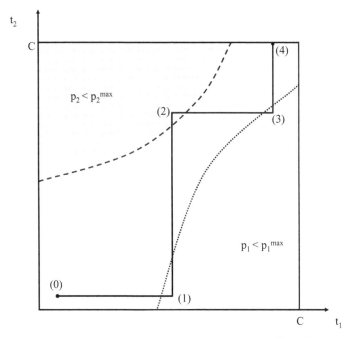

Figure 11.3 Graphical traces of a `solveMFGC` run, $\lambda_3^T > \lambda_{\max}^T$.

11.5 Numerical Evaluation

In this section we evaluate the numerical complexity of our algorithm. To this end we used the algorithm proposed by Heredia et al. in [6–8] as a reference. Henceforth we refer to this algorithm as HCO after its authors' initials.

The HCO algorithm requires the optimal *prioritization order* as input, i.e. a list of call types sorted by their relative priorities [8]. If *t* is the policy setting for which the maximum capacity is achieved, the optimal prioritization order is the permutation $\sigma^* \in \Sigma$, $\Sigma := \{(\sigma_i, \ldots, \sigma_{2N}) : \sigma_i \in \mathbb{N}, 1 \leq \sigma_i \leq 2N\}$, such that $t(\sigma_1^*) \leq t(\sigma_2^*) \leq \ldots \leq t(\sigma_{2N}^*) = C$. Selecting the optimal prioritization order is a complicated task as it depends on both QoS constraints and system characteristics as pointed out in [8]. In general there are a total of $(2N)!$ different prioritization orders. In [8] the authors give some guidelines to construct a partially sorted list of prioritization orders according to their likelihood of being the optimal ones. Then a trial and error process is followed using successive elements of the list until the optimal prioritization order is

Table 11.1 Comparison of the HCO algorithm (with known prioritization order) and our algorithm with and without speed-up technique (figures in Mflops).

	HCO		Our algorithm	
C	–	speed-up	–	speed-up
5	5.70	2.00	1.17	0.39
10	60.20	20.00	13.80	4.53
20	438.00	156.00	145.00	46.60

found. For each element the HCO algorithm is run and if after a large number of iterations it did not converged, another prioritization order is tried.

Our algorithm does not require any *a priori* knowledge. Indeed, after obtaining the policy setting t for which the maximum capacity is achieved, the optimal prioritization order is automatically determined as a by-product of our algorithm. This constitutes by itself a significant advantage of our algorithm over the HCO algorithm. Moreover, in what follows we show through numerical examples that our algorithm is still more efficient than the HCO algorithm when the latter is provided with the optimal prioritization order.

For the numerical examples we considered a system with two services ($N = 2$). Unless otherwise indicated, the values of the parameters are $b = (1, 2)$, $f = (0.8, 0.2)$, $\mu_c = (1/180, 1/300)$, $\mu_r = (1/900, 1/1000)$, $B^n = (0.02, 0.02)$, $B^{ft} = (0.002, 0.002)$; all tolerances have been set to $\varepsilon = 10^{-2}$. By (11.3), $B^h \approx (0.01002, 0.00668)$ and then $p_{max} \approx (0.02, 0.02, 0.01002, 0.00668)$.

A comparison of the number of floating point operations (flops) required by the HCO algorithm and our algorithm is shown in Table 11.1 and in Figure 11.4. Both algorithms were tested with and without the speed-up technique (see Section 11.4.1) yielding a total of four cases. The speed-up technique divides the flop count by a factor of about three.

To asses the impact of mobility on computational complexity, different scenarios were considered with varying mobility factors (μ_i^r/μ_i^c) for each service. The rest of the parameters have the same values as the ones used in the previous example, except μ_i^r which is varied to obtain four different mobility factor combinations: A) $\mu_1^r = 0.2\mu_1^c$, $\mu_2^r = 0.2\mu_2^c$; B) $\mu_1^r = 0.2\mu_1^c$, $\mu_2^r = 1\mu_2^c$; C) $\mu_1^r = 1\mu_1^c$, $\mu_2^r = 0.2\mu_2^c$; D) $\mu_1^r = 1\mu_1^c$, $\mu_2^r = 1\mu_2^c$. Computational cost results are displayed in Table 11.2 and aggregated costs across scenarios are plotted in Figure 11.5. Again, our algorithm performs better

Table 11.2 Comparison of the HCO algorithm (with known prioritization order) and our algorithm with speed-up technique for different mobility factors (figures in Mflops).

		C						Total
		5	10	15	20	25	30	
	A	2.08	17.54	45.78	74.33	267.04	407.74	814.51
HCO	B	2.67	14.06	50.25	147.13	266.41	487.41	967.93
(speed-up)	C	1.12	24.54	54.56	110.41	309.38	410.93	910.94
	D	2.24	16.86	53.39	121.39	106.49	462.12	762.49
	Total	8.11	73.00	203.98	453.26	949.32	1768.20	3455.8
	A	0.35	4.42	18.51	53.64	119.46	199.69	396.07
Our algorithm	B	0.34	3.87	16.49	43.01	83.60	172.73	320.04
(speed-up)	C	0.38	3.93	17.59	47.95	119.33	191.66	380.84
	D	0.31	3.93	15.33	45.92	95.25	172.58	333.32
	Total	1.39	16.15	67.92	190.51	417.64	736.66	1430.3

than the HCO algorithm provided with the optimal prioritization order, and with the speed-up technique. The gain factor ranges from 1.4 to 7.8 with an average of 3.8, and in general it decreases when the number of resource units (C) increases.

It is worth noting that, as expected, the disagreement among the values obtained for the optimal capacity computed using the different methods was within tolerance in all tested cases. The same can be said for the policy setting t.

11.6 Conclusions

We proposed a new algorithm for computing the optimal parameters setting of the of Multiple Fractional Guard Channel (MFGC) admission policy in multiservice mobile wireless networks. The optimal parameters setting maximizes the offered traffic that the system can handle while meeting certain QoS requirements. Compared to a recently published algorithm (HCO) ours offers the advantage of not needing a call prioritization order as input. We observed that a further enhancement of both algorithms is possible by eliminating the iterative procedure for computing the handoff call arrival rates. Numerical examples show that our algorithm is faster than the HCO algorithm even if the latter is provided with the optimal prioritization order and is enhanced with the above mentioned observation.

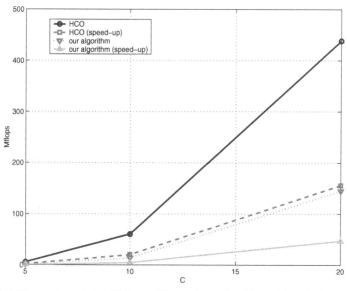

Figure 11.4 Comparison of the HCO algorithm and our algorithm with and without speed-up technique.

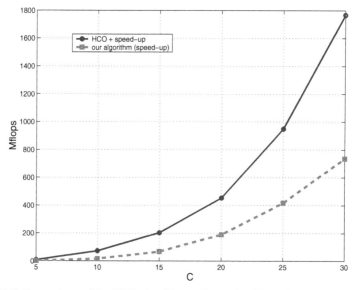

Figure 11.5 Comparison of the HCO algorithm and our algorithm with speed-up technique; aggregate values for scenarios A, B, C and D.

Acknowledgements

This work has been supported by the Spanish Ministry of Education and Science (30%) and by the European Commission (70% FEDER) through projects TSI2007-66869-C02-02 and TIN2008-06739-C04-02.

References

[1] D. Hong and S. S. Rappaport, Traffic model and performance analysis for cellular mobile radio telephone systems with prioritized and nonprioritized handoff procedures, *IEEE Transactions on Vehicular Technology*, vol. VT-35, no. 3, pp. 77–92, August 1986. (See also: CEAS Technical Report No. 773, June 1, 1999, College of Engineering and Applied Sciences, State University of New York, Stony Brook, NY 11794, USA.)

[2] Y.-B. Lin, S. Mohan and A. Noerpel, Queueing priority channel assignment strategies for PCS hand-off and initial access, *IEEE Transactions on Vehicular Technology*, vol. 43, no. 3, pp. 704–712, August 1994.

[3] F. Barceló, Comparison of handoff resource allocation strategies through the state dependent rejection scheme, in *Proceedings of ITC 17*, pp. 323–334, 2001.

[4] B. Li, C. Lin and S. T. Chanson, Analysis of a hybrid cutoff priority scheme for multiple classes of traffic in multimedia wireless networks, *Wireless Networks Journal (WINET)*, vol. 4, no. 4, pp. 279–290, 1998.

[5] N. Bartolini and I. Chlamtac, Call admission control in wireless multimedia networks, in *Proceedings of IEEE PIMRC*, vol. 1, pp. 285–289, 2002.

[6] H. Heredia-Ureta, F. A. Cruz-Pérez and L. Ortigoza-Guerrero, Multiple fractional channel reservation for optimum system capcity in multi-service cellular networks, *Electronics Letters*, vol. 39, no. 1, pp. 133–134, January 2003.

[7] H. Heredia-Ureta, F. A. Cruz-Pérez and L. Ortigoza-Guerrero, Multiple fractional channel reservation for multi-service cellular networks, in *Proceedings of IEEE ICC*, pp. 964–968, 2003.

[8] H. Heredia-Ureta, F. A. Cruz-Pérez and L. Ortigoza-Guerrero, Capacity optimization in multiservice mobile wireless networks with multiple fractional channel reservation, *IEEE Transactions on Vehicular Technology*, vol. 52, no. 6, pp. 1519–1539, November 2003.

[9] R. Ramjee, R. Nagarajan and D. Towsley, On optimal call admission control in cellular networks, *Wireless Networks Journal (WINET)*, vol. 3, no. 1, pp. 29–41, 1997.

[10] N. Bartolini, Handoff and optimal channel assignment in wireless networks, *Mobile Networks and Applications (MONET)*, vol. 6, no. 6, pp. 511–524, 2001.

[11] K. W. Ross, *Multiservice Loss Models for Broadband Telecommunication Networks*, Springer Verlag, 1995.

[12] V. Pla and V. Casares-Giner, Optimal admission control policies in multiservice cellular networks, in *Proceedings of the International Network Optimization Conference (INOC)*, October, pp. 466–471, 2003.

[13] M. M. Zonoozi and P. Dassanayake, User mobility modeling and characterization of mobility patterns, *IEEE Journal on Selected Areas in Communications*, vol. 15, no. 7, pp. 1239–1252, September 1997.

[14] F. Barceló and J. Jordán, Channel holding time distribution in public telephony systems (PAMR and PCS), *IEEE Transactions on Vehicular Technology*, vol. 49, no. 5, pp. 1615–1625, September 2000.

[15] Y. Fang and I. Chlamtac, Teletraffic analysis and mobility modeling of PCS networks, *IEEE Transactions on Communications*, vol. 47, no. 5, pp. 1062–1072, July 1999.

[16] M. Rajaratnam and F. Takawira, Handoff traffic characterization in cellular networks under nonclassical arrivals and service time distributions, *IEEE Transactions on Vehicular Technology*, vol. 50, no. 4, pp. 954–970, July 2001.

[17] M. Sidi and D. Starobinski, New call blocking versus handoff blocking in cellular networks, *Wireless Networks Journal (WINET)*, vol. 3, no. 1, pp. 15–27, March 1997.

[18] E. Chlebus and W. Ludwin, Is handoff traffic really Poissonian?, in *Proceedings of ICUPC'95*, pp. 348–353, 1995.

[19] M. Ruggieri, F. Graziosi and F. Santucci, Modeling of the handover dwell time in cellular mobile communications systems, *IEEE Transactions on Vehicular Technology*, vol. 47, no. 2, pp. 489–498, May 1998.

[20] V. Pla and V. Casares-Giner, Analytical-numerical study of the handoff area sojourn time, in *Proceedings of IEEE GLOBECOM*, November, pp. 886–890, 2002.

[21] M. Rajaratnam and F. Takawira, Nonclassical traffic modeling and performance analysis of cellular mobile networks with and without channel reservation, *IEEE Transactions on Vehicular Technology*, vol. 49, no. 3, pp. 817–834, May 2000.

[22] A. Alfa and W. Li, PCS networks with correlated arrival process and retrial phenomenon, *IEEE Transactions on Wireless*, vol. 1, no. 4, pp. 630–637, October 2002.

[23] S. Dharmaraja, K. Trivedi and D. Logothetis, Performance modeling of wireless networks with generally distributed handoff interarrival times, *Computer Communications*, vol. 26, pp. 1747–1755, 2003.

[24] F. Khan and D. Zeghlache, Effect of cell residence time distribution on the performance of cellular mobile networks, in *Proceedings of VTC'97*, IEEE, New York, pp. 949–953, 1997.

[25] V. Pla and V. Casares-Giner, Effect of the handoff area sojourn time distribution on the performance of cellular networks, in *Proceedings of IEEE MWCN*, September, pp. 401–405, 2002.

[26] P. V. Orlik and S. S. Rappaport, On the handoff arrival process in cellular communications, *Wireless Networks Journal (WINET)*, vol. 7, no. 2, pp. 147–157, March/April 2001.

12

Performance Modelling and Analysis of a 4G Handoff Priority Scheme for Cellular Networks

Demetres D. Kouvatsos[1], Yue Li[2] and Weixi Xing[2]

[1]*PERFORM – Networks and Performance Engineering Research Group, University of Bradford, Bradford BD7 1DP, West Yorkshire, U.K.; e-mail: d.d.kouvatsos@scm.bradford.ac.uk*
[2]*Institute of Advanced Telecommunications, Swansea University, Swansea SA2 8PP, Wales, U.K.; e-mail: {y.li, w.xing}@swansea.ac.uk*

Abstract

A novel handoff priority channel assignment protocol is proposed for a fourth generation (4G) wireless cellular network (4G-HoP) with bursty multimedia traffic flows consisting of all IP (Internet protocol) voice calls, streaming media and data packets. Under the auspices of an efficient medium access control (MAC), the 4G-HoP protocol employs the principles of channel reservation, extended channel sub-rating and priority/buffer threshold based queueing, subject to a generalized partial sharing (GPS) traffic handling scheme.

The impact of the 4G-HoP protocol on the performance of the 4G cellular network is assessed by devising an analytic framework, based on an open queueing network model (QNM) and the information theoretic principle of maximum entropy (ME). The QNM consists of five interacting multiple class generalized exponential (GE)-type queueing and delay systems with finite capacity and multiple servers under a GPS traffic handling scheme. The systems consist of (i) a loss system for new and handoff IP voice calls, (ii) a queueing system for handoff IP voice calls with guard channels under a head-of-line (HoL) priority rule and a complete buffer sharing (CBS) management scheme, (iii) a queueing system for new and handoff IP streaming media

D. D. Kouvatsos (ed.), Performance Modelling and Analysis of Heterogeneous Networks, 215–243.

packets with low and high buffer thresholds under a first-come-first-served (FCFS) rule and CBS, (iv) a queueing system for handoff IP streaming media packets with dedicated channels, HoL rule and CBS, and (v) a delay system for IP data packets under a discriminatory processor share (DPS) transfer rule.

New ME analytic solutions are characterized for the state and blocking probability distributions of the queueing and delay systems, subject to appropriate GE-type theoretic mean value constraints. Typical numerical experiments are included to verify the credibility of the ME solutions against Java-based simulation results at 95% confidence intervals and, moreover, an investigation is undertaken into the effect of bursty new and handoff multiple class traffic flows upon the performance of the wireless cell.

Keywords: Wireless cellular networks, internet protocol (IP), medium access control (MAC) protocol, quality-of-service (QoS), loss system, first-come-first-served (FCFS) rule, head-of-line (HoL) priority rule, discriminatory processor share (DPS) rule, generalized partial sharing (GPS) traffic handling scheme, performance evaluation, queueing network model (QNM), generalized exponential (GE) distribution, maximum entropy (ME) principle.

12.1 Introduction

Efficient algorithms for the analysis of queueing and delay network models under various traffic handling schemes are widely recognized as powerful and realistic tools for the performance evaluation and prediction of complex wireless networks with ever increasing volumes of multimedia traffic with different quality-of-service (QoS) guarantees.

Earlier performance models for 2G wireless cellular networks are based on the Global System for Mobile (GSM) telecommunications for the transmission of voice calls (e.g., [1]) and its extension, based on the General Packet Radio Service (GPRS), which allows data communication with higher bit rates than those provided by a single GSM channel (e.g., [1–3]). The introduction of the 3G wireless multi-service Universal Mobile Telecommunication Systems (UMTS) for multimedia traffic flows, such as voice calls, streaming media and data packets, has been of major interest for mobile network providers and has received widespread attention in the literature (e.g., [4,5]). More recently, 4G wireless systems have been proposed, broadly as a collection of different radio networks, which have access to Internet Protocol (IP) based services and where roaming is envisaged to be seamless [6,7].

The general industrial trend to migrate the 4G wireless network towards an all IP based solution has received considerable attention in industry and academia worldwide. This allows easy and cost effective service creation through reuse of application software as well as simple interpretability with existing Internet services. IP is technology independent and, therefore, it can be compatible with any underlying access technology. Thus, it represents an efficient interface amongst different radio access networks. Moreover, to support multimedia packet traffic flows with improved QoS, wireless 4G cell architectures are based on medium access control (MAC) protocols composed on both time-division and code-division statistical multiplexing [6–8]. For multimedia communications, the MAC protocol is expected to accommodate packets with different QoS requirements. Although the mobile and wireless research community has adopted these views, nevertheless mobility management as well as performance modelling and prediction of 4G network architectures remain two most critical issues requiring further investigation [7].

Recent performance evaluation studies for GSM/GPRS/UMTS wireless cells often use simulation modelling and the numerical solution of Marcov chains [1, 2, 4, 5]. Simulation modelling is an efficient tool for studying detailed system behaviour but it becomes costly, particularly as the system size increases. Markov chain modelling on the other hand provide greater flexibility but associated numerical solutions may suffer from several drawbacks, such as restrictive assumptions of Poisson arrival processes and/or state space explosion, limiting the analysis to small size systems. Thus, analytic approximations with low error tolerance are required to provide simple, reliable and cost-effective alternative tools for the performance predictionof 4G Cellular networks. Note that an earlier application of the information theoretic principle of maximum entropy (ME) [9, 10] to the performance analysis of a GSM/GPRS wireless cells with bursty generalized exponential (GE) traffic, subject to discriminatory partial sharing (DPS) and generalized PS (GPS) traffic handling schemes, can be seen in [4, 5].

In this paper, a novel handoff priority channel assignment scheme is proposed, named 4G-HOP, for a 4G cellular network with GE-type bursty multimedia traffic flows consisting of all IP voice calls, streaming media and guaranteed/best effort data packets with differentiated QoS requirements under an efficient MAC protocol. The new scheme enables group handover and complies with the principles of channel reservation, extended channel subrating and priority queueing [11], subject to a buffer threshold based GPS traffic handling scheme. In essence, the GPS scheme utilizes the informa-

Hand off Classes

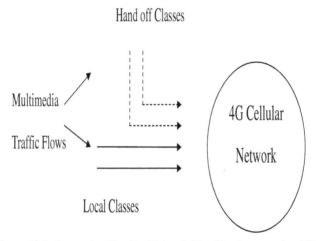

Figure 12.1 A generic 4G cell with handoff traffic classes under GPS.

tion of traffic rate distribution and QoS requirements to assign priorities to the users and employs appropriate buffer management rules, based on priority rules and buffer thresholds, for the efficient transmission of multimedia packets in the time slots of each transmission frame. To this end, channel efficiency can be maximized for multiple access media, subject to QoS constraints. A generic evisceration of a wireless multimedia 4G cell under GPS with new and handoff classes can be seen in Figure 12.1.

In this context, an open queueing network model (QNM) of finite capacity and multiple servers is devised to assess the impact of the 4G-HOP protocol on the performance of the wireless 4G cellular network. The QNM is decomposed into five interacting multiple class GE-type queueing and delay systems of new and handoff multimedia traffic flows under a GPS traffic handling scheme, namely

(i)   a $GE/GE/c_{11}/c_{11}$ loss system for new and handoff IP voice calls with c_{11} shared channels;

(ii) a $GE/GE/c_{12}/N_{12}/HoL/CBS$ queueing system for handoff IP voice calls with c_{12} dedicated channels, head-of-line (HoL) priorities and a complete buffer sharing (CBS) management scheme;

(iii) a $GE/GE/c_{21}/N_{21}/FCFS/CBS(T_l, T_h)$ queue for new and handoff IP streaming media packets with low (T_l) and high (T_h) buffer thresholds, c_{21} channels and first-come-first-served (FCFS) rule under complete buffer sharing (CBS) scheme;

(iv) a GE/GE/c_{22}/N_{22}/HOL/CBS queueing system for handoff IP streaming media packets with c_{22} dedicated channels, HoL rule under CBS; and

(v) a GE/GE/1/N_3/DPS delay system for IP data packets under a discriminatory processor share (DPS) transfer rule.

New analytic solutions for the aggregate and blocking state probabilities of the GE-type queueing and delay systems, as building blocks of the open QNM, are characterized, based on the information theoretic principle of maximum entropy (ME) and GE-type probabilistic algebra [10], subject to appropriate GE-type queueing theoretic constraints. Moreover, typical numerical experiments are presented to validate the credibility of the ME solutions for a 4G cellular network architecture under the 4G-HoP protocol against Java-based simulation results at 95% confidence intervals and also to study the effect of bursty multimedia class traffic flows upon the performance of the cell.

Some preliminary remarks on the GE-type distribution and the ME formalism are presented in Section 12.2. The performance modelling and evaluation of a wireless cell with handoff under the 4G-HOP and GPS schemes is described in Section 12.3. An overview of the ME solutions of the GE/GE/c/N/FCFS/CBS queue (a generalization of the GE/GE/c/c loss system) and GE/GE/1/N/DPS delay system as well as the GE/GE/c/N/HoL/CBS and GE/GE/c/N/FCFS/CBS(T_l, T_h) queueing systems are highlighted in Section 12.4. Typical experiments are devised in Section 12.5. Concluding remarks follow in Section 12.6.

12.2 Preliminary Remarks

12.2.1 GE-type Distribution

The GE-type distribution is of the form [10]

$$F(t) = P(X \le t) = 1 - \tau e^{-\tau vt}, \ \tau = 2/(1 + C^2), \ t \ge 0 \qquad (12.1)$$

where X is an inter-event time random variable and $1/v$; C^2 are the mean and squared coefficient of variation (SCV) of the inter-event times, respectively.

The GE distribution has a counting compound Poisson process (CPP) with geometrically distributed batch sizes with mean $1 = 1/v$. It is a credible model of bursty inter-arrival times of multiple class mobile connections with different minimum capacity demands. It is widely known an IP packet length distribution is non-exponential and should at least be described by the mean, v, and SCV, C^2. This is because the underlying physical network, such as Ethernet and IP, restricts IP packets and thus, they have different packet lengths, typically 1500 bytes and 53 bytes, respectively.

The GE distribution may also be employed to model short-range dependence (SRD) traffic with small error. For example, an SRD process may be approximated by an ordinary GE distribution whose first two moments of the count distribution match the corresponding first two SRD moments. This approximation of a correlated arrival process by an uncorrelated GE traffic process may facilitate (under certain conditions) problem tractability with a tolerable accuracy and, thus, the understanding of the performance behaviour of external SRD traffic in the interior of the network. It can be further argued that, for a given buffer size, the shape of the autocorrelation curve, from a certain point onwards, does not influence system behaviour [2]. Thus, in the context of system performance evaluation, an SRD model may be used to approximate with a tolerable accuracy long-range dependence (LRD) real traffic. This is of particular relevance as experimental studies in wireless systems have shown that the guaranteed/best effort data packets traffic has been shown to have a self-similar or a LRD property [14]. Note that, in the context of 4G wireless cells, the GE-type traffic could to model the multi-connections in either one-time slot or in several time slots.

12.2.2 Maximum Entropy (ME) Formalism

The principle of ME [9] provides a self-consistent method of inference for characterizing, under general conditions, an unknown but true probability distribution, subject to known (or, known to exist) mean value constraints. The ME solution can be expressed in terms of a normalizing constant and a product of Lagrangian coefficients corresponding to the constraints. In an information theoretic context, the ME solution is associated with the maximum disorder of system states and, thus, is considered to be the least biased distribution estimate of all solutions satisfying the system's constraints. Major discrepancies between the ME distribution and an experimentally observed [9] or stochastically derived [10] distribution indicate that important physical or theoretical constraints have been overlooked. Conversely, experimental or theoretical agreement with the ME solution represents evidence that the constraints of the system have been properly identified.

In the field of systems modelling, expected values performance distributions, such as those relating to the marginal and joint state probabilities (i.e., number of jobs at an individual queue or a network) may be known to exist and thus, they can be used as constraints for the characterization of the form of the ME solution. Alternatively, these expectations may be often analytically established via classical queueing theory in terms of some moments of the inter-arrival time and service time distributions. An efficient analytic (exact or approximate) implementation of the ME solution clearly requires the a priori estimation of the Lagrangian coefficients via exact or asymptotic expressions involving basic system parameters and performance metrics, as appropriate. Hence, the principle of ME may be applied to characterize useful information theoretic approximations of performance distributions of queueing/delay systems and related networks. The ME solution for the joint state probability distribution of an arbitrary open network with queueing and delay stations, subject to marginal mean value constraints, can be interpreted as a product-form approximation. Thus, entropy maximization implies a decomposition of a complex network into individual station each of which can be analyzed separately with revised inter-arrival and service times. Moreover, the marginal ME state probability of a single station, in conjunction with suitable formulae for the estimation of flow moments in the network, can play the role of a cost-effective analytic building block towards the computation of the performance metrics of the entire network. Further information on ME formalism and queueing networks can be found in [10].

12.3 A Wireless 4G Cellular Network under the 4G-HoP Scheme

An open QNM of a wireless 4G cellular network under the 4G-HoP channel assignment scheme is displayed in Figure 12.2. The multiple class arrival processes of all new and handoff IP voice calls, steaming media and guaranteed/best effort data packets are represented by CPPs with geometrically distributed batch sizes (or, equivalently, GE-type interarrival times [10]). Moreover, GE-type distributions are employed to describe the IP voice traffic durations as well as the channel transmission times of streaming media/data packets. In the context of this paper, only uplink traffic is considered. Note that the proposed QNM is of robust nature enabling the inclusion of some essential functionalities envisaged to be present in 4G mobile networks in the future towards a cost-effective bandwidth management. More specifically, it does not only support all IP based multimedia traffic flows but also incorporates efficient handoff service mechanisms in favour of handoff IP voice calls and streaming media under a GPS traffic handling scheme.

The decomposition of an open QNM into five interacting GE-type building block queueing and delay systems is described below. A number of common bufferless channels may be allocated to the transmission of multiple

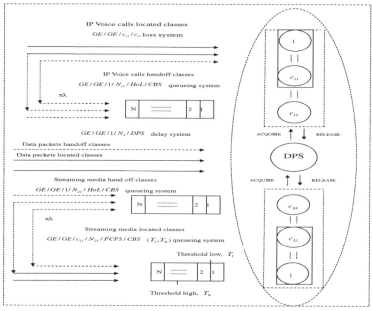

Figure 12.2 An open QNM of a wireless 4G cellular network under the 4G-HoP channel assignment scheme.

IP voice call classes consisting of new and high priority handoff calls. High-speed moving mobile terminals, each of which requires the assignment of a single channel for its entire duration, can generate handoff calls belonging to higher priority classes. Clearly, this transmission resource for IP voice calls can be modelled by a GE-type $GE/GE/c_{11}/c_{11}$ loss system with c_{11} common channels. If all c_{11} channels are busy, arriving new voice calls will be lost. However, blocked handoff voice calls are transferred, as calls of higher priority, to a GE-type $GE/GE/c_{12}/N_{12}/HoL/CBS$ queue with c_{12} guard channels, finite capacity, N_{12}, HoL priorities and CBS buffer management scheme. The HoL priority queue may also accommodate handoff voice calls of lower priority classes, which are generated by lower speed moving mobile terminals. A handoff IP voice call may be lost on arrival when the GE-type queue is full or, may be dropped out of the waiting line when a randomly distributed timeout period ends. Thus, the handoff $GE/GE/c_{12}/N_{12}/HoL/CBS$ queue serves multiple classes of handoff voice calls generated by mobile terminals moving at different speeds.

Moreover, another GE-type $GE/GE/c_{22}/N_{22}/HoL/CBS$ priority queue with c_{22} guard channels, finite capacity, N_{22}, HoL priorities, and CBS scheme is employed to serve a multiple class of handoff IP streaming media packets, where the priority of the handoff IP packets clearly depends on the moving speed of associated mobile terminals. Moreover, in a similar fashion to the handoff IP voice calls, the handoff IP streaming media packets either drop out of the waiting line at the end of a time out period or get blocked on arrival when the GE-type HoL priority queue is full. Consequently, in the latter case, the IP streaming media packets migrate to another transmission resource represented by a $GE/GE/c_{21}/N_{21}/FCFS/CBS(T_l, T_h)$ queueing system. The latter has finite capacity, N_{21}, FCFS rule and c_{21} common channels for the transmission of new and blocked (by the HoL queue) handoff IP streaming media packets with low and high buffer thresholds T_l and T_h, respectively, under CBS buffer management scheme. This buffer threshold scheme can be seen as a congestion control contribution for the uplink traffic scheduling, stipulating efficient channel sharing under the GPS traffic-handling scheme. Note that the streaming media packets of the $GE/GE/c_{21}/N_{21}/FCFS/CBS(T_l, T_h)$ queue may either be dropped out of the waiting line at the end of a time out period or blocked/lost on arrival when the GE-type HoL priority queue is full.

Finally, all admitted guaranteed/best effort IP data packet connections can share the available bandwidth according to the priority for each data class. Physically, a wireless 4G cell should be capable of allocating all available channels to one connection (subject to some battery restrictions) [12]. Thus, the transmission resource for the IP data packets can be clearly modelled by a GE/GE/1/N3/DPS delay system with guaranteed/best effort classes of data packets under discriminatory PS (DPS) rule. This delay system is clearly complying with the extended channel sub-rating principle [12] and incorporates according to assigned priorities differentiated transmission rates according to packet priorities and finite capacity N3, which represents the maximum number of connections sharing simultaneously the available data bandwidth.

Under the 4G-HOP protocol (based on the open QNM of Figure 12.2 and the GPS traffic-handling scheme), the transmission capacities of all IP voice calls, streaming media and data packets are dynamically evolving and they can be determined as follows.

At any given time, free IP dedicated voice call channels belonging to the $GE/GE/c_{11}/c_{11}$ loss system may be 'acquired' by the IP data partition to increase the transmission capacity of the $GE/GE/1/N_3/DPS$ delay system. However, new arrivals of new and handoff IP voice calls will cause the immediate 'release' to the GE-type loss system of some or all of the IP voice channels 'acquired' by the data partition, as appropriate. This is because

IP voice calls have higher pre-emptive priority in relation to the guaranteed/best effort IP data packets. The transmission of the guaranteed/best effort IP data traffic is enhanced further by adopting another type of 'acquired-released' mechanism associated with the $GE/GE/c_{21}/N_{21}/FCFS/CBS(T_l, T_h)$ queueing system. This process employs an auxiliary function $C(n_{21})$, where n_{21} is the aggregate queue length n_{21} $(0 \leq n_{21} \leq N_{21})$. The function defines the variable number of operational IP streaming media channels at the $GE/GE/C(n_{21})/N_{21}/FCFS/CBS(T_l, T_h)$ queue $(C(n_{21}) < c_{21})$ or, alternatively, the number of 'acquired' streaming media channels $c_{21} - C(n_{21})$ by the data partition. The evaluation of $C(n_{21})$ depends not only on n_{21} but also on the buffer thresholds (T_l, T_h). Consequently, at any given time, the number of available streaming media channels, $C(n_{21})$, of the $GE/GE/C(n_{21})/N_{21}/FCFS/CBS(T_l, T_h)$ queue, is determined for different sets of values of n_{21}. More details with regard the evaluation of function $C(n_{21})$ can be seen in Section 12.4.2.1.

12.4 ME Analysis of the Open QNM under the 4G-HOP Protocol

This section presents an overview of the ME analysis of the multiple class GE/GE/c/N/FCFS/CBS queue – a generalization of the GE/GE/c/c loss system - the GE/GE/1/N/DPS delay system, the $GE/GE/c/N/FCFS/CBS(T_l, T_h)$ queue with thresholds and the GE/GE/c/N/HOL/CBS priority queue, which play the role of building blocks in the decomposition process of the open QNM of Figure 12.2. Note that without loss of generality, c is used as generic parameter to represent the number of multi channels, when appropriate. A more extensive analysis of these systems, based on the principle of ME and GE-type probabilistic algebra, can be seen in [13]. As the focus of this paper is the evaluation of the performance impact of the proposed 4G-HoP channel assignment mechanism with priority handoff, it is assumed that no dropping of IP handoff packets takes place from the queues of the QNM due to time out periods. The study of finite capacity queues with time out periods adds substantial analytic complexity and will be the subject of a sequel paper [14].

The open QNM under 4G-HOP (see Figure 12.2) can be clearly decomposed into individual multiple class queueing and delay systems, respectively, by making use of the generic first two moments of the GE-type splitting streams of handoff IP voice calls and streaming media. The mean arrival rates and the squared coefficient of variations (SCVs) of the GE-type interarrival times of blocked handoff voice calls and streaming media classes of higher (h) priorities, respectively, can be analytically established via the generic

GE-type flow formulae relating to splitting streams [10], namely,

$$\lambda_{h, \text{ handoff, blocked}} = \lambda_{h, \text{ handoff}} * \pi_{h, \text{ handoff}} \tag{12.2}$$

$$Ca^2_{h, \text{ handoff, blocked}} = Ca^2_{h, \text{ handoff}}(1 - \pi_{h, \text{ handoff}}) + \pi_{h, \text{ handoff}} \tag{12.3}$$

where $\lambda_{h, \text{ handoff, blocked}}$ and $Ca^2_{h, \text{ handoff, blocked}}$ be mean arrival rate and inter-arrival time SCV of higher priority blocked handoff classes. And $\lambda_{h, \text{ handoff}}$, $Ca_{h, \text{ handoff}}$ and $\pi_{h, \text{ handoff}}$ be mean arrival rate, inter-arrival time SCV and blocking probability for higher priority handoff classes.

12.4.1 Analysis of GE/GE/c/N/FCFS/CBS and GE/GE/1/N/DPS Delay Systems

Notation
For each class i ($i = 1, 2, \ldots, R, R > 1$) let $\{1/\lambda_i, C^2_{ai}\}$, $\{1/\mu_i, C^2_{si}\}$ be the mean and squared coefficient of variation (SCV) of the interarrival and service time distributions, respectively. Note that μ_i ($i = 1, 2, \ldots, R$) under DPS rule is defined, subject to discriminatory weights. Moreover, for either a GE/GE/c/N/FCFS/CBS queueing system or a GE/GE/1/N/DPS delay system, let at any given time.

$\mathbf{S} = (c, c_1, c_2, \ldots, c_R)$, $\sum_R^{i=1} c_i \leq N$, where c_i ($i = 1, 2, \ldots, R$) is the class of the ith job in either system with c being the class of the job in service under the FCFS rule.

$\mathbf{Q}$ is the set of all feasible states of $\mathbf{S}$;

$P(S)$ $\mathbf{S} \in \mathbf{Q}$ is the stationary state probability;

n_i is the number of entities of class i ($i = 1, 2, \ldots, R$);

π_i is the blocking probability that an arrival of class i will find the system at full capacity.

For each state $\mathbf{S}$, $S \in Q$, and class i, $i = 1, 2, \ldots, R$, the following auxiliary functions are defined:

$n_i(S)$ = the number of class i jobs present in either the GE/GE/1/N/DPS or the GE/GE/c/N/FCFS/CBS system

$$s_i(S) = \begin{cases} 1, & \text{if the job in service is of class } i \\ 0, & \text{otherwise} \end{cases}$$

$$s_{ik}(S) = \begin{cases} 1, & \text{if } n_i \geq k \text{ and } k \leq c - \sum_{j=1}^{i-1} n_j \\ 0, & \text{otherwise} \end{cases}$$

$$f_i(S) = \begin{cases} 1, & \text{if } \sum_{i=1}^{R} n_i(S) = N; \text{ and } s_i(S) = 1 \\ 0, & \text{otherwise} \end{cases}$$

$$f_{ik}(S) = \begin{cases} 1, & \text{if } \sum_{i=1}^{R} n_i(S) = N, \text{ and } s_{ik}(S) = 1 \\ 0, & \text{otherwise} \end{cases}$$

Note that the random variables of the server state are $s_i(S)$ and $s_{ik}(S)$, which are related to the utilization of the bandwidth for the GE/GE/c/N/FCFS/CBS and GE/GE/1/N/DPS systems, respectively. On the other hand, the random variables of the queue capacity are designed as $f_i(S)$ or $f_{ik}(S)$.

Suppose that the following mean value constraints about the state probability $P(S)$ are known to exist:

(i) Normalization

$$\sum_{S \in Q} P(S) = 1 \tag{12.4}$$

(ii) Probabilities $\{U_i, i = 1, 2, \ldots, R\}$, $\{U_{ik}, i = 1, 2, \ldots, R; k = 1, 2, \ldots, c\}$

$$\begin{cases} \sum_{S \in Q} s_i(S) P(S) = U_i, \quad 0 < U_i < 1, \\ \quad i = 1, 2, \ldots, R, \hspace{3cm} \text{GE/GE/1/N/DPS} \\ \sum_{S \in Q} s_{ik}(S) P(S) = U_{ik}, \quad 0 < U_{ik} < 1, \\ \quad i = 1, 2, \ldots, R; k = 1, 2, \ldots, c, \hspace{1cm} \text{GE/GE/c/N/FCFS/CBS} \end{cases} \tag{12.5}$$

(iii) Average number in the system $\{L_i, i = 1, 2, \ldots, R\}$,

$$\sum_{S \in Q} n_i(S) P(S) = L_i, \quad i = 1, 2, \ldots, R \tag{12.6}$$

with $U_i < L_i < N$ for GE/GE/1/N/DPS and $U_{i1} < L_i < N$ for GE/GE/c/N/ FCFS/CBS.

(iv) Full buffer state probabilities $\{\phi_i, \quad \phi_{ik}, \quad i = 1, 2, \ldots, R; \quad k = 1, 2, \ldots, c\}$

$$
\begin{cases}
\sum_{\mathbf{S} \in \mathbf{Q}} f_i(\mathbf{S}) P(\mathbf{S}) = \phi_i, \quad 0 < \phi_i < 1, \quad \text{GE/GE/1/N/DPS} \\[2em]
\sum_{\mathbf{S} \in \mathbf{Q}} f_{ik}(\mathbf{S}) P(\mathbf{S}) = \phi_{ik}, \quad 0 < \phi_{ik} < 1, \quad \text{GE/GE/c/N/FCFS/CBS}
\end{cases}
$$

(12.7)

satisfying the class flow balance equations, namely

$$
\lambda_i(1 - \pi_i) = \mu_i U_i, \quad i = 1, \ldots, R \tag{12.8}
$$

The form of the ME joint state probability distribution $\{P(\mathbf{S}), \mathbf{S} \in \mathbf{Q}\}$, can be characterized by maximizing the entropy functional $H(\mathbf{P}) = -\sum_{\mathbf{S} \in \mathbf{Q}} P(\mathbf{S}) \log P(\mathbf{S})$, subject to prior information expressed by mean value constraints (12.4)–(12.7). By employing Lagrange's method of undetermined multipliers, the following solutions are obtained [13]:

$$
P(\mathbf{S}) =
\begin{cases}
\dfrac{1}{z} \displaystyle\prod_{i=1}^{R} g_i^{s_i(\mathbf{S})} x_i^{n_i(\mathbf{S})} y_i^{f_i(\mathbf{S})}, \forall \mathbf{S} \in \mathbf{Q}, \quad \text{GE/GE/1/N/DPS;} \\[2em]
\dfrac{1}{z} \displaystyle\prod_{i=1}^{R} \left(\prod_{k=1}^{c} g_{ik}^{s_{ik}(\mathbf{S})} \right) x_i^{n_i(\mathbf{S})} y_{ik}^{f_{ik}(\mathbf{S})}, \forall \mathbf{S} \in \mathbf{Q}, \quad \text{GE/GE/c/N/FCFS/CBS,}
\end{cases}
$$

(12.9)

where $Z = 1/P(0)$, is the normalizing constant, $\{g_i, x_i, y_i, i = 1, 2, \ldots, R\}$ and $\{g_{ik}, x_i, y_{ik}, i = 1, 2, \ldots, R; \quad k = 1, 2, \ldots, c\}$ are the Lagrangian coefficients corresponding to constraints (12.4)–(12.6) per class, respectively.

12.4.1.1 The Aggregate ME Probability Distribution

By using equations (12.8), the aggregate state probabilities $\{P(n), n = 0, 1, \ldots, N\}$ are given by

$$P(n) = \begin{cases} \dfrac{1}{Z} \left(\displaystyle\sum_{i=1}^{R} g_i x_i y_i^{f_{ik}(\mathbf{n})} \right) X^m, & n \in [1, N], \ \text{GE/GE/1/N/DPS} \\[6mm] \dfrac{1}{Z} \displaystyle\prod_{i=1}^{R} \left\{ \displaystyle\prod_{k=1}^{c} g_{ik}^{h_{ik}(\mathbf{n})} x_i y_{ik}^{f_{ik}(\mathbf{n})} \right\} X^m, & n \in [1, N], \ \text{GE/GE/c/N/FCFS} \end{cases}$$

$$(12.10)$$

where

$$Z = \frac{1}{P(0)}, \quad X = \sum_{R}^{i=1} x_i \ \text{and} \ \sum_{i=1}^{R} n_i = n, \quad m = \begin{cases} n - 1, & \text{if } n \leq N - 1 \\ N, & \text{if } m = N \end{cases}$$

12.4.1.2 The Blocking Probability

A universal expression for the marginal blocking probabilities $\{\pi_i, i = 1, 2, \ldots, R\}$ of a stable multiple class GE/GE/1/N/DPS delay system and GE/GE/c/N/FCFS/ CBS queueing systems, respectively, can be approximated by focusing on a tagged job within an arriving bulk and making use of GE-type probabilistic arguments.

To this end, let $\sigma_i = 2/Ca_i^2 + 1$ and $\gamma_i = 2/Cs_i^2 + 1$ be the GE-type interarrival and service time non-zero stage selection probabilities associated with the GE/GE/1/N/PS and GE/GE/c/N/FCFS/CBS systems, respectively. Given that the system is at state $\mathbf{n} = (n_1, n_2, \ldots, n_R)$ with $n = \sum_{i=1}^{R} n_i$, the number of available buffer spaces equal to $N - n$. By focusing on a tagged job within an arriving bulk of class i ($i = 1, 2, \ldots, R$), the following blocking probability can be clearly determined:

(i) A class i tagged job is blocked and its bulk finds the queue at state

$$\mathbf{0} = (0, 0, \ldots, 0))$$

$$= \begin{cases} \delta_i(0)(1 - \sigma_i)^N P(0), & \text{GE/GE/1/N/DPS} \\ \delta_i^c(0)(1 - \sigma_i)^N P(0), & \text{GE/GE/c/N/FCFS/CBS} \end{cases}$$

$$(12.11)$$

where

$$\delta_i(0) = \frac{r_i}{r_i(1 - \sigma_i) + \sigma_i}$$

(ii) A class i tagged job is blocked and its bulk finds the queue at state

$$\mathbf{n} = (n_1, n_2, \ldots, n_R)), \quad \sum_{i=1}^{R} n_i \leq N \tag{12.12}$$

$$= \begin{cases} (1 - \sigma_i)^N P(\mathbf{n}), & \text{GE/GE/1/N/DPS} \\ \begin{cases} \delta_i^{c-n}(0)(1 - \sigma_i)^{N-n} P(\mathbf{n}), \ 0 < n < c \\ (1 - \sigma_i)^{N-n} P(\mathbf{n}), \ c \leq n \leq N \end{cases} & \text{GE/GE/c/N/FCFS/CBS} \end{cases}$$

where

$$\delta_i(0) = \frac{r_i}{r_i(1 - \sigma_i) + \sigma_i}.$$

Combining equations (12.11)–(12.12), after some manipulation, the blocking probabilities $\{\pi_i, i = 1, 2, \ldots, R\}$ can be expressed by

$$\pi_i = \begin{cases} \sum_{n=0}^{N} \delta_i(n)(1 - \sigma_i)^{N-n} P(n), & \text{GE/GE/1/N/DPS} \\ \sum_{j=1}^{R} \sum_{n=0}^{N} \delta_i^L(0)(1 - \sigma_i)^{N-n} P(n), & \text{GE/GE/c/N/FCFS/CBS} \end{cases} \tag{12.13}$$

where

$$\delta_i(0) = \frac{r_i}{r_i(1 - \sigma_i) + \sigma_i}, \quad \delta_i(n) = 1 \quad \forall n > 0$$

and

$$L = \begin{cases} c & \text{if } n = 0 \\ \max(0, c - n) & \text{if } 0 < n < c \\ 0 & \text{if } c \leq n \leq N \end{cases}$$

12.4.2 ME Analysis of GE/GE/c/N/FCFS/CBS(T_l, T_h) and GE/GE/c/N/HoL/CBS Queues

Notation
For each class i ($i = 1, 2, \ldots, R, R > 1$) let $\{1/\lambda_i, C_{ai}^2\}$, $\{1/\mu_i, C_{si}^2\}$ be the mean and squared coefficient of variation (SCV) of the interarrival and service time distributions, respectively. Let c be a generic number of servers for either queueing or delay systems. Moreover, let

n_i be the number of entities of class i ($i = 1, 2, \ldots, R$);

$\mathbf{n} = (n_1, n_2, \ldots, n_R)$ be a joint system state (n.b., $\mathbf{0} = (0, \ldots, 0)$));

Ω be the set of all feasible joint states $\{\mathbf{n}\}$;

$P(\mathbf{n})$, $\mathbf{n} \in \Omega$, be the joint state probability;

π_i be the blocking probability that an arrival of class i will find the system at the full capacity.

12.4.2.1 ME Solutions of GE/GE/c/N/FCFS/CBS(T_l,T_h)

The form of the ME joint state probabilities $\{P(\mathbf{n}), \forall \mathbf{n} \in \Omega\}$ of the GE/GE/c/N/ FCFS/CBS(T_l,T_h) queueing system, where c is the maximum number of servers, n_i is the number of class i $(1 < i < R)$ in the system, can be determined, subject to appropriate marginal (class) mean value constraints. The latter are based on the class utilization $\{U_{ik} = 1 - \sum_{n=1}^{k-1} P_i(n), i = 1, 2, \ldots, R; k = 1, 2, \ldots, c\}$ for GE/GE/c/N/FCFS(T_l, T_h) queueing system, and mean number of entities per class $\{\phi_{ik}, i = 1, 2, \ldots, R; k = 1, 2, \ldots, c\}$ for GE/GE/c/N/FCFS(T_l, T_h) queueing system with a class i entity in service satisfying the flow balance equations

$$\lambda_i(1 - \pi_i) = \mu_i U_i, \quad i = 1, \ldots, R, \tag{12.14}$$

where U_i is the marginal utilization of class i, $i = 1, 2, \ldots, R$. By employing Lagrange's method of undetermined multipliers, the ME steady state joint probability $P(\mathbf{n})$ can be expressed by [10]

$$P(\mathbf{n}) = \frac{1}{Z}\Gamma_1\Gamma_2\Gamma_3\Gamma_4, \, GE/GE/c/N/FCFS/CBS(T_l, T_h) \tag{12.15}$$

with

$$\Gamma_1 = \frac{\prod_{l=0}^{R-1}\binom{n-n_l}{n_{l+1}}}{R^{n-c}} \tag{12.16}$$

$$\Gamma_2 = \sum_{i=1}^{R}\left\{\sum_{k_i=1}^{E(n)}\frac{E(n)\prod_{l=1}^{R-1}\binom{E(n)}{k_i}\binom{n-E(n)}{n_i-k_i}}{\prod_{l=0}^{R-1}\binom{n-n_l}{n_{l+1}}}\right\} \tag{12.17}$$

$$\Gamma_3 = \prod_{i=1}^{R}\left(\prod_{k=1}^{k_i}g_{ik}^{h_{ik}(n)}y_{ik}^{f_{ik}(n)}\right) \tag{12.18}$$

$$\Gamma_4 = \prod_{n=0}^{R-1} x_i^{n_i - k_i} x_R^{n_R - E(n) + \sum_{i=1}^{R} k_i} \tag{12.19}$$

where $Z = 1/P(0)$, $\{h_{ik}(n), f_i(n), f_{ik}(n) n \in \Omega\}$ are suitable auxiliary functions and $\{x_i, g_{ik}, y_{ik}, \forall i, n\}$ are Lagrangian coefficients for GE/GE/c/N/FCFS/CBS(T_h, T_l) queueing system corresponding to the aforementioned constraints (U_{ik}, L_i, ϕ_{ik}, $i = 1, 2, \ldots, R$). These Lagrangian coefficients can be determined by making asymptotic connections to the corresponding infinite capacity system and using flow balance equations (12.14) [10]. Moreover, as mentioned in Section 12.3, the auxiliary function ($C(n)$ $0 \leq C(n) < c$), where n ($0 \leq n \leq N$). $n = \sum_{i=1}^{R} n_i$ represents, at any given time, the number of available streaming media channels at the GE/GE/c/N/FCFS/CBS(T_h, T_l) queueing system. Consequently, ($c - C(n)$) is the number of channels which can be 'acquired' by the IP best effort data partition. In the context of the analytic ME solution (12.15) of the GE/GE/c/N/FCFS /CBS(T_h, T_l) queue, accordance with the proposed 4G-HoP mechanism, namely

$$C(n) = \begin{cases} n, & \text{for } n \leq c_1 \\ c_1, & \text{for } n \leq T_l \\ \frac{c_1 + c}{2}, & \text{for } T_l < n < T_h \\ c, & \text{for } n \geq T_h \end{cases} \tag{12.20}$$

where c_1 ($1 \leq c_1 < c$) is the minimum number of servers associated, in support of the quality of service, with the GE/GE/c/N/FCFS/CBS(T_l, T_h) queueing system. More specifically, note that if $T_l < n < T_h$ and the traffic flow shifts, eventually, the aggregate number of streaming media packets, n, from T_l towards T_h, then $C(n) = c_1$ and from T_h towards T_l, then $C(n) = c$. Thus, if n is between T_l and T_h, $C(n)$ can be estimated by the average number of available servers, $(c_1 + c)/2$. Finally, if $T_h \leq n$, then $C(n) = c$ (i.e., the GE/GE/c/N/FCFS/CBS(T_l, T_h) queue operates with the maximum number of servers, c. Note that the credibility of the analytic ME solution for the GE/GE/c/N/FCFS/CBS(T_l, T_h) queue (12.15) against simulation is included in Section 12.5.

Moreover, the aggregate probability $P(n)$ can be obtained by unconditioning the joint probability $P(n)$ over all classes and is determined by

$$P(n) = \frac{1}{Z} \sum_{n \subseteq A_n} \prod_{i=1}^{R} \left\{ \prod_{k=1}^{C(n)} g_{ik}^{h_{ik}(n)} y_{ik}^{f_{ik}(n)} \right\} X^{n-1} \qquad (12.21)$$

where $Z = 1/P(0)$, $X = \sum_{i=1}^{R} x_i$, $A_n = \{n : n = \sum_{i=1}^{R} n_i, 0 \le n_i \le N\}$.

12.4.2.2 A ME Solution of the GE/GE/c/N/HoL/CBS Queueing System

For each handoff class i ($i = 1, 2, \ldots, R, R > 1$) of the GE/GE/c/N/HoL/CBS queue with R priority classes and HoL rule, let $\{1/\lambda_i, C_{ai}^2\}$, $\{1/\mu, C_{si}^2\}$, be the mean and SCV of the interarrival and service time, respectively. Let c be a generic number of servers for the GE/GE/c/N/HoL/CBS queueing system. The form of the ME joint state probabilities $P(n)$ is the joint state probability and $P(n)$ is the aggregate number of entities of class i ($i = 1, 2, \ldots, R$). Note that $\{P(n), \forall n \in \Omega\}$ of the GE/GE/c/N/HoL/CBS building block queueing systems, where n_i is the number of class i entities representing mobility in the system, as appropriate, can be determined, subject to appropriate marginal (class) mean value constraints. These constraints are based on the class utilizations $\{U_{ik} = 1 - \sum_{n=1}^{k-1} P_i(n), i = 1, 2, \ldots, R; k = 1, 2, \ldots, c\}$ for GE/GE/c/N/HoL/CBS queueing system, mean number of entities per class $\{\phi_{ik}, i = 1, 2, \ldots, R; k = 1, 2, \ldots, c\}$ for GE/GE/c/N/HoL/CBS queueing system, with a class i entity in service satisfying the flow balance equations.

By employing Lagrange's method of undetermined multipliers, the form of the ME joint state probability $P(\mathbf{n})$ can be characterized at equilibrium

$$P(n) = \frac{1}{Z} \prod_{i=1}^{R} \left(\prod_{k=1}^{\min(n_i, c)} g_{ik} y_{ik}^{\delta_i 0} \right) \prod_{i=1}^{R} x_i^{n_i} \qquad (12.22)$$

Subsequently, aggregating over all relevant joint states, the steady state aggregate probability $P(n)$ of GE/GE/c/N/HOL/CBS queueing system can be expressed by

$$P(n) = \frac{1}{Z} \sum_{i=1 \wedge n = \sum_{i=1}^{R} n_i} \prod_{i=1}^{R} \left(\prod_{k=1}^{\min(n_i, c)} g_{ik} y_{ik}^{\delta_i(0)} \right) \prod_{i=1}^{R} x_i^{n_i} \qquad (12.23)$$

where $Z = 1/P(0)$, $\{h_{ik}(n), f_i(n), f_{ik}(n) n \in \Omega\}$ are suitable auxiliary functions and $\{x_i, g_{ik}, y_{ik}, \forall i, n\}$ are Lagrangian coefficients for GE/GE/c/N/HoL/CBS queueing system and $(U_{ik}, L_i, \phi_{ik}, i = 1, 2, \ldots, R)$.

Note that the Lagrangian coefficients can be determined by making asymptotic connections to the corresponding infinite capacity system and using flow balance equations (12.14) [6].

12.4.2.3 A Unified Blocking Probability of GE/GE/c/N/FCFS/CBS(T_l, T_h) and GE/GE/c/N/HoL/CBS Queueing System

Focusing on a tagged IP handoff either streaming media or voice call at either GE/GE/c/N/FCFS/CBS (T_l, T_h) or GE/GE/c/N/HoL/CBS queueing systems, respectively, within an arriving bulk and applying GE-type probabilistic arguments, the blocking probability per class π_i ($i = 1, 2, \ldots, R$) can be approximated [10, 13] by

$$\pi_i = \sum_{j=1}^{R} \sum_{n=0}^{N} \delta_i^L(0)(1 - \sigma_i)^{N-n} P(n) \qquad (12.24)$$

where

$$\delta_i(0) = \frac{r_i}{r_i(1 - \sigma_i) + \sigma_i}, \qquad \delta_i(n) = 1 \quad \forall n > 0$$

and

$$L = \begin{cases} c & \text{if } n = 0, \\ \max(0, c - n) & \text{if } 0 < n < c \\ 0 & \text{if } c \le n \le N. \end{cases}$$

12.4.3 Weighted Performance Measures under GPS

The mean performance metrics of the GE/GE/1/N/DPS delay system serving the IP guaranteed/best effort data packets (DP) (see Figure 12.2) can be determined by making use of weighted averages. This can be achieved by taking probabilistically into account the dynamic increase/decrease of the transmission capacities of the IP voice calls (VC) at the GE/GE/c_{11}/c_{11} loss system and IP streaming media (SM) packets at the GE/GE/c_{21}/N_{21}/FCFS/CBS(T_l, T_h) queue under the GPS traffic handling scheme. More specifically, the guaranteed/best effort DP will receive under GPS variable transmission capacity depending on the availability of free IP VC and SM channels. To this end, a class i ($i = 1, 2, \ldots, R$) average performance statistic for IP guaranteed/best effort DP under DPS rule, say S'_{DP}, can be defined by the weighted average measure:

$$S'_{DP} = \sum_{l_{11}=0}^{c_{11}} \sum_{l_{21}=0}^{c_{21}(n_{21})} \Xi_{11} \Xi_{21}, \quad i = 1, 2, \ldots, R \qquad (12.25)$$

$$C_{21}(n_{21}) = \begin{cases} n_{21}, & \text{for } n_{21} \leq c_1 \\ c_{21,l}, & \text{for } c_{21,l} < n_{21} \leq T_l \\ \frac{c_{21,l}+c_{21}}{2}, & \text{for } T_l < n < T_h \\ c_{21}, & \text{for } n_{21} \geq T_h \end{cases} \tag{12.26}$$

where $\Xi_{11} = S_{DP}^{l_{11},l_{21},i} P_{VC}(n_{11} = l_{11})$, $\Xi_{21} = P_{SM}(n_{21} = l_{21})$ and $c_{21,l}$ is the minimum number of available servers at the GE/GE/c_{21}/N_{21}/FCFS/CBS(T_l, T_h) queue. In addition, $P_{VC}(.)$ and $P_{SM}(.)$ are the aggregate VC and SM steady state probabilities of IP voice calls of the GE/GE/c_{11}/c_{11} loss system and the streaming media of the GE/GE/c_{21}/N_{21}/FCFS/CBS(T_l, T_h) queue, respectively and $\{S_{DP}^{l_{11},l_{21},i}, l_{11} = 0, 1, 2, \ldots, c_{11}, l_{21} = 0, 1, 2, \ldots, C_{21}\}$, are the estimated values of statistic S'_{DP} corresponding at each stage to the maximum available mean DP transfer rates. Clearly, these rates (under the GPS traffic handling scheme) are given by $\{\mu_{DP}+c_{11} - l_{11}\mu_{VC}+(c_{21}-C_{21}(n_{21}))\mu_{SM}\}$. Moreover, $C_{21}(n_{21})$ is the auxiliary function (see Section 12.4.2.1), which estimates the number of the streaming media channels $\{c - C_{21}(n_{21})\}$ acquired by the data partition when the aggregate number of streaming media at the GE/GE/c_{21}/N_{21}/FCFS/CBS(T_l, T_h) queue is n_{21}. Finally, $\{\mu_{VC}, \mu_{SM}, \mu_{DP}^i\}$ are the initially allocated aggregate transfer rates to the IP voice calls and streaming media systems and the class i ($i = 1, 2, \ldots, R$) transfer rates for data packets system, respectively.

12.5 Numerical Results

This section presents typical numerical experiments (see Figures 12.3–12.10) with reference to the proposed open QNM of Figure 12.2 in order to evaluate the credibility of the GE-type performance approximations of the ME methodology against Java based simulation results at 95% confidence intervals and assess the adverse effect of traffic variability on system performance.

The numerical results are presented in Figures 12.3–12.9 and involve:

- Three traffic classes of IP voice calls at the GE/GE/c_{11}/c_{11} loss system, namely, two new (local) classes and one handoff class with mean transmission rates 9.6 Mbytes, 6.0 Mbytes and 6.0 Mbytes, respectively;
- Three handoff traffic classes for IP voice calls at the GE/GE/1/N_{12}/HoL/CBS priority queue, namely, one highest priority handoff class consisting of those calls blocked at the GE/GE/c_{11}/c_{11} loss

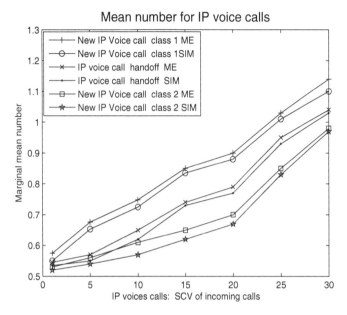

Figure 12.3 Effect of traffic variability on the marginal mean number of IP voice calls for a multiple class GE/GE/c/c loss system.

system and two lower priority handoff classes with mean transmission rates 6.0 Mbytes, 9.6 Mbytes and 12.5 Mbytes, respectively;

- Three handoff traffic classes of IP streaming media at the $GE/GE/1/N_{22}/HoL/CBS$ priority queue, namely one highest priority handoff class and two lower priority handoff classes with mean transmission rates 62.5 Mbytes, 300 Mbytes and 62.5 Mbytes, respectively.
- Three traffic classes of IP streaming media at the $GE/GE/c_{21}/N_{21}/FCFS/CBS(T_l, T_h)$ queue, namely, the highest priority handoff class, which is blocked at the $GE/GE/c_{22}/N_{22}/HoL/CBS$ priority queue and diverted towards the FCFS threshold queue and two new classes with mean transmission rates 300 Mbytes, 62.5 Mbytes and 62.5 Mbytes, respectively.
- Three classes of IP data packets at the $GE/GE/1/N_3/DPS$ delay system, namely two new classes (e.g., voice mail and web browsing) and a handoff class with mean transmission rates 12.5 Mbytes, 15 Mbytes and 12.5 Mbytes, respectively.

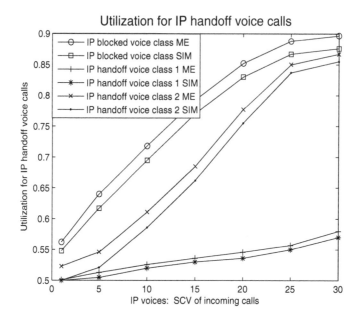

Figure 12.4 Effect of traffic variability on the marginal utilisation of IP voice calls for a multiple class GE/GE/1/N/HoL/CBS queueing system.

It is assumed that the partitions of the IP voice calls, streaming media and data packets consist of multi-bandwidth based on time-slot providing total capacity of 1 Gbps per second for IP voice calls, streaming media and data packets (the sum of the all mean transmission rates).

For each traffic class r ($r = 1, 2 \ldots, R$), let $\lambda_{ij,r}$ and $\mu_{ij,r}$ symbolize the mean arrival/transmission rates relating to IP voice calls, steaming media and data packets at the 'ij' queue or delay system, as appropriate. Moreover, let $\{Ca^2_{ij,r}, Cs^2_{ij,r}\}$, be the square coefficients variation of the interarrival and transmission times per class r ($r = 1, 2, \ldots, R$), respectively, at each queueing and delay systems, as appropriate, with mean input/output batch packets size given by $\{\frac{Ca^2+1}{2}, \frac{Cs^2+1}{2}\}$ [3]. Moreover, let c_{ij} and N_{ij} be the maximum number of servers and buffer capacity at the 'ij' queueing or delay system, as appropriate.

The following fixed input data (Figures 12.3–12.8) which are consistent with the aforementioned IP packet length) are used: $\{\lambda_{11,1} = 1.0, \lambda_{11,2,\text{handoff}} = 1.0, \lambda_{11,3} = 0.8, \mu_{11,1} = 0.96, \mu_{11,2,\text{handoff}} = 0.96, \mu_{11,3} = 0.6, c_{11} = 6, Ca^2_{11,2,\text{handoff}} = 2, Ca^2_{11,3} = 2, Cs^2_{11,1} = 4, Cs^2_{11,2,\text{handoff}} = 2,$

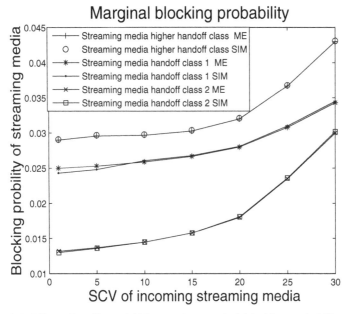

Figure 12.5 Effect of traffic variability on the marginal blocking probability of steaming media for a multiple class GE/GE/1/N/HoL/CBS queueing system.

$Cs_{11,3}^2 = 2\}$ for IP voice calls, $\{\lambda_{12,1} = 1.0, \lambda_{12,2} = 2.0, \lambda_{12,3} = 3.0, \mu_{12,1} = 0.6, \mu_{12,2} = 0.96, \mu_{12,3} = 1.25, c_{12} = 2, Ca_{12,2}^2 = 2, Ca_{12,3}^2 = 2, Cs_{12,1}^2 = 2, Cs_{12,2}^2 = 2, Cs_{12,3}^2 = 2\}$ for IP handoff voice calls (head of line), $\{\lambda_{21,1} = 4.0, \lambda_{21,2} = 1.0, \lambda_{21,3,\text{handoff}} = 2.5, \mu_{21,1} = 3.0, \mu_{21,2} = 0.625, \mu_{21,3,\text{handoff}} = 0.625, Ca_{21,1}^2 = Cs_{21,2}^2 = 3, Cs_{21,2}^2 = 3, Ca_{21,3,\text{handoff}} = Cs_{21,3,\text{handoff}} = 3, c_2 = 6, N_{21} = 10\}$ for streaming media packets, $\{\lambda_{22,1} = 2.5, \lambda_{22,2} = 1.0, \lambda_{21,3,\text{handoff}} = 0.5, \mu_{22,1} = 3.0, \mu_{22,2} = 0.625, \mu_{22,3,\text{handoff}} = 0.625, Ca_{22,1}^2 = Cs_{22,2}^2 = 3, Cs_{22,2}^2 = 3, Ca_{22,3,\text{handoff}} = Cs_{22,3,\text{handoff}} = 3, c_2 = 2, N_{21} = 20\}$ for handoff streaming media packets (head of line) and $\{\lambda_{31,1} = 0.5, \lambda_{31,2} = 1.0, \lambda_{31,3,\text{handoff}} = 0.5, \mu_3 = 4.0,$ PS service discrimination weight $\mu_{31,1} = 1/6\mu_3, \mu_{31,2} = 5/12\mu_3, \mu_{31,3\text{handoff}} = 5/12\mu_3, Ca_{31,1}^2 = 5, Ca_{31,1}^2 = 2, Ca_{31,3,\text{handoff}}^2 = 2, Cs_{31,1}^2 = 5, Cs_{31,2}^2 = 5, Cs_{31,3,\text{handoff}}^2 = 5, N_3 = 20\}$ for data packets. Without loss of generality, the comparative study focuses on marginal performance metrics of mean number of streaming media and data packets per class as well as the aggregate blocking probabilities of voice calls, streaming media packets and data packets,

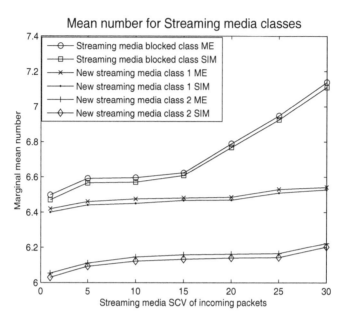

Figure 12.6 Effect of traffic variability on the marginal mean number of streaming media for a multiple class GE/GE/c/N/FCFS/CBS(T_l, T_h) queueing system.

respectively. Note that the numerical tests make use of varying traffic values of the interarrival time SCVs, Ca_{111}^2, Ca_{121}^2 Ca_{212}^2 and Ca_{222}^2.

Figure 12.3 illustrates the variability of the marginal mean IP voice calls per class versus the SCV of their interarrival times at the GE/GE/c/c. It can be seen that, as the interarrival time SCVs of IP voice calls progressively increases, the mean number of voice calls per class as their SCV of the IP voice calls increases too. In Figure 12.4, an assumed blocked stream of IP handoff calls at the GE/GE/c/c loss system forms the highest priority traffic class at the GE/GE/1/N/HoL/CBS queueing system with HoL priority rule and, as expected, it obtains the best overall performance. Focusing on the GE/GE/1/N/HoL/CBS priority queue for streaming media, it can be observed in Figure 12.5 that the blocking probability of the highest priority handoff class is the largest in magnitude when compared to the blocking probabilities of lower priority handoff classes. Moreover, the adverse effect of traffic variability on the marginal multiple class mean numbers of the IP streaming media of the GE/GE/c/N/FCFS/CBS(T_l, T_h) and the IP guaranteed/best effort data packets GE/GE/1/N/DPS delay system displayed in Figures 12.6 and 12.7, respectively. Note that the weighted formula (12.17)

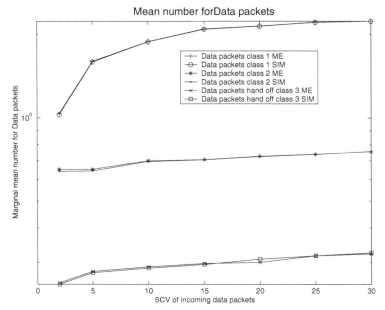

Figure 12.7 Effect of traffic variability on the marginal mean number of data packets for a multiple class GE/GE/1/N/DPS delay system.

is used for the calculation of the mean number of IP data packets of the DPS system. Finally, it can be seen in Figures 12.8–12.10 that the analytically established mean number of streaming media and guaranteed/best effort data packets deteriorate rapidly with different interarrival-time SCVs and a variable of the buffer size. In particular, Figure 12.10 illustrates that the GE-type GE/GE/c/N/FCFS/CBS(T_l, T_h) queue achieves higher performance than the corresponding GE/GE/c/N/FCFS/CBS queueing system without thresholds.

More specifically, it can be observed in Figures 12.8 and 12.9 that the analytically established mean number of streaming media and data packets deteriorate rapidly with increasing external interarrival-time SCVs (or, equivalently, average batch sizes) beyond a specific critical value of the buffer size which corresponds to the same mean number of data packets for two different SCV values. It is interesting to note, however, that for smaller buffer sizes in relation to the critical buffer size and increasing mean batch sizes, the mean number of data packets steadily improves with increasing values of the corresponding SCVs. This 'buffer size anomaly' can be attributed to the fact that, for a given arrival rate, the mean batch size of arriving bulks becomes larger as the SCV of the interarrival time increases, resulting in a greater

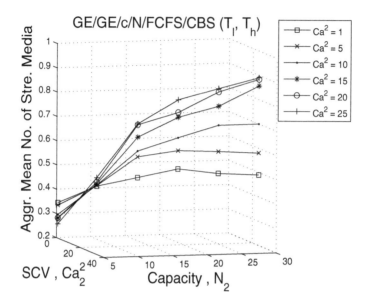

Figure 12.8 Effect of traffic variability on the aggregate mean number of streaming media for a multiple class GE/GE/c/N/FCFS/CBS (T_l, T_h) queueing system under PSS (Experiment 11 – Input Data: $\{\lambda_{21} = 2.0, \lambda_{22} = 1.2, \lambda_{23} = 0.6, \mu_{21} = \mu_{22} = \mu_{23} = 6.0, Ca_{21}^2 = Cs_{21}^2 = 5, Cs_{22}^2 = 5, Ca_{23} = Cs_{23} = 2\ c_2 = 2\}$ for streaming media packets).

proportion of arrivals being blocked (lost) and, thus, a lower mean effective arrival rate; this influence has much greater impact on smaller buffer sizes.

For additional numerical experiments, we refer to [14].

12.6 Conclusions

This work focuses on the performance modelling and evaluation of a wireless 4G cell architecture with bursty multimedia traffic flows of all IP voice calls, streaming media and data packets, subject to the novel 4G-HOP handoff priority channel assignment protocol. This protocol utilizes the principles of channel reservation, extended channel sub-rating and priority queueing, subject to a buffer threshold based generalized partial sharing (GPS) traffic handling scheme.

The impact of the 4G-HOP protocol on the performance of the 4G wireless cellular network is assessed via an open QNM consisting of five GE-type queueing and delay systems. In this context, new analytic closed form expres-

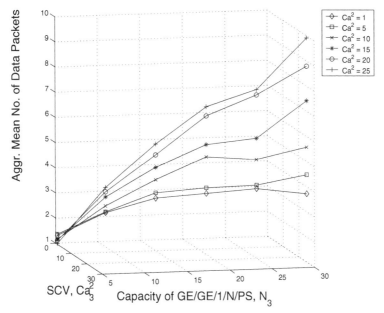

Figure 12.9 Effect of traffic variability on the aggregate mean number of data packets for a multiple class GE/GE/1/N/PS(T) queueing system under PSS (Experiment 12 – Input Data: $\{\lambda_{31} = 1.0, \lambda_{32} = 0.5, \lambda_{33} = 0.2, \mu_3 = 9.0$, PS service discrimination weight $= 1/3, (\mu_{31} = 1/3\mu_3, \mu_{32} = 1/3\mu_3), \mu_{33} = 1/3\mu_3, Ca_{31}^2 = 5, Ca_{32}^2 = 2, Ca_{33}^2 = 2, Cs_{31}^2 = 5, Cs_{32}^2 = 5, Cs_{33}^2 = 5\}$ for data packets).

sions for the aggregate and blocking state probabilities are highlighted, based on the information theoretic principle of maximum entropy (ME), subject to appropriate GE-type queueing theoretic constraints. Moreover, typical numerical experiments are used (i) to validate the credibility of the ME solutions for a 4G cellular network architecture under the 4G-HOP protocol against Java-based simulation results and (ii) to study the adverse effect of bursty multimedia class traffic flows upon the performance of the cell.

The numerical experiments of Figures 12.3–12.10 indicate that the ME analytic results are very comparable to those obtained via simulation and that traffic burstiness indicated by the SCVs of the interarrival times has an adverse effect on the performance metrics of all service classes. In this context, it is shown that the 4G-HoP handoff channel assignment protocol enables performance improvements for the priority and buffer threshold based handoff IP voice calls and streaming media, as appropriate. Morever, it provides addi-

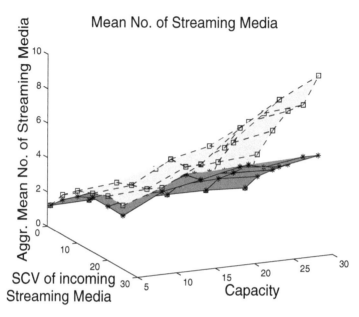

Figure 12.10 Effect of traffic variability on the aggregate mean number of streaming media for a multiple class GE/GE/c/N/FCFS/CBS(T_l, T_h) queueing system under 4G-HoF.

tional transmission capacity to the IP data partition whilst maintains fairness for new voice calls and streaming media.

Further work will extend the proposed QNM to include busty and self-similar multimedia traffic flows, subject to randomly distributed time out periods. Some initial analytic GE-type results can be seen in [14].

Acknowledgements

This work is supported in part by the EC IST project VITAL (IST-034284 STREP) and in part by the EC NoE Euro-FGI (NoE 028022).

References

[1] Y. R. Haung, Y. B. Lin and J. M. Ho, Performance analysis for voice/data integration on a finite-buffer mobile system, *IEEE Transactions on Vehicular Technology*, vol. 49, no. 2, pp. 367–378, March 2000.

[2] C. H. Foh, B. Meini, B. Wydrowski and M. Zuerman, Modelling and performance evaluation of GPRS, in *Proceedings of IEEE VCT*, Rhodes, Greece, pp. 2108–2112, 2001.

[3] D. D. Kouvatsos, I. Awan and K. Al-Begain, Performance modelling of GPRS with bursty multi-class traffic, *IEE Proc. Computers and Digital Techniques*, vol. 150, no. 2, pp. 75–85, March 2003.

[4] D. D. Kouvatsos and I. Y. R. Haung, Performance analysis for UMTS networks with queued radio access bearers, *IEEE Transactions on Vehicular Technology*, vol. 51, no. 6, pp. 1330–1337, 2002.

[5] V. Huang and W. Zhuang, QoS-oriented access control for 4G mobile multimedia CDMA communication, *IEEE Communications*, vol. 40, pp. 118–125, 2002.

[6] F. Paint, P. Engelstad, E. Vanerm, T. Haslestad, A. M. Nordvic, K. Myksvoll and S. Svaet, Mobility aspects in 4G networks, White Paper, IEEE Telecommunications Society, 2003.

[7] F. H. P. Fitzek, D. Angelini, G. Mazzini and M. Zorzi, Design and performance of an enhanced IEEE 802.11 MAC protocol for multihop coverage extension, *IEEE Wireless Communications*, vol. 10, pp. 30–39, 2003.

[8] E. T. Jaynes, Prior probabilities, *IEEE Transactions on Systems Science and Cybernetics*, vol. 4, pp. 227–241, 1968.

[9] D. D. Kouvatsos, Entropy maximisation and queueing network models, *Annals of Operation Research*, vol. 48, pp. 63–126, 1994.

[10] N. P. Shah, I. Awan and D. D. Kouvatsos, Modelling handover-priority channel assignment scheme in mobile networks, in *Proceedings of Workshop Proceeding of the Sixth Informatics for Research Students*, D. Rigas (Ed.), School of Informatics, University of Bradford, March 2005, pp. 161–164, 2005.

[11] T. S. Rappaport, *Wireless Communications*, Prentice Hall, NJ, 1996.

[12] Y. Li and D. D. Kouvatsos, Recent analytic results for multiclass GE-type queueing and delay systems, Research Report RR-03-08, PERFORM – Networks and Performance Engineering Research Unit, University of Bradford, March 2008.

[13] D. D. Kouvatsos, S. Tantos and Y. Li, GE-type probabilistic algebra for the analysis of finite capacity queues with time out periods, Research Report RR-07-08, PERFORM – Networks and Performance Engineering Research Unit, University of Bradford, Research Work in Progress, July 2008.

[14] Y. Li, J. He and W. Xing, Bandwidth management of WiMAX systems and performance modelling, *KSII Transactions on Internet and Information Systems*, vol. 2, no. 2, pp. 63–81, April 2008.

13

On the Estimation of Signalling Performance and Efficiency in UMTS and Beyond Radio Access Architectures

Peter Schefczik[1] and Anja Wiedemann[2]

[1]*Alcatel-Lucent, Bell Labs Germany, Lorenzstraße 10, 70435 Stuttgart, Germany; e-mail: pschefczik@alcatel-lucent.com*
[2]*Department of Experimental Mathematics, University of Duisburg-Essen, 45326 Essen, Germany; e-mail: wiedem@exp-math.uni-essen.de*

Abstract

Current standardization activities of 3rd Generation Partnership Project (3GPP) Long Term Evolution (LTE) working groups show that hierarchical UMTS networks influenced by GSM and ISDN history are changing towards decentralized and distributed network architectures based on IP principles. Equipment manufacturers submitted various proposals for UMTS R6. Subsequently 3GPP ordered a feasibility study and requested an evaluation of the different architecture proposals with regard to performance. In this paper we present a simulation modeling approach, which provides an environment for the easy assessment of different evolution scenarios of the UMTS architecture with regard to signalling performance. We apply the presented simulation concept to a case study, which compares a new distributed radio access network (DRAN) architecture for UMTS and the current Universal Terrestrial Radio Access Network (UTRAN). For both RANs under study a Central Processing Unit (CPU) capacity planning is performed aiming at optimization of the signalling performance. Furthermore the developed simulation concept also allows for an estimation of the allocated CPU capacity for the RANs under consideration.

D. D. Kouvatsos (ed.), Performance Modelling and Analysis of Heterogeneous Networks, 245–275.

Keywords: UMTS, architecture, simulation, signalling traffic, next generation networks, long term evolution.

13.1 Introduction

In today's competitive environment, service providers are offering more and more value-added services. Thus the architectures capable of delivering such services become a critical success factor for generating new revenue streams. Next-generation networks will depend on the ability to deliver those services in a non-costly and easy applicable way. Thereby it must be ensured that the costly network is utilized optimally. Therefore equipment manufacturers must proactively define the optimal network architecture to ensure quality and optimal use of network resources and easy day-to-day operations.

Next generation services such as Voice over IP (VoIP) or Virtual Private Network (VPN) are often delivered over a magnitude of layers and subnetworks. For UMTS R5 the ALL-IP option and the IP Multimedia Subsystem (IMS) start the convergence of wireline and wireless networks. Those so-called Next Generation Networks will combine the best of two worlds: the hierarchical 5 nines telecom network will be merged with the Internet Engineering Task Force (IETF) IP packet network. The race to find a good suitable architecture has already begun. In 3GPP R6 the study item of the evolution of the Universal Terrestrial Radio Access Network (UTRAN) architecture [1] was created. Several proposals all separating user and control plane were made so far. For UTRAN R7 the requirements for the evolved UTRAN are studied in [2]. The work on the next generation wireless architecture was recently identified in the Long Term Evolution Workshop in 3GPP [3]. The final architecture shall obey the so-called "ilities". That means architecture aspects are conceived according to functionality, reliability, manageability, scalability, operability, flexibility, software complexity, performance and cost. Thus the functionality must be carefully distributed over the various network elements. To find an optimal distribution is one aim of this paper. Two network architectures containing similar functional entities but combined into different network elements are compared in a simulative approach.

The simulation methodology applied is described in full detail in [4, 5]. The modeling methodology uses Message Sequence Charts (MSCs) and is capable of incorporating arbitrary signalling flows and also easily supports changes in the system architecture. Thereby we follow a use case approach and construct a general event driven signalling protocol performance model using the OPNET Modeler tool. Within the simulation environment we analyze the alternative network architectures and protocol stacks with respect to their signalling performance.

The work presented here is part of the German research project "IPon-Air". The IPonAir project [6] focuses on the seamless integration of the wireless access network with the existing and future IP-based backbone network.

The focus of this paper is to present results provided by an experimental comparison of two different UTRAN architectures. The paper is organized as follows. In Section 13.2 we describe the architectures under consideration. These are the current UTRAN architecture as defined by 3GPP [7] and the enhanced Distributed Radio Access Network (DRAN) architecture [8]. In Section 13.3 we give an overview of the modeling and simulation concept. In Section 13.4 we present the experimental setup of the simulation studies performed and we discuss and interpret the results of the architecture analysis and also provide some initial cost accounting calculations for both architectures. Finally we provide an outlook towards future work of interest.

13.2 Radio Access Network Architectures

In this section we present two different architectures that are analyzed with our simulation methodology.

The UMTS architecture, which is taken as the reference architecture is depicted in Figure 13.1. The UTRAN is a hierarchical network architecture highly influenced by GSM and ISDN history. It consists of base transceiver stations (Node Bs) and centralized Radio Network Controllers (RNCs). ATM is used as transport technology in the UTRAN. In this radio access architecture the main intelligence is concentrated in the RNC, which manages the provisioning of all necessary bearer services for control and user traffic in order to establish the Radio Access Bearer (RAB) between User Equipment (UE) and Core Network (CN). The UE complies with the mobile terminal handling the user traffic in the cellular system. The CN consists of a Circuit Switched (CS) and a Packet Switched (PS) domain. The CS domain of the CN consists of a Media Gateway (MGW) for voice user data, the Mobile Switching Center (MSC) server for signalling purposes and the Gateway MSC (GMSC) server for connection to external networks. It must be noted that unfortunately inside the CN of the UMTS system the term MSC denotes Mobile Switching Center, but in the rest of this paper an MSC denotes a Message Sequence Chart. The PS domain within the CN comprises the Serving

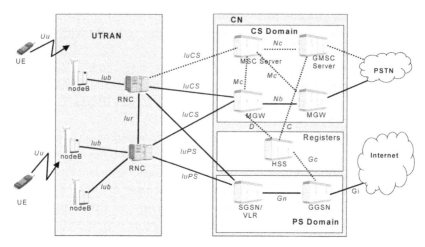

Figure 13.1 UTRAN reference network architecture.

GPRS Support Node (SGSN), which is comparable to the Mobile Switching Center but for PS connections and the Gateway GPRS Support Node (GGSN) for connection to external packet data networks. The CS domain handles all circuit switched sessions and connects the UTRAN to the Public Switched Telephone Network (PSTN) while the PS domain handles the packet switched sessions and connects the UTRAN to the public internet. Also depicted is the Home Subscriber Server (HSS) storing the relevant user data. There are some more network elements performing call control and other signalling functions as well as multimedia resource functions. Those are not shown here for shortness sake. In Figure 13.1 dotted lines denote pure signalling links. Solid lines can transport signalling as well as user traffic.

The UTMS evolution scenario that will be compared to the reference architecture, is the DRAN architecture (cf. Fig. 13.2), which was a contribution of Lucent Technologies to the 3GPP standardization discussion on R6 in [1].

The DRAN architecture is mainly motivated by the ever increasing need for IP services which is already required by UMTS R5, but the current UMTS R5 architecture will probably suffer from drawbacks resulting from IP transport between the Node B and RNC. This stems from the fact that the synchronization between these network elements requires high Quality of Service (QoS) which cannot be provided by IP as a best-effort transport mechanism without application of specific QoS differentiation strategies.

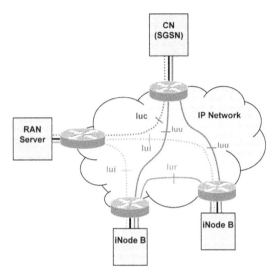

Figure 13.2 DRAN network architecture.

Therefore modifications of this radio access network (RAN) architecture were investigated and incorporated into the evolved DRAN architecture.

In DRAN, the complete radio specific processing is relocated from the RNC into the Node B, which is called "intelligent Node B" (iNode B) in this architecture proposal. Thereby the Iub interface formerly located between the Node B and RNC becomes node internal to the iNode B in DRAN. Because of transferring all radio-specific user data handling and related control plane functionality from the RNC to the iNode B, the RNC loses its centralized functionality for the RAN. However, in DRAN, the RNC does not become completely obsolete but remains with some control functions with regard to the management of bearer services for the RAN and is called "RAN Server" in this architecture proposal. The cell-specific Radio Resource Management (RRM) located in the iNode B is connected via the new Iui interface to the RRM functions within the RAN Server. User traffic is directly routed to the destination iNode B via the Iuu interface, i.e. user plane traffic bypasses the RAN Server. Furthermore the RAN Server keeps track of all users in its UTRAN and manages micro mobility inside the RAN.

In the DRAN the user and control plane of the former UTRAN can be completely separated from each other, logically as well as physically. Both planes can be optimized independently from each other. This eases the scalability of the network architecture and the integration of new services. The

DRAN can easily handle increasing user plane traffic volume by expanding the IP network and no extensions of UMTS nodes are necessary. iNode Bs can be added as long as the RAN server is capable of handling it. By co-location of user traffic functions at the edge of the network the architecture is simplified for the incorporation of new services within next generation networks and at the same time the architecture is inherently scalable for user traffic.

In DRAN ATM as the transport technology in the UTRAN is eliminated in favor of an IP-routed network, which is not disadvantageous in DRAN, because the time critical radio resource processing is completely relocated into the iNode B and does not require any synchronization with the RAN Server. Thereby the DRAN does not need a complex and costly ATM network to be set up for the Iub and Iur interfaces. The same holds for the Iu interface, which is now split into the Iuu interface for user data traffic and Iuc for control data traffic. Furthermore QoS can be provided in the transport network, which is now capable of service differentiation down to the iNode B. Moreover appropriate IETF protocols for resource allocation and traffic engineering may be used, for example Multi-Protocol Label Switching (MPLS) with Resource Reservation Protocol with Traffic Engineering (RSVP-TE). The Resource Reservation Protocol (RSVP) is an internet scheme that allows the reservation of resources for "flows" of packets thereby managing application traffic's QoS. However, RSVP can reserve a path but cannot establish a path. To facilitate this RSVP-TE was conceived. RSVP-TE includes all the necessary extensions to establish label-switched paths (LSPs) in MPLS networks. In this manner all available transport resources within the UTRAN are under control of the RAN Server, although the RAN Server is not concerned with user plane traffic.

The new iNode B internal Iub interface can be optimized without taking into account the defined standards for the Iub interface. This approach simplifies call control and also yields shorter delays for user traffic due to less protocol overhead for the transport of user traffic. Regarding user traffic the iNode B is directly connected to the SGSN. Thus the RNC hop has gone for user traffic in the access network and this further reduces the delay. Moreover the MAC scheduler for the RLC packets is now located in the iNode B and is aware of IP packets arriving into the node as well as the current wireless channel conditions. This is why the scheduler can base scheduling decisions on both incoming user traffic and radio conditions and better quality and scheduling decisions can be made. The MAC scheduler placement inside the iNode B is also needed for new radio technologies like High Speed Downlink Packet Access (HSDPA). Thus DRAN is already prepared for HSDPA.

By avoiding changes to the Uu and Iu interfaces towards the mobile terminal and the CN, the DRAN architecture proposal ensures interworking with existing mobile terminals and backward compatibility with legacy UTRAN of previous releases.

Handover in the proposed distributed RAN architecture differs from the handling in the legacy RAN environment. In the DRAN soft and softer handover are autonomously performed by the involved iNode Bs via Iur interface. Interaction with RAN Server is necessary only in case of hard handover in a way similar to SRNS relocation. Except for soft handover between legacy NodeBs and iNode Bs, interworking with the legacy RAN is seamlessly possible as the SGSN sees a standard Iu interface presented by the iNode B as Iuu. In addition, hard handover can easily be performed by redirecting the user traffic at the Iuu interface to the new iNode B.

While UTRAN is a hierarchical network, DRAN is a step forward towards a flat network and towards increased network reliability. Nevertheless in DRAN the RAN Server centralizes the main signalling functionality. In more detail this means that in case of an RAN Server failure the connected UEs still can receive/send data as long as no mobility management or call management signalling is needed. This is not possible in UTRAN due to the fact that the RNC centralizes the main intelligence of the RAN so that those parts of the network served by a particular RNC will be completely down and inaccessible, if this RNC fails. Nevertheless in case of a RAN Server failure all call handling procedures towards the RAN Server cannot be initiated any more und thus those portions of the network served by this particular RAN Server will also be unusable for users intending to perform certain network transactions (e.g. call setup).

Another disadvantage of the DRAN is that the CN has to handle many more network elements directly because the iNode Bs are directly connected to the SGSN. It may well be that this can be eased by increasing the CPU capacity of the SGSN. However the mobility handling can be more complex, e.g. each hard handover is handled similarly to SRNS relocation. This may result in unwanted load (e.g. in the SGSN) and new user traffic protocols may be needed to handle such handovers. Moreover the handover functionality needs more bandwidth on the last mile to the iNode B. This results from the fact that with relocation of the Iur interface down to the iNode B each iNode B is (logically) connected to its neighbours. In case of a handover the serving iNode B performs the necessary radio specific processing to serve its own Uu

Table 13.1 Pros and cons of the UTRAN and the DRAN architecture.

Feature	Conventional UTRAN	Proposed DRAN
Scalability on network element basis	no	yes for u-plane
Decreasing delay due to user plane processing closer to the air interface	no	yes
O&M changed	no	yes
ATM paths needed to set up a network	yes	no
Reliability for u-plane	Single point of failure	Distributed into iNode Bs
Reliability for c-plane	Single point of failure	Still single point of failure
Split of user and control plane	Partly inside network elements	fully
Optimization of transport costs	Preconfigured ATM paths	IP with MPLS and RSVP-TE can be used
Easy to adapt to other radio technologies (e.g. HSDPA)	difficult	yes
Easy to adapt to new IP based services	difficult	yes
Independent scaling of user and control plane	Not at all	yes
Easy to adapt to varying traffic (using Traffic Engineering for traffic flow optimization)	difficult	MPLS using RSVP-TE can be used
SRNC Relocation frequency	low	high
Easy to adjust to varying traffic (using scheduler closer at the air interface)	difficult	yes
More bandwidth needed on the last mile for handover	no	yes
Easier to manage due to less boxes	no	yes
Optimized scheduling done in NodeB	no	yes

interface and transfers the RLC/MAC PDUs simultaneously to the drift iNode Bs. The user-traffic in question is thus transferred twice on the link between the last access router and the serving iNode B: Once as GTP-u (IP) packet and once after processing as stream of RLC/MAC PDUs, cf. [9]. However, we do not regard the occurrence of more complex handovers as too critical, because the tendency within the UMTS standardization discussion is towards an all IP based network architecture so that the portion of circuit switched network traffic can be assumed to decrease in the light of VoIP.

Moreover additional Operations & Maintenance (O&M) interfaces may be needed to configure and operate the iNode B. For example the PDCP, RRC, RLC and MAC layer configurations are now maintained in the iNode B.

We summarize the above mentioned thoughts in Table 13.1. For more transparency the disadvantages for DRAN and UTRAN are underlined.

Overall despite the disadvantages for the DRAN, the DRAN looks promising to be one candidate for a future evolved UTRAN. To underpin this statement we take a deeper look at the signalling performance both of UTRAN and DRAN in the rest of this paper.

13.3 Modeling and Simulation Methodology

In this section we present the detailed simulation methodology and the developed tool chain. We explain the relevant inputs, give an overview on the generation of the performance simulation model and we comment on the outputs of the event driven simulation. Because our goal is to assess alternative network architectures and protocol stacks with respect to signalling performance the developed tool chain is extremely flexible and can deal with different architectures, signalling flows, traffic models and resources.

The modeling concept focuses on the signalling traffic in the network architectures as described in Section 13.2. Each signalling load use case is modeled in the form of a Message Sequence Chart (MSC) [7, 8] and forms one "System Function" (SF). Examples for signalling sequences are the Mobile Originated (MO) voice call setup and release procedures necessary for initiating and terminating voice calls by a mobile terminal. The MSCs describing the signalling message flow use cases in UTRAN are taken from standards documents [11]. The signalling sequences in DRAN did not exist and had to be specified newly. For our specification we only need a small subset of MSC-2000 [10], but we need performance extensions to MSC-2000 for specification of the cost of message processing and message transfer.

Each signalling message is annotated with a number tuple consisting of a complexity class factor and the message length specified in bytes. The complexity class factor is related to the message processing operation. It basically assigns the amount of service (in milliseconds) a signalling message receives within a Central Processing Unit (CPU) module. The higher the complexity class factor, the higher is the amount of service for a particular message and the more processing time is needed to process the message in the CPU. The message length is used to derive the cost for link traversal of the message between network nodes. Therefore on the one hand the response time (End-to-End (E2E) Delay) of a SF is composed of the propagation and transmission delay resulting from link traversal. On the other hand processing and queuing delay resulting from processing times while being in service within a certain CPU as well as those times waiting for a free CPU to get serviced add to the E2E delay of a SF.

As many information elements of signalling messages are optional in the standards, we often must use an average message length or one that is measured frequently in a real implementation. Also complexity classes can be partly derived from measurements inside the real network elements. Where

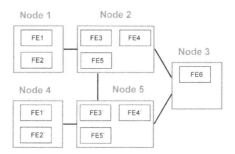

Figure 13.3 Abstract allocation of FEs to network elements.

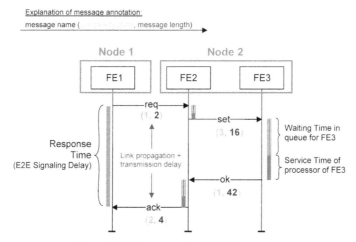

Figure 13.4 Short annotated MSC sequence.

this is not possible they have to be guessed. The set of MSCs together with the complexity class annotations is called the *Load Model*.

The messages of a SF are exchanged between stateless "Functional Entities" (FEs). Those FEs are architecture subentities that reside inside the foreseen network elements. FEs can be relevant protocol entities or entities that contain a specific network functionality. Figure 13.3 shows an abstract allocation of FEs to network elements in which FEs with similar functionality are denoted with a prime.

Figure 13.4 depicts a short annotated MSC sequence in which the FEs exchange signalling messages. Figure 13.5 shows the CPU model, which is applied, i.e. a CPU in our modeling concept consists of a single First-In First-Out (FIFO) queue and one or more servers that process incoming messages.

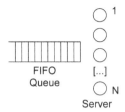

Figure 13.5 CPU definition.

Table 13.2 Resource consumption table for one single resource (CPU).

Complexity Class	Service Time
1	3 ms
2	10 ms
3	30 ms

FEs which logically process incoming messages (i.e. convert them to another message type) are physically located on CPUs. One or more FEs can be allocated to one single CPU. The messages logically processed by the FEs are afterwards processed by the CPU with regard to their complexity class factor. The bindings of FEs and resources to form specific node architectures as well as packet forwarding and message forwarding are explained in [4].

Remember that the annotation tuple of each message depicted in Figure 13.4 denotes the complexity class and the length of the message. The Complexity Class leads to a service time needed to process that message on a processor resource. A Complexity Class to Service Time mapping is shown exemplarily in Table 13.2 below. From Table 13.2 it can be derived that a message of complexity type 1 gets a service amount of 3 ms, a message denoted with complexity class 2 gets a service amount of 10 ms and a message of complexity type 3 becomes very costly with 30 ms service amount.

To simulate a change of the speed of a processor means to adjust the values in the table; e.g. doubling the CPU speed would introduce the factor 0.5 for the entries of Table 13.2. In the implementation this translates into the configuration of a scale factor for each CPU denoting the speed of the CPU. Hence the simulation of faster CPUs is easily accomplished. The resources and their processing properties and the interconnecting communication channels with their channel capacities are called the *Resource Model*.

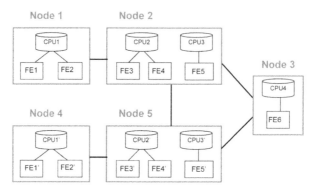

Figure 13.6 Example of a system configuration model.

The mapping of the FEs onto the resources is called the *System Configuration Model*. Thus the System Configuration Model defines which FE runs on which processor. Together with the Load Model this determines which message is processed on which FE and CPU. Note that two or more FEs can be mapped onto one CPU. Figure 13.6 exemplifies a System Configuration Model of the abstract FE allocation scenario illustrated in Figure 13.3.

To impose a traffic load onto the system a *Traffic Model* is specified. The Traffic Model mainly contains the use case call rates, i.e. the number of instantiations per time unit for each MSC. Moreover an appropriate traffic distribution is needed to specify the Traffic Model. A typical assumption is to choose the interarrival times as independent and identically exponential distributed thereby forming a Poisson process for the traffic. Further a correlated traffic model is used to inject traffic into the modeled architecture. The correlated traffic model approach works as follows. The call setup MSC is triggered independently using a Poisson process. The call release MSC however is connected to the call setup process in the following way: On the one hand the number of call releases is proportional to the number of active calls in the system. On the other hand the number of active calls is delimited by an upper limit, which complies with the maximum cell throughput. This upper boundary represents the Call Admission Control Function (CAC). The CAC is used to keep a certain QoS in the cell. In theory the number of users in a Code Division Multiple Access (CDMA) system can grow infinitely, but each new user adds interference to the cell thereby making the conditions worse for all existing users. Thus to keep a certain speech quality the number of users has to be limited. With this correlated approach state information like

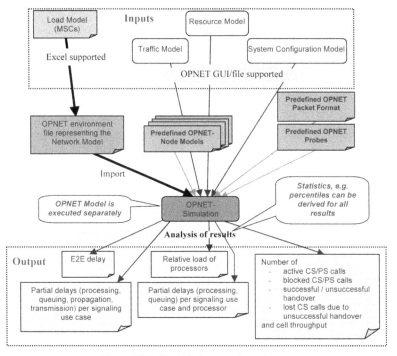

Figure 13.7 Tool chain overview picture.

the number of active users in the cell or the number of signalling and traffic channels currently used in the network can easily be derived.

A tool chain supports the methodology described in this section. An overview of the tool chain described so far is depicted in Figure 13.7.

The Load Model, Traffic Model, Resource Model and the System Configuration Model are specified within Microsoft Excel. By application of a Visual Basic for Applications (VBA) algorithm, input necessary for simulation execution within the OPNET Modeler [12] simulation environment is derived. The implementation of the OPNET model necessary for evaluation of various radio network architectures under study is a generic implementation, which can be applied to various architecture variants without the necessity for changes to the implementation. Hence an easy assessment of the architectures under study is enabled. The output of the simulation provides information about signalling E2E delays for the signalling SFs under study. Furthermore information about the composition of the E2E delays can be derived (e.g. if the signalling E2E delay for MO Call Setup is 5 ms, it might be composed out of 3 ms processing time, 1.8 ms queuing time, 0.15 ms transmission delay and 0.05 ms propagation delay). Moreover the simulation results obtain information about the utilization of the processors inside the network elements and the network designer can get information about the amount of time a signalling SF spent within a single processor, which is of interest when dimensioning CPUs in network elements (e.g. if the signalling E2E delay for MO Call Setup is 5 ms, whereas 3 ms are processing time and 1.8 ms are queuing time, the network designer can see that, for instance, 2 ms of processing were performed in CPU 1 and 1 ms in CPU 2 and 1 ms of queuing time was received in CPU 1 and 0.8 ms in CPU 2). Furthermore the simulation results provide information about the number of

- active circuit switched (CS) and packet switched (PS) calls and their throughput in the radio cell and in the network;
- blocked CS and PS calls in the radio cell/in the network;
- successful and unsuccessful handovers;
- lost CS calls due to unsuccessful handovers.

The simulation settings for the CPUs inside particular network elements also allow for conclusions and estimations regarding the costs arising from the construction of such CPUs and their allocation to network elements.

Subsequently we go into detail with regard to the FEs we identified within the two architectures. Table 13.3 provides a survey of the FEs we identified to be modeled. Furthermore a short description of each FE is given. Figures 13.6 and 13.7 show the allocation of FEs to network elements within UTRAN and DRAN control planes. By looking at the allocation of FEs to network elements illustrated in Figures 13.8 and 13.9 the relocation of FEs between the two architectures is visible. On the one hand the elimination of FEs which are needed in UTRAN but no longer needed in DRAN becomes apparent, i.e. protocol entities necessary for signalling between the Node B and the RNC (e.g. NBAP) or with regard to the ATM transport network connection (e.g. ALCAP) are eliminated in DRAN, because DRAN abandons ATM as transport technology in favor of IP. On the other hand there are new FEs emerging in DRAN, i.e. entitites with regard to the newly integrated IP transport technology (e.g. RSVP-TE).

In Figures 13.10 and 13.11 we illustrate the MSC signalling flow in UTRAN and DRAN in order to establish a Radio Resource Control (RRC) signalling connection for a mobile terminal, i.e. we take the RRC Connection Establishment procedure as an example. The RRC Connection Setup establishes the radio connection between the UE and the Node B and further

Table 13.3 Survey of FEs with short description.

#	FE Name	Short Description
1	Uu	Air interface
2	Iub	Interface between RNC and Node B
3	RRC Connection Management	Handles the radio resource connection between the UE and the UTRAN and is the termination point for control messages
4	TN Resource Management	Reserves the physical resources in the transport network
5	RSVP-TE	RSVP-TE agent on the iNode B for traffic engineering in the transport network
6	Iur RSVP-TE	Is responsible for the resource allocation on the Iur Interface
7	Iuc RSVP-TE	Is responsible for the resource allocation on the Iuc Interface
8	Iui RSVP-TE	Is responsible for the resource allocation on the Iui Interface
9	Iui	Interface between RAN Server and iNode B
10	Bearer Management	Manages bearers on Iuc, Iui and Iuu. Talks to TN Resource Management
11	Radio Bearer Management	Sets up and releases radio links
12	Bearer Mapping Unit	Coordinates the radio bearer and the transport bearer towards the CN
13	Mobility Management	Manages the mobility of the users in case of hard handover
14	Micro Mobility Management	Manages soft and softer handover inside UTRAN
15	Broadcast/Multicast Management	Manages broadcast channels
16	User Management	Holds the users data and states
17	Paging Entity	Executes paging co-ordination
18	NBAP	Handles the radio network signaling over the Iub interface
19	ALCAP	Establishes and tears down data transport bearers on the Iu, Iur and Iub interfaces
20	ALCAP_Iu	ALCAP unit for Iu interface
21	ALCAP_Iub	ALCAP unit for Iub interface
22	ALCAP_Iur	ALCAP unit for Iur interface
23	RANAP	Manages the Iu interface between CN and UTRAN
24	RNSAP	Interface for the Iur for handover between RNCs

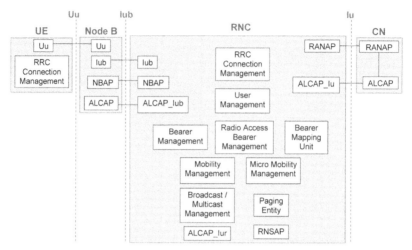

Figure 13.8 Allocation of FEs to UTRAN control plane.

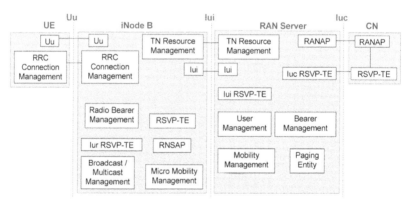

Figure 13.9 Allocation of FEs to DRAN control plane.

connects the UE to the RNC. This procedure is used in voice and data call setups to establish a signalling channel between a mobile terminal (UE) and the corresponding RNC. We observe that this signalling sequence is shorter in DRAN than in UTRAN, because the signalling exchange between the iNode B and the RAN Server is reduced due to the fact that the functionality of the iNode B was significantly enhanced by the allocation of radio specific processing so that interaction with the RAN Server can be considerably reduced.

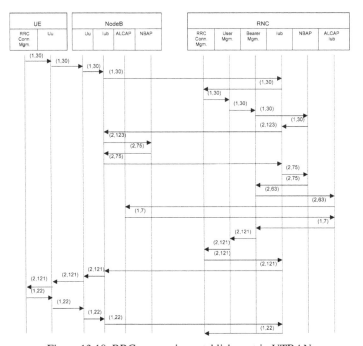

Figure 13.10 RRC connection establishment in UTRAN.

For a clear understanding of Figures 13.10 and 13.11 we recall that each message of the RRC signalling sequence is annotated with a number tuple in the form of (complexity class, message length).

13.4 Experimental Studies

In this section the experimental setup of two of our simulation studies is described. The first experiment is a simple experiment, which is included to clarify the developed modeling methodology, whereas the second experiment provides relevant conclusions about the radio access network architectures under study.

13.4.1 Initial Simulation Study

In the UTRAN reference architecture as well as in the DRAN architecture evolution scenario of the first experiment we consider a topology consisting of 180 Node Bs or iNode Bs respectively. Those are not modeled explicitly

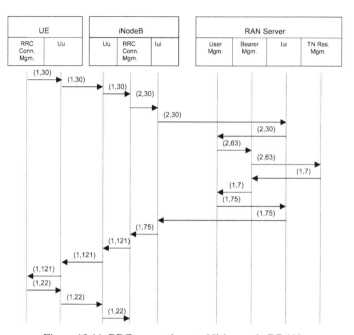

Figure 13.11 RRC connection establishment in DRAN.

but by application of a scaled down version of the network under study considering only nine (i)Node Bs. More precisely the Small Scale High-fidelity Reproduction of Network Kinetics (SHRiNK) modeling technique [13] is used. The basic idea of SHRiNK is to expose a sample of the traffic onto the scaled down system and then extrapolate from the scaled down version to the original system.

In the presented simulation study we consider a pure voice call network in which the voice calls are initiated and released by the mobile terminals. We assume a mean call holding time of 100 seconds for each of these MO voice calls. Also we assume a traffic value of 16.67 mErlang per user. This could for example be derived from a call holding time of 100 s and a Busy Hour Call Attempt (BHCA) rate of 0.6 per user per hour. For QoS purposes a load control mechanism is applied within the radio cells restricting the upper limit for the number of active voice calls to not more than 80 for each radio cell. As we are mainly interested in the comparison of the UTRAN and the DRAN network architecture, we can use a fictitious traffic model to generate the load needed in the simulation study.

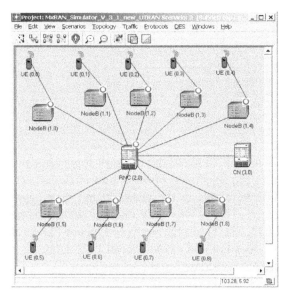

Figure 13.12 UMTS R5 reference architecture.

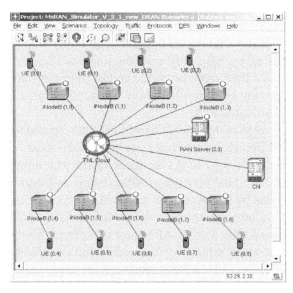

Figure 13.13 DRAN evolution scenario.

Table 13.4 Interarrival rates.

# User	SF	Total Rate [1/sec.]	Mean Interarrival Time [sec.]
100,000	MO Voice Call Setup	11.67	15.424
400,000	MO Voice Call Setup	46.67	3.857
700,000	MO Voice Call Setup	81.67	2.204
1,000,000	MO Voice Call Setup	116.67	1.543
1,060,000	MO Voice Call Setup	123.67	1.455
1,300,000	MO Voice Call Setup	151.67	1.187

Figures 13.12 and 13.13 show the network topologies under study. The mobile terminals (UEs) illustrated in these figures do not act in accordance with single mobile terminals in the network, but they aggregate all users within one radio cell. This is why only one UE per cell is depicted in Figures 13.12 and 13.13. The Transport Network Latency (TNL) cloud in DRAN (cf. Figure 13.13) does not add to the overall E2E delay, i.e. delay times in the IP network are not considered in this experiment.

The basic idea of the presented simulation experiment is as follows: We progressively increase the number of users in the network from 100,000 up to 1,300,000. Using the Erlang formula with the traffic model parameters above we calculate that blocking increases over 2% for 1,060,000 users and above. That means with 180 cells and 80 channels per cell this is the number of total voice users that can be satisfied in the whole RAN. Thus the operating point is chosen there. These total number of users seem rather high to be processed by the RAN. But as we have only call setup and call release implemented those two must be called more often to produce relevant traffic. We expect the number of users to be somewhat lower when more than 20 use cases are to be processed by the network.

Now we can calculate the mean interarrival rates needed to generate that traffic. Table 13.4 lists the interarrival rates used in this experiment.

We annotate that in order to get the mean interarrival times per cell from the total rates for the whole network provided by the assumed user model we divide the total rates by the number of assumed UE network elements which is 180 and then we take the inverse value of the division operation which is the mean interarrival time. For the MO voice call release SF the rate is made proportional to the rate for the MO voice call setup signalling sequence and to the number of active calls in the system.

In our simulation study we observe the influence of the increasing network load on the

Table 13.5 CPU parameters in UTRAN and DRAN.

	# Server	Speed Factor	Amount of Service (ms)		
			CC = 1	CC = 2	CC = 3
UE CPU	100	10	1	1	1
Node B / iNode B CPU	4	6	1	2	3
RNC BSC / RAN Server	4	663	1	2	36
RNC CPU 2	100	672	1	2	40
RNC CPU 3	100	672	1	2	40
RNC CPU 4	100	672	1	2	40
CN CPU 1	1000	100	1	1	1

- E2E delays of MO voice call setup and release signalling sequences;
- number of active MO voice calls within the network;
- CPU utilizations within network elements of interest (Node B and iNode B, RNC.BSC and RAN Server).

The simulation is performed for UTRAN and for DRAN whereas in DRAN the same CPU parameters as in UTRAN are applied. Table 13.5 lists the CPU parameters of UTRAN and DRAN and obtains information about

- the number of servers located within single network elements;
- the capacity of these servers in terms of their speed;
- a small table which assigns service times in milliseconds (ms) to each complexity class factor.

As we are only interested in the effects caused by the Node B, iNode B, RNC and RAN Server, all other network elements are equipped with many parallel CPUs. The same is true for the user traffic related CPUs 2 to 4 inside the RNC. That means the CPUs within the UE and CN as well as the CPUs except the Base Station Controller (BSC) CPU in the RNC which is involved in the signalling between Node B and RNC as well as iNode B and RAN Server are modeled as infinite server resources so that their contribution to the overall signalling E2E delay is rather small. The amount of service table within the RNC.BSC and RAN Server CPU states that messages of complexity class 1 will get an amount of service of 1 ms, whereas messages of complexity class 2 get an amount of service of 2 ms while messages of complexity class 3 become very costly with 36 ms as amount of service.

Table 13.6 displays a numerical representation of the SFs MO voice call setup and release in UTRAN and DRAN with regard to their distribution

Table 13.6 Numerical representation of SF MO voice call setup (SF 1) and release (SF 2).

		UE			Node B / iNode B			RNC / RAN Server												CN						Σ			
		CPU			CPU			BSC			CPU 2			CPU 3			CPU 4			CPU 1			CPU 2			Σ			
		CC			CC			CC			CC			CC			CC			CC			CC			CC			
		1	2	3	1	2	3	1	2	3	1	2	3	1	2	3	1	2	3	1	2	3	1	2	3	1	2	3	Σ
UTRAN	SF 1	12	9	0	17	20	0	8	18	0	5	10	0	10	13	0	0	0	0	1	0	0	2	5	0	55	75	0	130
UTRAN	SF 2	5	6	0	12	11	0	9	10	0	3	6	0	10	6	0	0	0	0	1	0	0	1	2	0	41	41	0	82
DRAN	SF 1	14	9	0	15	27	0	14	22	0										0	0	0	2	5	0	45	63	0	108
DRAN	SF 2	5	6	0	6	14	0	10	15	0										0	0	0	1	2	0	22	37	0	59

of messages within the complexity classes 1 to 3 while passing particular resources.

Table 13.6, for instance, obtains the information about SF 1 (MO voice call setup) in UTRAN that 17 messages of complexity class 1 and 20 messages of complexity class 2 pass the CPU located in the Node B network element. Table 13.6 furthermore provides the information that SF 1 in UTRAN consists of 55 messages of complexity class 1 and 75 more costly messages of complexity type 2, altogether it needs 130 signalling messages to perform MO voice call setup in UTRAN.

13.4.1.1 Results and Discussion of Initial Simulation Study

As a first plausibility test the theoretical results from the Erlang B traffic considerations are compared with the results from the simulation model. For this reason in Figure 13.14 the number of active MO voice calls in the UTRAN and DRAN network vs. increasing number of users is illustrated together with the theoretical curve.

We observe that the number of active MO voice calls in the network equals the number of active voice calls derived from Erlang B. The operating point of 2% blocking is near the point where the curves become nonlinear, i.e. at 1,060,000 users.

Note that the total number of users that the network can carry is rather high. This is due to the fact that only 2 SFs are considered at the moment. In the future we will add data traffic and mobility MSCs. This is why then this number will fall considerably.

Next we look at various server utilizations of interest when examining potential bottlenecks in the two network architectures. In Figure 13.15 we display the utilizations of the RNC.BSC and RAN Server CPUs and the utilization of the Node B and iNode B CPUs.

In the UTRAN we see that the RNC.BSC CPU is loaded at 83% if the number of users in the network increases up to the operating point of

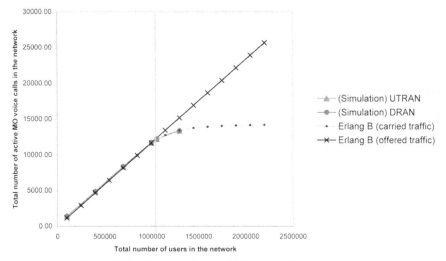

Figure 13.14 Number of active MO voice calls in the network.

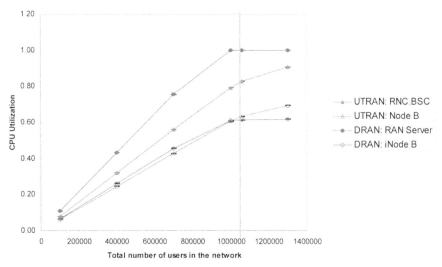

Figure 13.15 RNC.BSC and RAN server, Node B and iNode B CPU load.

1,060,000 users whereas in DRAN the RAN Server CPU is already fully utilized at that value (cf. Figure 13.15).

Comparing the utilization of the Node B and iNode B CPU we observe that in UTRAN the Node B CPU is working at 63% if there are 1,060,000

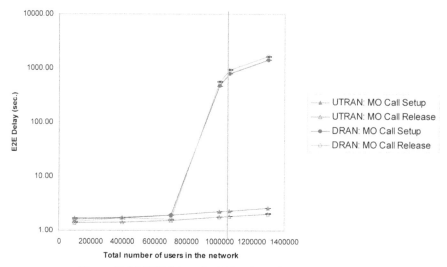

Figure 13.16 E2E delays for MO voice call setup and release.

users in the network. In DRAN the iNode B CPU is working at 62% load when the RAN Server CPU is already fully utilized.

Thus we conclude that in UTRAN the limiting factor within the simulated network is the RNC.BSC CPU whereas in DRAN the RAN Server CPU is identified as bottleneck.

The perpendicular line denotes the operating point of the network, i.e. the user traffic boundary of the total number of users in the RAN that can be satisfied with 2% blocking. Figure 13.16 shows the E2E signalling delays for MO voice call setup and release in UTRAN and DRAN.

We observe that the mean E2E delay for setting up MO voice calls is marginally higher in UTRAN than in DRAN whereas the mean E2E delay for releasing MO voice calls is slightly lower in UTRAN than in DRAN. By saying so, we only consider the results as long as the RAN Server CPU within DRAN is not congested.

Moreover we observe that the E2E delay signalling traffic in the DRAN explodes with increasing number of users in the RAN. Thus, if we want to operate in an acceptable delay range and with up to 1,060,000 users we must increase the involved CPU resources. This is done in a further simulation experiment where we increase the capacity of the CPUs located in the Node B and iNode B as well as the RNC.BSC and RAN Server CPU by adjusting the speed factor of the CPUs. Then we observe the influence of this capacity modification on the E2E signalling delays. For this experiment we assume 1,060,000 users as expected for the 2% blocking value in the network. We aim at serving these users by providing signalling delays, which comply with or are less than 2 sec.

Subsequently we provide a preliminary estimation of CPU costs based on the CPU speed factors of the resources allocated to network elements arising from the usage of UTRAN and DRAN. For this estimation we consider the speed parameters after capacity planning in UTRAN and DRAN so that the E2E signalling delays for MO Call Setup and Release are equal to or less than 2 sec.

Considering 180 Node Bs in UTRAN and 180 iNode Bs in DRAN respectively with a speed factor of 11 and 4 server we calculate a total CPU capacity of 7,920 ((speed factor * number of server) * number of network elements assumed) for the Node Bs and iNode Bs in UTRAN and DRAN.

In UTRAN we assume a speed factor of 663 for the RNC.BSC resource which occupies 4 server and a speed factor of 672 for three more resources (100 server each) inside the RNC necessary for handling user plane traffic. This means a CPU speed capacity of 204,252 in total for one single RNC. As far as the RNC is the central network element within UTRAN we must assure a high reliability of this network element, because if the RNC fails the complete part of the network served by this particular RNC will be down. To ensure the maintenance of operation of the network in case of an RNC failure, we assume a second RNC kept in hot standby mode which is perfectly synchronized so that it can take over the functionality of the failed RNC. By doing so we also have to consider the CPU capacity of the RNC twice so that we assume a total CPU capacity of 408,504 for two RNCs.

In DRAN we only consider one single CPU within the RAN Server with a speed factor of 763 and 4 server that we also double to ensure a reliable network so that we calculate a RAN Server CPU cpacity of 6,104. The assumption of one single CPU is sufficient, because the RAN Server is not concerned with user plane traffic.

Summing up the CPU capacity in UTRAN and DRAN we get a total capacity of 416,424 in UTRAN (180 Node Bs: 7,920 + RNC capacity: 408,504) and 14,024 in DRAN (180 iNode Bs: 7,920 + RAN Server capacity: 6,104). We realize that in DRAN we only need a fraction of the CPU capacity which is necessary in UTRAN. Thus it will be significantly cheaper to set up a DRAN with regard to the CPU capacity needed for the network elements within DRAN. However, we can also provide the same CPU capacity of UTRAN to DRAN and study potential signalling delay gains. This experiment is subject of the next section. The experiment also considers a full set of signalling MSCs (mobility MSCs included).

13.4.2 Advanced Architecture Evaluation

In the UTRAN reference architecture as well as in the DRAN architecture evolution scenario of the second experiment we consider a topology consisting of 360 Node Bs or iNode Bs respectively. Thereby we consider 180 Node Bs per RNC and 180 iNode Bs per RAN Server so that we consider a network topology with two RNCs and two RAN Servers, respectively. Again, these nodes are not modeled explicitly, i.e. the SHRiNK modeling approach is applied again.

We again assume a mean call holding time of 100 seconds for MO and MT voice calls and a holding time of 200 seconds for packet calls. The portion of MO to MT calls is 70% : 30%. For voice calls we assume a traffic value of 16.67 mErlang per user which leads to a data rate per user of 12.2 kbps, whereas for packet calls an equivalent traffic value of 14.44 mErlang per user is assumed. The blocking probability is 2% and the maximum cell throughput is 1000 kbps.

We consider a full set of signalling SFs as illustrated in Table 13.7 and an overall CPU capacity, which is equal in UTRAN and DRAN.

The simulation experiments and results, which will be described subsequently, were performed to answer the following questions:

1. Is it possible to operate the DRAN evolution scenario with the same processing capacity as the UTRAN reference architecture at comparable signalling performance?
2. What about the performance in terms of signalling E2E delays of UTRAN and DRAN in case the portion of data (PS) traffic increases under consideration of the fact that a higher portion of PS traffic can be assumed in future mobile networks?
3. What happens to the handover signalling sequences in DRAN, where handover signalling was identified to become more complex than handovers in UTRAN?

In order to answer these questions, the UTRAN reference simulator was calibrated first. The calibration was performed by distributing the processing capacity so that the utilization in the main processor of the Node B is about 55% and the utilization in the RNC.BSC processor, which is mainly involved in signalling between the Node B and RNC, is about 75% at the operating

Table 13.7 Full set of signalling SFs.

#	UTRAN Signaling SF
1	MO Voice / CS Data Call Establishment
2	MO Voice / CS Data Call Release
3	MT Voice / CS Data Call Establishment
4	MT Voice / CS Data Call Establishment
5	PS Data Transfer Establishment (with follow on: UE PS attaches and automatically gets PDP context, then goes to URA_PCH)
6	PS Detach (UE initiated)
7	Transition from URA_PCH to CELL_DCH due to Downlink Data Transfer
8	Transition from CELL_DCH to URA_PCH due to RRC inactivity timer expiring
9	MO PDP Context Activation (from PMM_CONNECTED+Cell_DCH)
10	MO PDP Context Deactivation (from PMM_CONNECTED+Cell_DCH)
11	IMSI Detach Signaling Flow
12	Location Updating Signaling Flow
13	UTRAN Registration Area (URA) Update (URAU) Signaling Flow
14	Routing Area (RA) Update (RAU) Signaling Flow
15	Intra-RNC Softer Handover
16	Intra-RNC Soft Handover
17	Inter-RNC Soft Handover
18	Intra-RNC Hard Handover
19	Inter-RNC Hard Handover
20	Serving Radio Network Subsystem (SRNS) Relocation
21	Paging in idle mode
22	Paging in connected mode
23	Paging unsuccessful
24	Measurement Report

point of the network, which is assumed to be 1 million of users in the 50/50 traffic mix (i.e. the portion of voice (CS) and data (PS) traffic is assumed to be 50% each). When distributing the processing capacity in UTRAN, knowledge about already existing physical equipment was considered. The CPUs inside the mobile terminal (UE) and the CN were again modeled as infinite server resources, because the focus is on the RAN part of the network architecture.

Thus, the overall processing capacity available in UTRAN was derived. This processing capacity can be distributed among the processors inside the network elements of the DRAN. As far as the network elements of DRAN (i.e. iNode Bs and RAN Server) have not been built, the network designer has the freedom to vary the assignment of processing capacity to processors inside the network elements. Furthermore the network designer can vary the number of processors inside the network elements and the allocation of FEs to these processors so that the simulation toolkit is also helpful when dimensioning new network nodes, which are to be physically produced by manufacturers.

Signalling E2E delays can be derived for all SFs listed in Table 13.7. In this publication only a subset of results is presented which is sufficient to answer the questions of interest that were raised above. In the experiments performed three traffic mixes imposing various portions of voice and data

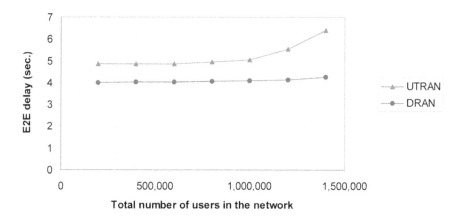

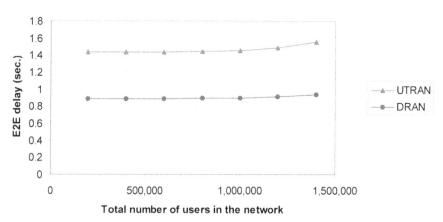

Figure 13.17 UTRAN/DRAN evaluation, MO voice call establishment, MO PDP context activation in UTRAN and DRAN, 20/80 traffic mix.

traffic onto the network were considered: 50% voice / 50% data (50/50), 80% voice / 20% data (80/20) and 20% voice / 80% data (20/80). The number of users in both networks is varied from 200,000 up to 1,400,000. We recall that the assumed processing capacity is the same in both architectures.

The simulation results illustrated in Figure 13.17 are suitable to answer the questions raised above. Figure 13.17 shows that it is possible to operate DRAN and UTRAN with the same processing capacity at comparable signalling performance. It is considerably important to point out that DRAN also performs well under consideration of higher portions of data traffic in the network . This is extremely important because for future networks higher portions of data traffic are expected.

Further simulations on Soft Handover as well as on SRNS Relocation in UTRAN and DRAN also show that it is also possible to operate DRAN and UTRAN with the same processing capacity at comparable signalling performance and this also applies for the more complex handover handling in DRAN.

13.5 Conclusion and Future Work

In this paper we presented an architecture comparison of a conceivable UMTS evolution scenario submitted for 3GPP standardization discussion with the current UMTS radio access reference network. An event driven protocol simulation model was used to investigate the signalling load in the proposed architectures. The results of this architecture evaluation and their implications with regard to the assessment of the presented UTRAN evolution scenario compared to the reference architecture were explained.

Summarizing the results we can say that by carefully selecting the CPU resources in the analyzed architectures we get comparable signalling performance results in UTRAN and DRAN for various portions of voice and data traffic in the network. The same is true even for the signalling of the handover handling which was identified to be more complex in distributed network architectures. Potentially it is even more cost-saving as DRAN applies to an IP routed network instead of a costly ATM network that needs to be established within UTRAN.

In the future we will extend the evaluation study of the presented radio access network architecture variants by setting up further evolution scenarios for comparison. We already compared another RAN evolution scenario to the UTRAN reference architecture, which is the Merged RAN (MRAN). MRAN combines the complete functionality of the Node B and the RNC into one single network element, which is the merged Node B. Initial evaluation results show that this evolution variant is also efficient provided that the CPU capacity is carefully allocated within the merged Node B (cf. [14]). However, we also consider the evaluation of further evolution scenarios like for example the Base Station Router (BSR, cf. [15]). The BSR is a network element that combines Node B, RNC, SGSN and GGSN functionalities in one single box, thereby enabling a flat, IP-based network architecture. The BSR can dramatically simplify and reduce the cost of operating IP-based mobile networks and services. We are confident that the proposed modeling and simulation concept based on MSCs as well as the simulation environment is applicable for the optimization and evaluation of future RAN architectures and facilitates future architecture decisions in the area of 3G/4G network architectures. One such an architecture candidate is the BSR, another is the DRAN architecture described above.

Acknowledgements

The research reported in this paper has been partly conducted within the IPonAir program funded by the German Ministry of Education and Research (BMBF) under grant 01BU161. The responsibility for contents of this publication is with the authors.

The work reported in this paper results from the cooperation of the Universities of Essen and Alcatel-Lucent. The authors also want to acknowledge the contributions of Wilfried Speltacker and Michael Soellner both with Alcatel-Lucent and Andreas Mitschele-Thiel from the University of Ilmenau. All assisted in the overall modeling concept and the ideas, which have been provided within the IPonAir project.

References

[1] 3GPP TR 25.897 V0.4.0 (2003-11), Feasibility study on the evolution of UTRAN architecture (Release 6), 3GPP TSG-RAN WG3 Meeting #38, Draft 3GPP TR 25.897 V0.3.1 (2003-08), Sophia-Antipolis, France, October 6–10, 2003.

[2] 3GPP TR 25.913 V2.1.0 (2005-05), Requirements for evolved UTRA and UTRAN (Release 7), 3GPP TSG RAN; 6-10, 2003.

[3] 3GPP RAN WG2, *Long Term Evolution Adhoc Workshop*, Sophia-Antipolis, France, June 20–21, 2005.

[4] N. Frangiadakis, G. Nikolaidis, P. Schefczik and A. Wiedemann, MxRAN functional architecture performance modeling, in *Proceedings of the OPNETWORK 2002 Conference*, Washington DC, USA, August 26–30, CD-Rom, 5 pp., 2002.

[5] A. Mitschele-Thiel, P. Schefczik and A. Wiedemann, Architecture optimization of 3rd generation wireless systems based on use cases, in *Proceedings of the Applied Telecommunication Symposium (ATS'04)*, Arlington, Virginia, USA, April 18–22, CD-Rom, 6 pp., 2004.

[6] IPonAir homepage, http://www.iponair.de

[7] 3GPP Technical Specification Group Services and Systems Aspects, Network architecture, TS 23.002 V6.5.0, 2004–2006.

[8] M. Bauer, P. Schefczik, M. Soellner and W. Speltacker, Evolution of the UTRAN architecture, in *Proceedings of 3G Mobile Communication Technologies (3G2003)*, London, UK, June 25–27, pp. 244–248, 2003.

[9] Lucent Technologies, Architectural proposal for UTRAN evolution study item, Tdoc R3-031379, 3GPP TSG-RAN WG3 Meeting #38, Sophia Antipolis, France, October 6–10, 2003.

[10] MSC-2000 Message Sequence Chart (MSC) (revised in 2001), SDL forum version of Z.120 (11/99) rev. 1(14/11/01).

[11] 3GPP TSG RAN, UTRAN Functions, examples on signaling procedures (Release 1999), TS 25.931, version 3.6.0, March 2002.

[12] OPNET Technologies, Inc., homepage: http://www.opnet.com

[13] R. Pan, B. Prabhakar, K. Psounis and D. Wischik, SHRINK: A method for scalable performance prediction and efficient network simulation, in *Proceedings of IEEE INFOCOM 2003*, San Francisco, USA, March 30–April 3, pp. 1943–1953, 2003.

[14] A. Mitschele-Thiel, P. Schefczik and A. Wiedemann, Comparison of signaling performance in the UMTS radio access network and a merged radio access network architecture, in *Proceedings of the IEEE Wireless Communications and Networking Conference (WCNC) 2005*, New Orleans, LA, USA, March 13–17, vol. 3, pp. 1767–1772, 2005.

[15] M. Bauer, P. Bosch, N. Khrais, L. G. Samuel and P. Schefczik, The UMTS Base Station Router (BSR), *Bell Labs Technical Journal*, vol. 11, no. 4, pp. 93–111, 2007.

14

An Analytical Model for Elastic Service-Classes in W-CDMA Networks

Vassilios G. Vassilakis, Georgios A. Kallos, Ioannis D. Moscholios
and Michael D. Logothetis*

*WCL, Department of Electrical & Computer Engineering, University of Patras,
265 04 Patras, Greece; e-mail: m-logo@wcl.ee.upatras.gr*

Abstract

In this article we present a teletraffic model, named Wideband Threshold Model, for the analysis of W-CDMA networks supporting elastic traffic. Mobile users generate Poisson arriving calls that compete for the acceptance to a W-CDMA cell under the complete sharing (CS) policy. A newly arriving call can be accepted with one of several possible Quality-of-Service (QoS) requirements, depending on the resource availability in the cell that is indicated by thresholds. Aiming at the efficient calculation of the system state probabilities and mainly the call blocking probabilities (CBP), for the uplink connection, we provide recurrent formulas. The accuracy of the analytical CBP results is verified by simulation results and is found to be absolutely satisfactory.

Keywords: Quality-of-Service, W-CDMA, elastic traffic, call blocking probability, recurrent formula.

---

* Author for correspondence.

D. D. Kouvatsos (ed.), Performance Modelling and Analysis of Heterogeneous Networks, 277–299.

14.1 Introduction

Modern wireless networks support applications with different Quality-of-Service (QoS) requirements, while offering wide range of voice and data services. In the case of Universal Mobile Telecommunications System (UMTS) networks, the air interface is the Wideband Code Division Multiple Access (W-CDMA) [1, 2]. In W-CDMA networks, the call-level performance modelling is complicated not only because of the presence of inter- and intra-cell interference, but also due to the heterogeneous traffic they support.

We distinguish three types of traffic: stream, elastic and adaptive. Stream traffic is generated by calls that have fixed resource and service time requirements, which cannot be reduced at any time (e.g. real-time audio or video). Elastic and adaptive traffic is generated by calls that may have different possible resource requirements depending on the resource availability. The service time of an elastic call is directly related to the resources allocated to this call, whereas the service time of an adaptive call is always constant and independent of the resources allocated to this call.

For the analysis of a multi-rate loss system in the presence of Poisson stream traffic, only, the famous Erlang Multi-rate Loss Model (EMLM) is widely used [3, 4]. A recurrent formula, known as Kaufman-Roberts (K-R) recursion, facilitates the calculation of Call Blocking Probabilities (CBP) at most. Since then, several modifications of this model were proposed either for wired or mobile networks [5–8]. In [5], a blocked call can retry many times, requesting for less resources each time. In [6], calls arrive to the link with several possible contingency resource requirements and select one resource requirement according to the total occupied resources which are indicated by thresholds, common to all service-classes. The Connection-Dependent Threshold Model (CDTM) [7] generalizes the retry and threshold models (as well as the EMLM) by individualizing the thresholds among different service-classes. The above-mentioned models have been proposed for wired connection-oriented networks with elastic traffic and they do not consider the resource allocation scheme of W-CDMA networks.

We study W-CDMA networks in the uplink direction, where a single base station (BS) controlling a cell can be modelled as a system of certain bandwidth capacity. The bandwidth unit (b.u.) can be an equivalent bandwidth defined by the load factor introduced e.g. by a lower rate service-class (e.g. voice). The load factor is determined by the signal-to-noise ratio, data rate and activity factor (probability that a call is transmitting/active) of the associated service-class [1]. As far as the inter-cell interference is concerned, it is assumed log-normally distributed and independent of the cell load. A call is accepted for service, as long as there are enough resources available in the cell. More precisely, the call admission control (CAC) policy is based on the estimation of the increase of the total interference (intra- and inter-cell interference plus thermal noise) caused by the new call's acceptance. After call acceptance, the signal-to-noise ratio of all in-service calls deteriorates; because of this, W-CDMA systems have no hard limits on call capacity. A call should not be accepted if it increases the noise of all in-service calls above a tolerable level, given that, according to the W-CDMA principle, a call is noise for all other calls. This call admission policy corresponds to the CS policy in wire networks, since no restriction per service-class is set. Under this realistic consideration of W-CDMA networks, while supporting stream traffic only, the applicability of the EMLM has been studied in [8], where the K-R recursion is successfully applied for the CBP calculation in the uplink direction.

In this article, we present an efficient teletraffic model with recurrent formulas, and having been challenged by the presence of elastic traffic in contemporary networks, we study the applicability of the CDTM in the uplink direction of W-CDMA networks under the realistic consideration mentioned above. The new model is named Wideband Threshold Model (WTM). A W-CDMA cell can accommodate not only elastic service-classes but also adaptive and stream service-classes. Each service-class (other than stream) is associated to an individual set of thresholds which indicate the occupied cell resources. Poisson arriving calls to a cell may have several contingency resource/QoS requirements. Upon their arrival, calls select one resource requirement according to the thresholds; in this way, the notion of elastic traffic is covered in a great extent at the call setup phase (while in-service, calls do not alter the assigned resources). A service time requirement corresponds to each resource requirement so that the total offered traffic-load [9] remains constant for every resource-and-service-time pair. This is valid for elastic traffic only. For adaptive traffic, the service time requirement is constant. For stream traffic both resource and service time requirements are constant. The WTM does not have a product form solution (PFS) [10, 11]. By considering the so called macro system, where a state represents the total amount of the occupied cell resources, and introducing appropriate approximations, a one-dimensional reversible Markov chain is obtained. Based on it, we present an approximate recurrent formula for the calculation of macro state probabilities and consequently CBP as well as other performance metrics.

14.2 The Erlang Multi-rate Loss Model (EMLM)

14.2.1 Model Description

In the EMLM, a system of bandwidth capacity C b.u. accommodates Poisson arriving calls of K independent service-classes. Calls compete for the available bandwidth under the CS policy [10]. Each service-class k ($k = 1, \ldots, K$) call requests upon arrival b_k b.u. If they are available, the call is accepted in the system and occupies the b_k b.u. for a time, exponentially distributed with mean μ_k^{-1}. Otherwise the call is blocked and lost. The arrival rate of service-class k calls is denoted by λ_k. The offered traffic-load of service-class k calls is calculated by $a_k = \lambda_k \mu_k^{-1}$. The system state j ($j = 0, \ldots, C$) is defined as the total number of occupied b.u. The probability that the system is in state j is denoted by $q(j)$.

14.2.2 Local Balance Equation and Bandwidth Share

In the EMLM the following local balance exists between adjacent system states [3]:

$$a_k q(j - b_k) = Y_k(j) q(j) \tag{14.1}$$

where $Y_k(j)$ is the average number of service-class k calls in state j. We can calculate the bandwidth share, $P_k(j)$, (proportion of the total occupied bandwidth j that corresponds to calls of a specific service-class) of service-class k calls in state j ($j > 0$) as follows:

$$\frac{a_k b_k q(j - b_k)}{j q(j)} = \frac{Y_k(j) b_k}{j} \iff \frac{a_k b_k q(j - b_k)}{j q(j)} = P_k(j) \tag{14.2}$$

14.2.3 A Tutorial Example

As an abstract example consider a link of capacity $C = 5$, accommodating $K = 3$ service-classes, with fixed bandwidth per call requirements: $b_1 = 1$, $b_2 = 2$ and $b_3 = 3$. In Figure 14.1 we present the (macro-) state transition diagram of this example, where each (macro-) state is defined by the occupied link bandwidth j by all calls that are present in the link. Clearly, the EMLM has a PFS because of the reversibility in each state. The state $j = 5$ is a blocking state for all service-classes. The state $j = 4$ is a blocking state for the second and third service-class, while the state $j = 3$ is a blocking state for the third service-class only.

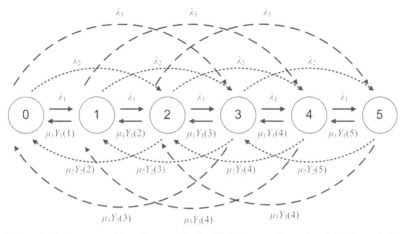

Figure 14.1 State transition diagram for the EMLM: one dimensional Markov chain.

14.2.4 State Probabilities and Call Blocking Probabilities

The state probabilities, $q(j)$, are calculated by the Kaufman–Roberts recursion [3,4]:

$$jq(j) = \sum_{k=1}^{K} a_k b_k q(j - b_k) \text{ for } j = 1, \ldots, C \text{ and } q(j) = 0 \text{ for } j < 0,$$

$$\text{under the normalization condition } \sum_{j=0}^{C} q(j) = 1. \qquad (14.3)$$

A new arriving call with requirement b_k b.u. is accepted in the system if and only if $j \leq C - b_k$. Thus, for the service-class k the states $j = C - b_k + 1$, $\ldots$, C are blocking states, while the states $j = 0, 1, \ldots, C - b_k$ are non-blocking states. The CBP of service-class k is determined by adding the state probabilities of all blocking states:

$$B_k = \sum_{j=C-b_k+1}^{C} q(j) \qquad (14.4)$$

14.3 The Connection-Dependent Threshold Model (CDTM)

14.3.1 Model Description

The CDTM is an extension of the EMLM that can be used for wired, connection-oriented networks supporting elastic traffic [7]. Assume a link of capacity C b.u. commonly shared among Poisson arriving calls of K independent service-classes. The arrival rate of service-class k calls is λ_k ($k = 1, \ldots, K$) and an arriving service-class k call has S_k bandwidth and mean service time requirements with values $b_{k,l}$ and $\mu_{k,l}^{-1}$, ($l = 1, \ldots, S_k$) respectively, where $b_{k,S_k} < \ldots < b_{k,l} < \ldots < b_{k,2} < b_{k,1}$ and $\mu_{k,S_k}^{-1} > \ldots > \mu_{k,l}^{-1} > \ldots > \mu_{k,2}^{-1} > \mu_{k,1}^{-1}$. The service time of service-class k calls is assumed to be exponentially distributed. The pair $(b_{k,1}, \mu_{k,1}^{-1})$ is used from service-class k calls when the number of occupied b.u. at the call arrival is $j \le J_{k,1}$, where $J_{k,1}$ is the lowest threshold of the service-class k. The pair $(b_{k,l}, \mu_{k,l}^{-1})$, (with $l > 1$), is used from service-class k calls when $J_{k,l} < j \le J_{k,l+1}$, where $J_{k,l}$ and $J_{k,l+1}$ are two successive thresholds of the service-class k. The pair $(b_{k,S_k}, \mu_{k,S_k}^{-1})$ is used from service-class k calls when $J_{k,S_k-1} < j \le C - b_{k,S_k}$. Finally, a service-class k call is blocked when $C - b_{k,S_k} < j \le C$. The offered traffic-load of service-class k calls with resource requirement $b_{k,l}$ is defined as: $a_{k,l} = \lambda_k \mu_{k,l}^{-1}$. The total offered traffic load is equal for every pair $(b_{k,l}, \mu_{k,l}^{-1})$ and is defined as: "the product of the offered traffic load by the required bandwidth per call", $a_{k,l}b_{k,l}$ [9]. Similarly to the EMLM, the parameter j ($j = 0, \ldots, C$) represents the system state and the state probabilities are denoted by $q(j)$. Note that in the CDTM after the initial bandwidth allocation, calls do not change their allocated bandwidth while in service.

14.3.2 A Tutorial Example

We demonstrate the basic principles of the CDTM with the aid of the example of the previous subsection. The first service-class has no thresholds. The second service-class has one threshold ($J_{2,1} = 2$) and two contingency bandwidth requirements ($b_{2,0} = 2, b_{2,1} = 1$), whereas the third service-class has two thresholds ($J_{3,1} = 2, J_{3,2} = 3$) and three contingency bandwidth requirements ($b_{3,0} = 3, b_{3,1} = 2, b_{3,2} = 1$), as shown in Figure 14.2. In this specific example, a call is blocked and lost whenever the link bandwidth is fully occupied; this is valid for all three service-classes because their minimum bandwidth per call requirement is 1.

In Figure 14.3 we show the state transition diagram of the above example, for the second service-class only, in order to simplify the presentation. By

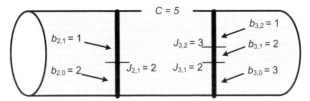

Figure 14.2 Principles of the CDTM.

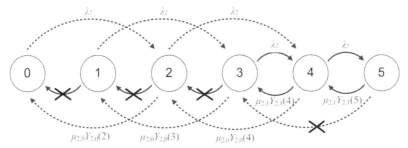

Figure 14.3 State transition diagram in the CDTM (second service-class only).

$Y_{2,0}(j)$ and $Y_{2,1}(j)$, we denote the mean number of the second service-class calls with bandwidth requirements $b_{2,0}$ and $b_{2,1}$, respectively, in state j. The system does not have a PFS because it is irreversible. This is true either when we consider the number of in-service calls as system state or when the system state is the (macro-) state j; in the latter case irreversibility is obvious from Figure 14.3 (where all arrows are valid – ignore the X's on the arrows, for the moment).

14.3.3 Basic Approximations

By introducing the following approximations, we modify the state transition diagram of Figure 14.3 so that it becomes reversible and an approximate but recurrent formula for the calculation of the state probabilities, $q(j)$, can be obtained:

1. Local balance between adjacent system states (rate up = rate down).
2. Migration approximation: Calls accepted with other than the maximum bandwidth requirement are negligible within a space, called migration space. More precisely, the population of calls with bandwidth requirement (with $l \neq 0$) in state j is negligible when $0 < j < J_{k,l} + b_{k,l}$ and $J_{k,l+1} + b_{k,l} \leq j \leq C$.

3. Upward migration approximation: Calls accepted in the system with their maximum bandwidth requirement are negligible within a space, called upward migration space. More precisely, the population of calls with bandwidth requirement $b_{k,0}$ in state $j = 0$ is negligible when $J_1 + b_{k,0} < j \leq C$.

In the example of Figure 14.2, according to the migration approximation, the population of calls with requirement $b_{2,1}$ in states $j = 1, 2$ and 3 is assumed negligible. Thus corresponding transitions $3 \to 2, 2 \to 1$ and $1 \to 0$ are ignored, as it is shown in Figure 14.3 with the X's on the arrows. Similarly, due to the upward migration approximation, the transition $5 \to 3$ is ignored for calls with requirement $b_{2,0}$. In this way and due to the local balance assumption, the resultant Markov chain becomes reversible.

14.3.4 Local Balance Equation and Bandwidth Share

Due to the approximations of the previous subsection, the following local balance exists between adjacent system states:

$$a_{k,l}\delta_{k,l}(j)q(j - b_{k,l}) = Y_{k,l}(j)\delta_{k,l}(j)q(j) \tag{14.5}$$

Consequently, the bandwidth share of service-class k calls with requirement $b_{k,l}$ in state j $(j > 0)$, $P_{k,l}(j)$, is given by

$$\frac{a_{k,l}b_{k,l}\delta_{k,l}(j)q(j - b_{k,l})}{jq(j)} = \frac{Y_{k,l}(j)b_{k,l}\delta_{k,l}(j)}{j}$$

$$\Longleftrightarrow \frac{a_{k,l}b_{k,l}\delta_{k,l}(j)q(j - b_{k,l})}{jq(j)} = P_{k,l}(j) \tag{14.6}$$

14.3.5 State Probabilities and Call Blocking Probabilities

Based on the basic approximations, the state probabilities, $q(j)$, are calculated by the following recurrent formula [7]:

$$jq(j) = \sum_{k=1}^{K}\sum_{l=0}^{S_k} a_{k,l}b_{k,l}\delta_{k,l}(j)q(j - b_{k,l}) \quad j = 1, \ldots, C$$

$$\text{and } q(j) = 0 \text{ for } j < 0, \text{ with } \sum_{j=0}^{C} q(j) = 1 \tag{14.7}$$

where the parameters $\delta_{k,0}(j)$ and $\delta_{k,l}(j)$ (with $l \neq 0$) are defined by the following expressions:

$$\delta_{k,0}(j) = \begin{cases} 1, & \text{for } 1 \leq j \leq C \text{ and } b_{k,1} = 0 \\ 1, & \text{for } j \leq J_{k,1} + b_{k,0} \text{ and } b_{k,1} > 0 \\ 0, & \text{otherwise} \end{cases} \qquad (14.8)$$

$$\delta_{k,l}(j) = \begin{cases} 1, & \text{for } J_{k,l+1} + b_{k,l} \geq j > J_{k,l} + b_{k,l} \text{ and } b_{k,l} > 0 \\ 0, & \text{otherwise} \end{cases} \qquad (14.9)$$

$\delta_{k,0}(j)$ is related to upward migration space, while $\delta_{k,l}(j)$ (with $l \neq 0$) to migration space of calls with requirement $b_{k,l}$.

A new arriving service-class k call is accepted in the system only whenever its minimum bandwidth requirement, b_{k,S_k}, can be satisfied – i.e. if and only if $j \leq C - b_{k,S_k}$. Thus for the service-class k the states $j = C - b_{k,S_k} + 1, \ldots, C$ are blocking states, while the states $j = 0, 1, \ldots, C - b_{k,S_k}$ are non-blocking.

The CBP of service-class k is determined by adding the state probabilities of all the blocking states:

$$B_k = \sum_{j=C-b_{k,S_k}+1}^{C} q(j) \qquad (14.10)$$

14.4 The Wideband Threshold Model (WTM)

14.4.1 Model Description

Consider a W-CDMA system that supports K independent service-classes. We examine a reference cell surrounded by neighbouring cells in the uplink direction only (calls from mobile users to the BS).

The QoS offered to each service-class k ($k = 1, \ldots, K$) belongs to one out of S_k alternative QoS levels that depend on the occupied cell resources. In the rest of the article a service-class k call of QoS level l ($l = 1, \ldots, S_k$) is referred to as service-class k, l call.

The following QoS parameters characterize a service-class k, l call:

(a) $R_{k,l}$: Transmission bit rate;
(b) $\mu_{k,l}^{-1}$: Mean service time (exponential);
(c) $(E_b/N_0)_{k,l}$: Bit error rate (BER) parameter.

We distinguish three types of service-classes:

- Stream type: service-classes that have only one QoS level ($S_k = 1$).
- Elastic type: service-classes that have more than one QoS levels ($S_k > 1$) and the calls' mean service time strongly depends on the QoS level (it holds: $\mu_{k,S_k}^{-1} > \ldots > \mu_{k,l}^{-1} > \ldots > \mu_{k,2}^{-1} > \mu_{k,1}^{-1}$).
- Adaptive type: service-classes that have more than one QoS levels ($S_k > 1$) and the calls' mean service time is the same for every QoS level (it holds: $\mu_{k,S_k}^{-1} = \ldots = \mu_{k,l}^{-1} = \ldots = \mu_{k,2}^{-1} = \mu_{k,1}^{-1}$).

The arrival rate of service-class k calls is Poisson with mean λ_k. The offered traffic-load of service-class k, l calls is defined as: $a_{k,l} = \lambda_k \mu_{k,l}^{-1}$. For the purposes of our analysis, we express later in the article the different service's QoS requirements as different resource/bandwidth requirements.

14.4.1.1 Interference and Call Admission Control

We assume perfect power control – i.e. at the BS, the received power from each service-class k, l call is the same and equal to $P_{k,l}$. Since in W-CDMA systems all users transmit within the same frequency band, a single user "sees" the signals generated by all other users as interference. We distinguish the intra-cell interference, I_{intra}, caused by users of the reference cell and the inter-cell interference, I_{inter}, caused by users of the neighbouring cells. We also consider the existence of the thermal noise, P_N, which corresponds to the interference of an empty system.

The CAC in W-CDMA is performed by measuring the noise rise, NR, which is defined as the ratio of the total received power at the BS, I_{total}, to the thermal noise power, P_N [1]:

$$NR = \frac{I_{\text{total}}}{P_N} = \frac{I_{\text{intra}} + I_{\text{inter}} + P_N}{P_N} \tag{14.11}$$

When a new call arrives, the CAC estimates the noise rise and if it exceeds a maximum value, $NR_{\max}$, the new call is blocked and lost.

14.4.1.2 User Activity

A service-class k call alternates between active (transmitting) and passive (silent) periods. This behavior is described by the activity factor, v_k, which represents the fraction of the call's active period over the entire service time ($0 < v_k \leq 1$). Users that at a time instant occupy system resources are referred to as active users.

14.4.1.3 Load Factor and Cell Load

The cell load, n, is defined as the ratio of the received power from all active users to the total received power:

$$n = \frac{I_{\text{intra}} + I_{\text{inter}}}{I_{\text{intra}} + I_{\text{inter}} + P_N} \qquad (14.12)$$

From (14.11) and (14.12) we can derive the relation between the noise rise and the cell load:

$$NR = \frac{1}{1 - n} \text{ and } n = \frac{NR - 1}{NR} \qquad (14.13)$$

We define the maximum value of the cell load, n_{max}, as the cell load that corresponds to the maximum noise rise, NR_{max}. A typical value of the maximum cell load is $n_{\text{max}} = 0.8$ and it can be considered as the shared system resource [8].

The load factor, $L_{k,l}$, given in (14.14) can be considered as the resource/bandwidth requirement of a service-class k, l call:

$$L_{k,l} = \frac{(E_b/N_0)_{k,l} R_{k,l}}{W + (E_b/N_0)_{k,l} R_{k,l}} \qquad (14.14)$$

where by W we denote the chip rate of the W-CDMA carrier which is 3.84 Mcps.

The cell load can be written as the sum of the intra-cell load, n_{intra} (cell load that derives from the active users of the reference cell), and the inter-cell load, n_{inter} (cell load that derives from the active users of the neighbouring cells). They are defined in (14.15) and (14.16), respectively:

$$n_{\text{intra}} = \sum_{k=1}^{K} \sum_{l=1}^{S_k} m_{k,l} L_{k,l} \qquad (14.15)$$

where $m_{k,l}$ is the number of active users of service-class k, l.

$$n_{\text{inter}} = (1 - n_{\text{max}}) \frac{I_{\text{inter}}}{P_N} \qquad (14.16)$$

Figure 14.4 helps us in understanding the concept of the cell load in W-CDMA systems. Two users of the reference cell (controlled by the BS-1) during their active periods occupy L_1 and L_2 resources which contribute to the intra-cell load, n_{intra}, as it is measured at the BS-1. The users from the two

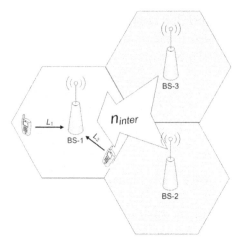

Figure 14.4 Intra-cell and inter-cell load in W-CDMA systems.

neighbouring cells (controlled by BS-2 and BS-3) contribute to the inter-cell load, n_{inter}.

According to the adopted CAC policy, the following condition is used at the BS in order to decide whether to accept a new service-class k, l call or not:

$$n + L_{k,l} \leq n_{\max} \tag{14.17}$$

14.4.2   A Tutorial Example

We demonstrate the basic principles of the WTM with the aid of the example of the previous subsections. We extend the example of subsection III.B by introducing the local blockings $LB_{k,l}(j)$. Due to the existence of local blockings, in the WTM there are no pure blocking and non-blocking states, but blocking of a service-class k, l call may occur in any state j with probability $LB_{k,l}(j)$, which we call Local blocking factor. In Figure 14.5 we show the (macro-) state transition diagram of this example for the second service-class. We see that, due to the local blockings, in Figure 14.5 the transition rates from lower states to higher, are reduced by the factor $1 - LB_{k,l}(j)$ in comparison to the CDTM (Figure 14.3). Another difference is that the system state j can exceed C. The maximum value of j is denoted by $j_{\max}$ and is the state at which $LB_{k,l}(j)$ approach 1 – i.e the higher states are unreachable due to the local blockings. This fact shows that in W-CDMA systems there is no hard limit on the system capacity and therefore we talk about soft capacity. The

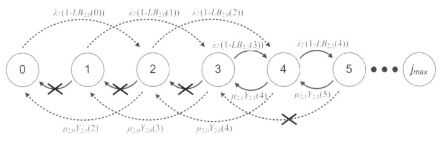

Figure 14.5 State transition diagram in the WTM.

Markov chain of Figure 14.5 does not have a PFS (all arrows are valid – ignore the X's on the arrows, at the moment) due to the existence of different contingency bandwidth requirements (as in the CDTM) and also due to the local blockings, since they destroy the local balance between adjacent system states.

14.4.3 Basic Approximations

By introducing the following approximations, we modify the one-dimensional Markov chain of the model, in order to make it reversible and consequently determine the state probabilities recursively:

1. Local balance between adjacent system states (rate up = rate down).
2. Migration approximation: calls accepted with other than the maximum bandwidth requirement are negligible within a space, called migration space. More precisely, the mean number of calls, $Y_{k,l}(j)$ $(l \neq 0)$, in state j with bandwidth requirement $b_{k,l}$ is negligible when $0 < j < J_{k,l} + b_{k,l}$ and $J_{k,l+1} + b_{k,l} \leq j \leq j_{\max}$.
3. Upward migration approximation: calls accepted in the system with their maximum bandwidth requirement are negligible within a space, called upward migration space. More precisely, the mean number of calls, $Y_{k,0}(j)$, in state j with bandwidth requirement $b_{k,0}$ is negligible when $J_{k,1} + b_{k,0} < j \leq j_{\max}$.

In the example of Figure 14.5, according to the migration approximation, the population of calls with requirement $b_{2,1}$ in states $j = 1, 2$ and 3 is assumed negligible. Thus corresponding transitions $3 \rightarrow 2, 2 \rightarrow 1$ and $1 \rightarrow 0$ are ignored, as is shown in Figure 14.5 with the X's on the arrows. Similarly, due to the upward migration approximation, the transition $5 \rightarrow 3$

is ignored for calls with requirement $b_{2,0}$. In this way and due to the local balance assumption, the resultant Markov chain becomes reversible.

14.4.4 Local Blocking Probabilities

Due to (14.17), the probability that a new service-class k, l call is blocked when arriving at an instant with intra-cell load, n_{intra}, is called local blocking probability (LBP) and can be calculated by:

$$\beta_{k,l}(n_{\text{intra}}) = P(n_{\text{intra}} + n_{\text{inter}} + L_{k,l} > n_{\text{max}}) \tag{14.18}$$

In order to calculate the LBP of (14.18) we can use (14.14)–(14.16). We notice that the only unknown parameter is the inter-cell interference, I_{inter}. We model I_{inter} as a lognormal random variable (with parameters μ_I and σ_I), that is independent of the intra-cell interference. Hence, the mean, $E[I_{\text{inter}}]$, and the variance, $\text{Var}[I_{\text{inter}}]$, of I_{inter} are calculated by:

$$E[I_{\text{inter}}] = \exp\left(\mu_I + \frac{\sigma_I^2}{2}\right) \tag{14.19}$$

$$\text{Var}[I_{\text{inter}}] = (e^{\sigma_I^2} - 1)e^{2\mu_I + \sigma_I^2} \tag{14.20}$$

Consequently, because of (14.16), the inter-cell load, n_{inter}, will also be a lognormal random variable. Its mean, $E[n_{\text{inter}}]$, and variance, $\text{Var}[n_{\text{inter}}]$, are calculated as follows:

$$E[n_{\text{inter}}] = \exp\left(\mu_n + \frac{\sigma_n^2}{2}\right) = \frac{1 - n_{\text{max}}}{P_N} E[I_{\text{inter}}] \tag{14.21}$$

$$\text{Var}[n_{\text{inter}}] = (e^{\sigma_n^2} - 1)e^{2\mu_n + \sigma_n^2} = \left(\frac{1 - n_{\text{max}}}{P_N}\right) \text{Var}[I_{\text{inter}}] \tag{14.22}$$

where μ_n and σ_n are the parameters of n_{inter}, which can be found by solving (14.21) and (14.22):

$$\mu_n = \ln(E[I_{\text{inter}}]) - \frac{\ln(1 + \text{CV}[I_{\text{inter}}]^2)}{2} + \ln(1 - n_{\text{max}} - \ln(P_N)) \tag{14.23}$$

$$\sigma_n = \sqrt{\ln(1 + \text{CV}[I_{\text{inter}}]^2)} \tag{14.24}$$

The coefficient of variation, $\text{CV}[I_{\text{inter}}]$, is defined as:

$$\text{CV}[I_{\text{inter}}] = \frac{\sqrt{\text{Var}[I_{\text{inter}}]}}{E[I_{\text{inter}}]} \tag{14.25}$$

Note that (14.18) can be rewritten as

$$1 - \beta_{k,l}(n_{\text{intra}}) = P(n_{\text{inter}} \leq n_{\text{max}} - n_{\text{intra}} - L_{k,l}) \qquad (14.26)$$

The right-hand side of (14.26), is the cumulative distribution function (CDF) of n_{inter}. It is denoted by $P(n_{\text{inter}} \leq x) = F_n(x)$ and can be calculated from

$$F_n(x) = \frac{1}{2}\left[1 + \text{erf}\left(\frac{\ln x - \mu_n}{\sigma_n\sqrt{2}}\right)\right] \qquad (14.27)$$

where erf( ) is the well-known error function.

Hence, if we substitute $x = n_{\text{max}} - n_{\text{intra}} - L_{k,l}$ into (14.27), from (14.26) we can calculate the LBP of service-class k, l calls as follows:

$$\beta_{k,l}(n_{\text{intra}}) = \begin{cases} 1 - F_n(x), & x \geq 0 \\ 1, & x < 0 \end{cases} \qquad (14.28)$$

14.4.5 State Probabilities

As we stated before, in W-CDMA networks the cell load can be considered as a shared system resource and the load factor as the resource requirement of a call. Thus, we can use a modification of (7) for the calculation of state probabilities in the WTM. Below we present five steps needed for the modification.

14.4.5.1 Resource Discretization

The discretization of the cell load, n, and the load factor, $L_{k,l}$, is necessary and is performed with the use of the basic cell load unit, g:

$$C = \frac{n_{\text{max}}}{g} \quad \text{and} \quad b_{k,l} = \frac{L_{k,l}}{g} \qquad (14.29)$$

The resulting discrete values of (14.29) can be considered as the system capacity and the service-class k, l bandwidth requirement, respectively.

14.4.5.2 Incorporation of the User Activity

The CDTM considers calls that are active during the entire service time and provides no way to distinguish active users from passive users. However, in W-CDMA networks we can exploit the fact that passive users do not consume any system resources. Hence, in the WTM a state j does not represent the total number of occupied b.u. as it happens in the CDTM. Instead, it represents the total number of b.u. that would be occupied if all users were active. We denote by c the total number of occupied b.u. at an instant. Note that in the CDTM, c is always equal to j, while in the WTM, we have $0 \le c \le j$. The case $c = 0$ corresponds to a situation when all users are passive, while when $c = j$, all users are active.

We denote by $q(j)$ the probability of the state j. The bandwidth occupancy, $\Lambda(c|j)$, is defined as the conditional probability that c b.u. are occupied in state j and can be calculated from (14.30) recursively:

$$\Lambda(c|j) = \sum_{k=1}^{K} \sum_{l=1}^{S_k} P_{k,l}(j)[v_k \Lambda(c - b_{k,l}|j - b_{k,l}) + (1 - v_k)\Lambda(c|j - b_{k,l}))],$$

for $j = 1, \ldots, j_{\max}$ and $c \le j$, with $\Lambda(0|0) = 1$ and $\Lambda(c|j) = 0$ for $c > j$

(14.30)

14.4.5.3 Incorporation of Local Blockings

In W-CDMA systems, due to the inter-cell interference, blocking of a service-class k, l call may occur at any state j with a probability $LB_{k,l}(j)$. This probability can be calculated by summing over c the LBPs multiplied by the corresponding bandwidth occupancies:

$$LB_{k,l}(j) = \sum_{c=0}^{j} \beta_{k,l}(c)\Lambda(c|j)$$ (14.31)

Note that when $j = 0$ we have $LB_{k,l}(0) = \beta_{k,l}(0)$, since in this case $j = c$ and $\Lambda(c|j) = 1$.

14.4.5.4 Determination of the Bandwidth Share

The service-class k, l bandwidth share in state j, is derived from (14.6) by incorporating the LBFs and the parameters Delta of (14.8) and (14.9):

$$P_{k,l}(j) = \frac{a_k(1 - LB_{k,l}(j - b_{k,l}))b_{k,l}\delta_{k,l}(j)q(j - b_{k,l})}{jq(j)}$$ (14.32)

14.4.5.5 Calculation of State Probabilities

The state probabilities are given by extending (14.7) due to the presence of local blockings:

$$jq(j) = \sum_{k=1}^{K} \sum_{l=1}^{S_k} a_{k,l}(1 - L_{k,l}(j - b_{k,l}))b_{k,l}\delta_{k,l}(j)q(j - b_{k,l}) \qquad (14.33)$$

$$\text{for } j = 1, \ldots, j_{\max} \text{ and } q(j) = 0 \text{ for } j < 0, \text{ with } \sum_{j=0}^{j_{\max}} q(j) = 1$$

14.5 Call Blocking Probabilities

The CBP of service-class k can be calculated by adding all the state probabilities multiplied by the corresponding LBFs:

$$B_k = \sum_{j=0}^{j_{\max}} \sum_{l=1}^{S_k} q(j)\gamma_{k,l}LB_k(j) \qquad (14.34)$$

However, due to contingency bandwidth requirements, $b_{k,l}$, we need also to sum over l in specific areas defined by thresholds. This is done with the aid of the parameters Gamma:

$$\gamma_{k,1}(j) = \begin{cases} 1, & \text{when } j \leq J_{k,l} \\ 0, & \text{otherwise} \end{cases} \qquad (14.35)$$

$$\gamma_{k,l}(j) = \begin{cases} 1, & \text{when } J_{k,l} < j \leq J_{k,l+1}, \text{ for } l \neq 1 \\ 0, & \text{otherwise} \end{cases} \qquad (14.36)$$

14.6 Evaluation

In this section, we present two application examples whereby we compare the analytical versus simulation CBP results of the WTM in order to show its accuracy. In the first application example we consider two service-classes and varying mean inter-cell interference. In the second application example we consider three service-classes and constant mean inter-cell interference. The simulation results have been obtained as mean values of six runs with

Table 14.1 Traffic parameters for the first application example.

	Voice	Data
Type	Adaptive	Elastic
Transmission rates (Kbps)	$R_{1,1} = 12.2$ and $R_{1,2} = 6.2$	$R_{2,1} = 64$ and $R_{2,2} = 32$
Thresholds	$J_{1,1} = 0.7$	$J_{2,1} = 0.6$
Activity factor	$v_1 = 0.67$	$v_2 = 0.8$
BER parameter (dB)	$(E_b/N_0)_1 = 5$	$(E_b/N_0)_2 = 4$

confidence interval of 95%. However, the resultant reliability ranges of our measurements are small enough and therefore we present only the mean CBP results.

14.6.1 Application Example 1

We consider a W-CDMA system with two service-classes: voice and data. The traffic parameters used for each service-class are as follows (see also Table 14.1):

- Voice: requires a transmission rate of $R_{1,1} = 12.2$ Kbps which can be reduced to $R_{1,2} = 6.2$ Kbps depending on the threshold $J_{1,1} = 0.7$. The activity factor is chosen to be $v_1 = 0.67$ and the required BER parameter is $(E_b/N_0)_1 = 5$ dB. This service-class is adaptive since the reduction of the transmission rate does not affect the service time.
- Data: requires a transmission rate of $R_{2,1} = 64$ Kbps which can be reduced to $R_{2,2} = 32$ Kbps depending on the threshold $J_{2,1} = 0.6$. The activity factor is chosen to be $v_2 = 0.8$ and the required BER parameter is $(E_b/N_0)_2 = 4$ dB. This service-class is elastic since the reduction of the transmission rate corresponds to the same increase of the service time.

We take measurements for eight different traffic-load points (x-axis of Figure 14.6). Each traffic-load point corresponds to some values of the offered traffic-load of both services-classes as is shown in Table 14.2.

In this example, we distinguish two cases of the inter-cell interference: Low interference with mean $E[I_{\text{inter}}] = 3\text{E-}18$ mW and high interference with mean $E[I_{\text{inter}}] = 5\text{E-}18$ mW.

Table 14.2 Offered traffic-load (erl) for the for the first application example.

	1	2	3	4	5	6	7	8
Voice (a_1)	5	10	15	20	25	30	35	40
Data (a_2)	2.5	4.0	5.5	7.0	8.5	10.0	11.5	13.0

In Figure 14.6 we show the analytical and simulation results of the WTM for two service-classes versus the offered traffic-load. The results show that the model's accuracy is absolutely satisfactory, especially for low offered traffic-load (points 1–6 in the x-axis of Figure 14.6). Furthermore, the increase in the inter-cell interference does not seem to affect the accuracy of the calculations.

In Figure 14.7 we show the analytical results of the Staehle–Mäder model (i.e. no thresholds are defined for the two service-classes) for the two service-classes versus the offered traffic-load. If we compare the results shown in Figures 14.6 and 14.7, we see that the WTM leads to lower CBP than the Staehle-Mäder model does, for both service-classes. The divergence is more obvious in the case of high offered traffic-load (points 6 to 8). This can be explained by the fact that with an increased offered traffic-load, more calls in average are present in the system and the arrival rate of new calls is increased. Since the Staehle–Mäder model assumes fixed bandwidth requirements for each call, it is apparent that under heavy traffic conditions, many new calls, instead of being accepted with reduced bandwidth (as it happens in the WTM), are blocked. On the other hand, under light traffic conditions, the system state j in the WTM rarely exceeds the thresholds and consequently only a few calls are accepted with reduced bandwidth. In the latter case the WTM performs pretty much the same with the Staehle–Mäder model. Moreover, the superiority of the WTM over the Staehle–Mäder model becomes clearer for low inter-cell interference. The reason is that, since both models deal with the inter-cell interference in the same way, an increased inter-cell interference is more important in the blocking of new calls than the call's bandwidth requirements; therefore, the call's bandwidth reduction introduced by the WTM has less impact during the CAC.

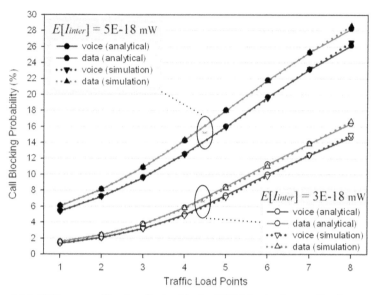

Figure 14.6 CBP versus offered traffic-load in the WTM for the first application example.

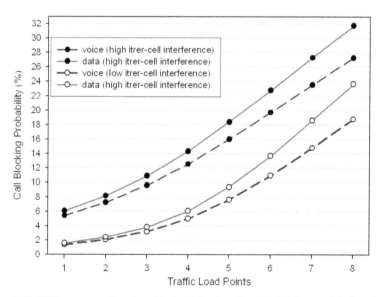

Figure 14.7 CBP versus offered traffic-load in the Staehle–Mäder model for the first application example.

Table 14.3 Traffic parameters for the second application example.

	Voice	Data	Video
Type	Adaptive	Elastic	Adaptive
Transmission rates (Kbps)	12.2 and 8.4	64 and 32	144, 128 and 112
Thresholds	0.7	0.6	0.4 and 0.6
Activity factor	0.5	1.0	0.3
BER parameter (dB)	5	4	3

14.6.2 Application Example 2

We consider a W-CDMA system with three service-classes: voice, data and video. The traffic parameters used for each service-class are as follows (see also Table 14.3):

- Voice: requires a transmission rate of $R_{1,1} = 12.2$ Kbps which can be reduced to $R_{1,2} = 8.4$ Kbps depending on the threshold $J_{1,1} = 0.7$. The activity factor is chosen to be $v_1 = 0.5$ and the required BER parameter is $(E_b/N_0)_1 = 5$ dB. This service-class is adaptive since the reduction of the transmission rate does not affect the service time.
- Data: requires a transmission rate of $R_{2,1} = 64$ Kbps which can be reduced to $R_{2,2} = 32$ Kbps depending on the threshold $J_{2,1} = 0.6$. The activity factor is chosen to be $v_2 = 1.0$ and the required BER parameter is $(E_b/N_0)_2 = 4$ dB. This service-class is elastic since the reduction of the transmission rate corresponds to the same increase of the service time.
- Video: requires a transmission rate of $R_{3,1} = 144$ Kbps which can be reduced to $R_{3,2} = 128$ Kbps and to $R_{3,3} = 112$ Kbps depending on the thresholds $J_{3,1} = 0.4$ and $J_{3,2} = 0.6$. The activity factor is chosen to be $v_2 = 0.3$ and the required BER parameter is $(E_b/N_0)_2 = 3$ dB. This service-class is adaptive since the reduction of the transmission rate does not affect the service time.

We take measurements for eight different traffic-load points (x-axis of Figure 14.8). Each traffic-load point corresponds to some values of the offered traffic-load of three considered service-classes as it shown in Table 14.4. In this example we consider low inter-cell interference with $E[I_{\text{inter}}] = 3\text{E-}18$ mW and $CV[I_{\text{inter}}] = 1$.

Table 14.4 Offered traffic-load (erl) for the for the second application example.

	1	2	3	4	5	6	7	8
Voice (a_1)	2.0	6.0	10.0	14.0	18.0	22.0	26.0	30.0
Data (a_2)	1.0	2.0	3.0	4.0	5.0	6.0	7.0	8.0
Video (a_3)	0.75	1.0	1.25	1.5	1.75	2.0	2.25	2.5

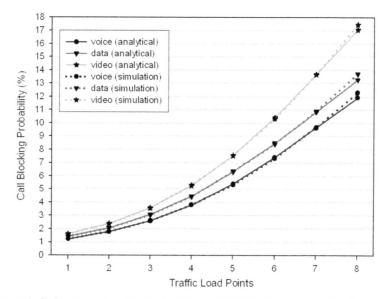

Figure 14.8 CBP versus offered traffic-load in the WTM for the second application example.

In Figure 14.8 we show the analytical and simulation results of the WTM for three service-classes versus the offered traffic-load. The results show that the model's accuracy is absolutely satisfactory, especially for low offered traffic-load.

14.7 Conclusion

We propose a new teletraffic model for the analysis of W-CDMA systems supporting elastic traffic. Poisson arriving calls, generated by mobile users, compete for the acceptance to a W-CDMA cell under the complete sharing policy. Each call upon arrival may have one out of several possible QoS requirements depending on the resource availability in the cell. We provide a recurrent formula for the calculation of call blocking probabilities. Simulations are used to verify the accuracy of the proposed calculations. We show by numerical examples that the accuracy of the new model is absolutely satisfactory.

Acknowledgements

This work was supported by the project ΠΕΝΕΔ-No.03ΕΔ420, which is funded for 75% by the European Union – European Social Fund and for 25% by the Greek State – General Secretariat for Research and Technology.

References

[1] H. Holma and A. Toskala (Eds.), *WCDMA for UMTS*, John Wiley & Sons, 2002.

[2] M. Schwartz, *Mobile Wireless Communications*, Cambridge University Press, 2005.

[3] J. S. Kaufman, Blocking in a shared resource environment, *IEEE Trans. Commun.*, vol. 29, no. 10, pp. 1474–1481, 1981.

[4] J. W. Roberts, A service system with heterogeneous user requirements, in *Performance of Data Communications Systems and Their Applications*, G. Pujolle (Ed.), North Holland, Amsterdam, pp. 423–431, 1981.

[5] J. S. Kaufman, Blocking in a completely shared resource environment with state dependent resource and residency requirements, in *Proceedings of IEEE INFOCOM'92*, pp. 2224–2232, 1992.

[6] J. S. Kaufman, Blocking with retrials in a completely shared resource environment, *Performance Evaluation*, vol. 15, pp. 99–113, 1992.

[7] I. D. Moscholios, M. D. Logothetis and G. K. Kokkinakis, Connection dependent threshold model: A generalization of the Erlang multiple rate loss model, *Performance Evaluation*, vol. 48, nos. 1–4, pp. 177–200, May 2002.

[8] D. Staehle and A. Mader, An analytic approximation of the uplink capacity in a UMTS network with heterogeneous traffic, in *Proceedings of 18th International Teletraffic Congress (ITC18)*, Berlin, Germany, September 2003.

[9] H. Akimaru and K. Kawashima, *Teletraffic – Theory and Applications*, Springer-Verlag, 1993.

[10] J. M. Aein, A multi-user-class, blocked-calls-cleared demand access model, *IEEE Trans. Commun.*, vol. 26, no. 3, pp. 378–385, 1978.

[11] K. W. Ross, *Multiservice Loss Models for Broadband Telecommunication Networks*, Springer, Berlin, 1995.

15

An Analytical Model of WCDMA Uplink Radio Interface Servicing Finite Number of Sources

Mariusz Głąbowski, Maciej Stasiak, Arkadiusz Wiśniewski
and Piotr Zwierzykowski

*Chair of Computer and Communication Networks, Poznań University of
Technology, ul. Polanka 3, 60-965 Poznań, Poland;
e-mail: mglabows@et.put.poznan.pl*

Abstract

The paper presents and describes the uplink blocking probability calculation
method in cellular systems with WCDMA radio interface and finite source
population. The method is based on the model of the full availability group
with multi-rate traffic streams. The proposed method can be easily applied to
3G network capacity calculations during planning stages of its design.

Keywords: WCDMA, blocking probability calculation, multiservice traffic.

15.1 Introduction

Universal Mobile Telecommunication System (UMTS) that uses WCDMA
radio interface is one of the standards proposed for third generation cellular
technologies (3G). The standard has been adopted in Europe and some Asian
countries. According to the ITU recommendations, the 3G system should
include services with circuit switching and packet switching, transmit data
with the speed of up to 2 Mbit/s, and ensure access to multimedia services [1].
In the GSM system, the maximal number of subscribers serviced by one cell
was unequivocally defined and depended on the number of used frequency

*D. D. Kouvatsos (ed.), Performance Modelling and Analysis of Heterogeneous
Networks,* 301–318.

channels. Because the GSM system was assumed to perform voice services only, all calculations of the network capacity could be done on the basis of the known dependencies worked out by Erlang. Capacity calculations of WCDMA radio interface, due to the possibility of resource allocation for different classes of traffic, are much more complex. Moreover, all users serviced by a given cell make use of the same frequency channel and a differentiation of the transmitted signals is possible only and exclusively by an application of orthogonal codes [2]. However, due to multipath propagation occurring in a radio channel, not all transmitted signals are orthogonal with respect to one another, and, consequently, are received by users of the system as interference adversely and considerably affecting the capacity of the system. Additionally, the increase in interference also arises from the users serviced by other cells of the system that make use of the same frequency channel as well as by the users making use of the adjacent radio channels. To ensure appropriate level of service of the UMTS network it is thus necessary to limit interference by decreasing the number of active users or the allocated resources employed to service them. It is estimated that the maximal usage of resources of radio interface without decreasing the level of service will be at 50–80% [3].

This paper presents the uplink blocking probability calculation method for cellular systems with WCDMA radio interface and finite source population. To achieve this, the Kaufman–Roberts recursion used for the calculation of blocking probability for a full-availability group has been adopted and appropriately modified. It is expected that the presented method, due to taking into account decreasing number of sources, can yield more accurate results than traditional Erlang approach.

The paper has been divided into five sections. Section 15.2 discusses basic dependencies describing radio interface load for the uplink direction. Section 15.3 presents an analytical model applied to blocking probability calculations for the uplink direction and finite source population. The next section, Section 15.4, includes a comparison of the results obtained in the calculation with the simulation data for a system comprising seven cells. Finally, Section 15.5 sums up the discussion.

15.2 Uplink Load Factor for WCDMA Radio Interface

WCDMA radio interface offers enormous possibilities in obtaining large capacities, however, imposes many limits as regards the acceptable level of interference in the frequency channel. In every cellular system with spread-

Table 15.1 Examples of E_b/N_0, v_j and L_j for different service classes.

Class of service (j)	Speech	Video call	Data	Data
W [Mchip/s]		3.84		
R_j [kb/s]	12.2	64	144	384
v_j	0.67	1	1	1
$(E_b/N_0)_j$ [db]	4	2	1.5	1
L_j	0.0053	0.0257	0.0503	0.1118

ing spectrum the radio interface capacity is drastically limited due to the occurrence of a few types of interference [4], namely:

- co-channel interference within a cell – coming from the concurrent users of a frequency channel within the area of a given cell,
- outer co-channel interference – from the concurrent users of the frequency channel working within the area of adjacent cells,
- adjacent channels interference – from the adjacent frequency channels of the same operator or other cellular telecommunication operators,
- all possible noise and interference coming from other systems and sources, both broadband and narrowband.

Accurate signal reception is possible only when the relation of energy per bit E_b to noise spectral density N_0 is appropriate [3]. A too low value of E_b/N_0 will cause the receiver to be unable to decode the received signal, while a too high value of the energy per bit in relation to noise will be perceived as interference for other users of the same radio channel. The relation E_b/N_0 for a user of the class j connection can be calculated as follows [3]:

$$\left(\frac{E_b}{N_0}\right)_j = \frac{W}{v_j \cdot R_j} \cdot \frac{P_j}{I_{\text{total}} - P_j}, \tag{15.1}$$

in which the following notation has been adopted:

- P_j – received signal power from a user of the class j connection,
- W – chip rate of spreading signal,
- v_j – activity factor of a user of the class j service,
- R_j – bit rate of a user of the class j service,
- I_{total} – total received wideband power including thermal noise power.

The mean power of a user of the class j service can be determined with the help of equation (15.2):

$$P_j = \frac{I_{\text{total}}}{1 + \frac{W}{\left(\frac{E_b}{N_0}\right)_j \cdot R_j \cdot v_j}} = L_j \cdot I_{\text{total}}, \qquad (15.2)$$

where L_j is the load factor for a user of j class connection:

$$L_j = \left(1 + \frac{W}{\left(\frac{E_b}{N_0}\right)_j R_j \cdot v_j}\right)^{-1}. \qquad (15.3)$$

Table 15.1 shows sample values E_b/N_0 for different traffic classes and the matching values of the load-factor L. Once L_j of a single user has been established, it is possible to determine the total load for the uplink connection:

$$\eta_{UL} = \sum_{j=1}^{M} N_j \cdot L_j, \qquad (15.4)$$

where m is the number of traffic classes (services), and N_j is the number of users of the class j service. Relation (15.4) is true only when we deal with a system that consists of a single cell. In fact, however, there are many cells available, in which the generated traffic influences the capacity of radio interface of other cells. Thus, relation (15.4) should be complemented with an element that would take into consideration interference coming from other cells [3]. To achieve this, a variable i is introduced, which is defined as other cell interference over own cell interference. The total load for the uplink can thus be rewritten as

$$\eta_{UL} = (1 + i) \sum_{j=1}^{M} N_j \cdot L_j. \qquad (15.5)$$

The bigger the load of a radio link, the higher the level of noise generated in the system. The increase in noise Δ_{nr} is defined as the relation of the total noise received in the system to the thermal noise and is described by the following equation:

$$\Delta_{nr} = I_{\text{total}}/P_N = (1 - \eta_{UL})^{-1}, \qquad (15.6)$$

where P_N is the thermal noise. When the load of the uplink approaches unity, the matching increase in noise tends towards infinity. Therefore, it is assumed that the actual maximal usage of the resources of a radio interface without lowering the level of the quality of service will amount to about 50–80% [4].

15.3 Model of the System

Before admitting a new connection in the systems with WCDMA radio interface, admission control needs to check whether the admittance will not sacrifice the quality of the existing connections. The admission control functionality is located in RNC (Radio Network Controller) where the load information from the surrounding cells can be obtained. The admission control algorithm estimates the load increase that the establishment of a new call would cause in the radio network [3]. This is done not only in the access cell, but also in the adjacent cells to take the inter-cell interference effect into account. The new call is rejected if the predicted load exceeds particular thresholds set by the radio network planning either in uplink or downlink [4]. Regarding voice source traffic modelling, it is well known that the process of a voice call transitioning an ON state to OFF state can be modelled as two-state Markov chain. The state transition occurs in such a way that the amount of time spent in each state is exponentially distributed and, given the present state of traffic source, the future is independent of the past. If we assume that OFF and ON rate from ON to OFF is μ, and from OFF to ON is λ, then the ON period endures for a random time with exponential distribution of parameter λ and then jumps to a silence state with an exponential distribution of parameter μ. The average length of the ON and OFF periods is $1/\mu$ and $1/\lambda$, respectively [5]. For data traffic such as Web traffic it has been shown that it can be effectively characterised by heavy-tailed models. One of the simplest heavy-tailed distribution is the Pareto distribution, which is power law over its entire range [5].

15.3.1 Basic Assumptions

Our model consists of seven cells with omni directional antennas, as it is shown in Figure 15.1. In the model we assume: perfect power control, no limitation due to: scrambling codes, Node B (base station) processing power (channel elements) and Iub resources. We apply load factor L_j for the user of j class connection to estimate whether a new call can be admitted or blocked. We assume, for example, that the new call, which generates L_j load in the access cell, will generate part of that value in the surrounding cells $(L_j < L'_j)$. The problem of estimating L'_j is not the subject of this paper.

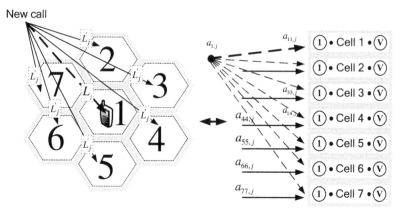

Figure 15.1 A set of full availability groups as the model of a wireless network.

The measurements or theoretical calculations of these values was presented in e.g. [6]. In our model we assume that each call which generates load L_j in access cell will cause load $L'_j = L_j/2$ in the neighbouring cells.

Model of full availability group servicing a mixture of the multi-rate traffic streams will approximate a radio interface which services the load generated in that cell and the load from the neighbouring cells (Figure 15.1) [7].

For Figure 15.1 the following notation is used: V is the cell capacity, $a_{zz,j}$ is the mean traffic of class j offered to the system by the users in the cell z and $a_{zr,j}$ is the mean traffic of class j offered in the cell r by the users of the cell z.

To determine the blocking probability of a new call appearing in the z cell, it is necessary to take into consideration the load generated by the call in the neighbouring cells. Therefore we propose a new analytical model, which can be used for the calculation of uplink blocking probability in UMTS network. In the model, we have used the idea of the model of the switching node servicing a mixture of multi-rate unicast and multicast traffic streams presented in [8].

15.3.2 Analytical Model

15.3.2.1 Full-Availability Group with Infinite Population of Traffic Sources

Let us consider the full-availability group with different multi-rate traffic streams. The full-availability group is a discrete link model that uses complete sharing policy [9]. This system is an example of a state-independent system in

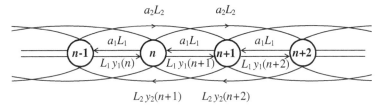

a_2L_2 a_2L_2

Figure 15.2 Fragment of a diagram of the one-dimensional Markov chain in a multi-rate system ($m = 2, L_1 = 1, L_2 = 2$).

which the passage between two adjacent states of the process associated with a given class stream does not depend on the number of busy bandwidth units in the system. Let us assume that the system services call demands having an integer number of the so-called BBUs (Basic Bandwidth Units) [9]. The total capacity of the group is equal to V BBUs. The assumed BBU in our model is, for example, 0.0001 of load factor. The idea of demanding resources by a call as the load factor was presented [10]. The group is offered m independent classes of Poisson traffic streams having the intensities: $\lambda_1, \lambda_2, \ldots, \lambda_m$. The class j call requires L_j BBU to set up a connection. The holding time for calls of particular classes has an exponential distribution with the parameters: $\mu_1, \mu_2, \ldots, \mu_m$. Thus the mean traffic offered to the system by the class j traffic stream is equal to

$$a_j = \frac{\lambda_j}{\mu_j}. \tag{15.7}$$

A multi-dimensional service process occurring in a system with different multi-rate traffic streams can be approximated by the one-dimensional Markov chain characterised by a product form solution. In the full-availability group model, Markov chain can be described by the Kaufman–Roberts recursion [11, 12]:

$$nP(n) = \sum_{j=1}^{m} a_j L_j P\left(n - L_j\right), \tag{15.8}$$

where $P(n)$ is the state probability in the multi-rate system.

The blocking probability b_k for the class j traffic stream can be written as follows:

$$b_j = \sum_{n=V-L_j+1}^{V} P(n). \tag{15.9}$$

The diagram presented in Figure 15.2 can be characterised by the Kaufman–Roberts recursion for the system with two call streams ($m = 2$, $L_1 = 1$, $L_2 = 2$). The $y_j(n)$ symbol denotes reverse transition rates of a class j service stream outgoing from state n. These transition rates for a class j stream are equal to the average number of class j calls serviced in state n. From equation (15.8) it results that the knowledge of the parameter $y_j(n)$ is not required for the determination of the occupancy distribution in the full-availability group with multi-rate traffic generated by infinite population of traffic sources. However, the value of this parameter in the given state of the group is the basis of the proposed method of the occupancy distribution calculation in the group with finite population of traffic sources. Accordingly, let us consider a part of the one-dimensional Markov chain diagram constructed for a system with multi-rate traffic (Figure 15.2). As stated earlier, the $y_k(n)$ symbol denotes reverse transition rates of a class j service stream outgoing from state n. This parameter determines – for class j traffic stream – the average number of class j calls serviced in state n and can be determined on the basis of the following reasoning [13]. Each state of the one-dimensional Markov chain in the multi-rate system (Figure 15.2) satisfies the following state equation:

$$P(n) \left[\sum_{j=1}^{m} a_j L_j + \sum_{j=1}^{m} L_j y_j(n) \right]$$

$$= \sum_{j=1}^{m} a_j L_j P(n - L_j) + \sum_{j-1}^{m} L_j y_j(n + L_j). \qquad (15.10)$$

From equation (15.8) it results that the sum of service streams outgoing from state n towards the lower states (states with lower number of busy BBUs) expressed in BBUs, is equal to n:

$$n = \sum_{j=1}^{m} L_j y_j(n). \qquad (15.11)$$

According to formulae (15.8) and (15.11), equation (15.10) can be expressed as follows:

$$\sum_{j=1}^{m} a_j L_j P(n) = \sum_{j=1}^{m} L_j y_j(n + L_j) P(n + L_j). \qquad (15.12)$$

Expression (15.12) is the equation of statistical equilibrium between the total stream outgoing from state n towards higher states and the total service stream entering state n from higher states. This equation holds only when local balance equations for call streams of individual traffic classes are satisfied:

$$a_j L_j P(n) = L_j y_j (n + L_j) P(n + L_j). \qquad (15.13)$$

Based upon equation (15.13), the reverse transition rate for class j calls in $(n+L_j)$ state is equal to:

$$y_j(n + L_j) = \begin{cases} a_j P(n)/P(n + L_j) & \text{for } n + L_j \leq V, \\ 0 & \text{for } n + L_j > V. \end{cases} \qquad (15.14)$$

Formula (15.14) determines the average number of class j calls serviced in the state $n + L_j$.

15.3.2.2 Full-Availability Group with Finite Population of Traffic Sources

Let us consider now the group servicing multi-rate traffic generated by finite population of sources [14]. Let us denote as N_j the number of sources of class j whose calls require L_j for service. The input traffic stream of class j is build by the superposition of N_j two-state traffic sources which can alternate between the active state ON (the source requires L_j) and the inactive state OFF (the source is idle). The class j traffic offered by idle source is equal to:

$$\alpha_j = \Lambda_j/\mu_j, \qquad (15.15)$$

where L_j is the mean arrival rate generated by an idle source of class j and $1/\mu_j$ is the mean holding (service) time of class j calls. In the considered model we have assumed that the holding time for the calls of particular classes has an exponential distribution. Thus, the mean traffic offered to the system by the idle class j traffic sources is equal to:

$$a_j = (N_j - n_j)\alpha_j, \qquad (15.16)$$

where n_j is the number of active (in-service) sources of class j. Interrelation between the offered traffic load and the number of in-service sources of a given class (equation (15.16)) makes the direct application of the Kaufman–Roberts recursion (15.8) for determination of occupancy distribution in the considered system impossible. The paper [14] proposes a new approximate method which enables us to make the mean value of traffic offered by class j

dependent on the occupancy state (the number of occupied BBUs) of the group, and thereby determination of the system with finite population of sources by the Kaufman–Roberts recursion. Let us notice that reverse transition rate $y_j(n)$ of class j determines the average number of class j calls serviced in the state of n busy BBUs. In the proposed method, it is assumed that the number of in-service n_j sources of class j in the state of n BBUs being busy is approximated by the parameter $y_j(n)$, i.e.

$$n_j(n) \approx y_j(n). \qquad (15.17)$$

Such an approach assumes that the average number of the given class calls being serviced in a given state of the group with infinite population of traffic sources is approximate to the average number of calls being serviced in the same state in the case of finite source population. Thereby, the parameter $y_j(n)$ can be determined by equation (15.14), where probabilities $P(n)$ are calculated by the Kaufman–Roberts recursion, under initial assumption that the offered traffic load is not dependent on the number of in-service sources and is equal to

$$a_j = N_j\alpha_j. \qquad (15.18)$$

The determined values $y_j(n)$ enable us to depend the mean value of the offered traffic on the occupancy state of the group in the following manner:

$$a'_j(n) = \left(N_j - y_j(n)\right)\alpha_j. \qquad (15.19)$$

Eventually, the approximated recurrent equation which determines the occupancy distribution in the full-availability group with multi-rate traffic and a finite population of sources can be calculated by

$$nP'(n) = \sum_{j=1}^{m} a'_j(n - L_j)L_j P'(n - L_j). \qquad (15.20)$$

In the proposed model, it is assumed that a new call is rejected when the increase in the load, both in the access cell and the neighbouring cells, will exceed acceptable thresholds. This means that servicing process for a new call occurring in the access cell and the adjacent cells, are mutually dependent. Therefore, the blocking probability $B_{zz,j}$ of the class j call occurring in the access cell z also depends on the blocking probability $B_{zh,j}$ of those calls in the adjacent cells (h). The blocking probability of class j call coming from

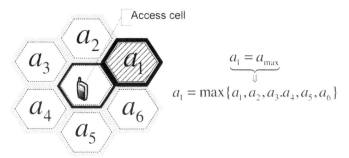

Figure 15.3 Blocking probability calculation in case of differently loaded neighbouring cells.

the cell z in the cell h ($h \neq z$) can be expressed by the following function:

$$B_{zh,j} = f\big((a_{hh,1}, L_{hh,1}), \ldots, (a_{hh,m}, L_{hh,m}), \ldots,$$
$$(a_{zh,1}, L'_{zh,1}), \ldots, (a_{zh,m}, L'_{zh,m})\big). \tag{15.21}$$

Thus, the blocking probability of the class j traffic streams coming from the z cell, in the h cell depends on "internal" traffic streams – generated in the cell h, i.e. the traffic streams $a_{hh,j}$ and on the contributing traffic streams from the cell z. For clarity, it is assumed the radio interface load of each cell h ($h \neq z$) is constant (i.e. 25%). This assumption makes it possible to express the probability $B_{zh,j}$ as only dependable on the traffic streams which influence the call h from the access cell z. Therefore we obtain:

$$B_{zh,j} = f\big((a_{zh,1}, L'_{zh,1}), \ldots, (a_{zh,m}, L'_{zh,m})\big). \tag{15.22}$$

The probabilities $B_{zh,j}$ can be determined on the basis of the full-availability group servicing a mixture of multi-rate traffic streams (equations (15.8) and (15.9)). To determine the value $a_{zh,j}$ influencing the load of cell h by the load generated in access cell z (for class j calls), we used the Fixed Point Methodology [8, 15]. According to the method, a given cell can be offered such a traffic which is not blocked in the neighbouring cells. This phenomenon leads to a decrease in the value of real traffic offered to a given cell and is called the thinning effect [15]. The classs j traffic stream decreased by thinning effect which is offered to the cell h by the calls occurring in the access cell z has been named effective traffic. The traffic can be determined on the basis of the following equation:

$$a_{zh,j} = a_{z,j} \prod_{\substack{r=1 \\ h \neq r}}^{u} \big(1 - B_{zr,j}\big), \tag{15.23}$$

where $a_{z,j}$ is the real traffic class j offered to the system by the users form the cell z and u is the number of cells in the considered system. Let us notice that to determine the value of the effective traffic $a_{zh,j}$ of class j it is indispensable to know the blocking probability $B_{zh,j}$ of the traffic of the same class in the neighbouring cells. Therefore, the iterative method is used to determine the values $a_{zh,j}$. The known values of the blocking probability $B_{zh,j}$ of class j in the cell h coming from cell z make it possible to determine the total blocking probability $B_{z,j}$ of calls in the access cell z:

$$B_{z,j} = 1 - \prod_{h=1}^{u} \left(1 - B_{zh,j}\right). \tag{15.24}$$

In case of differently loaded surrounding cells, to calculate blocking probability we can take into account only neighbouring cell with the highest blocking probability. Such an assumption is justified due to the way admission control works. A new call is rejected if the load in at least one cell would increase over particular threshold. In our model, each call causes the same load in the neighbouring cells. With this assumption in mind, the blocking of the connection from the access cell z can only occur in the adjacent cell h which is the most overloaded. Then to calculate the blocking probability we have to consider only two cells: the most loaded cell and the access cell (Figure 15.4). Thus the uplink blocking probability can be expressed in the following form (on the basis of equation (15.24)):

$$B_{z,j} = 1 - (1 - B_{zz,j}) \cdot (1 - B_{zh,j}), \tag{15.25}$$

where z is the access cell and h ($h \neq z$) is the most loaded neighbouring cell.

15.4 Numerical Results and Simulations

In order to confirm the proposed calculation method of uplink blocking probability in cellular system with WCDMA radio interface and with finite number of traffic sources, the results of analytical calculations have been compared with the results of simulation experiments. The simulations were performed with help of event scheduling method on the calls level [19]. The study has been carried out for subscribers demanding a set of services. The study was carried out for users demanding a set of services (Table 15.1) and it was assumed that

- a call of the particular services demanded $L_1 = 53$, $L_2 = 257$, $L_3 = 503$ and $L_4 = 1118$ BBUs,

- the class j call generated in an access cell with requirements equal to L_j BBUs requires constant value BBUs in all neighbouring cells ($L'_j = 0.3L_j$ or $L'_j = 0.5L_j$),
- the services were demanded in the following exemplary proportions: $a_1L_1 : a_2L_2 : a_3L_3 : a_4L_4 = 1 : 1 : 1 : 1$,
- for each cell, the maximum theoretical load capacity in uplink direction, was equal to 10000 BBUs,
- the L_{BBU} in the model was equal to 0.0001 of the interface capacity,
- the maximum uplink load of access cell (η_{UL}) was set to 80% of the theoretical capacity (i.e. $V = 8000$ BBUs for access cell),
- the maximum uplink load of neighboring cells was equal, for example, 25, 35, 42, 45, 48 and 50% of the theoretical capacity, respectively.

Figures 15.4–15.6 show the mean blocking probability as a function of the offered traffic for four call classes serviced by the system that consist of seven cells with radius of 1200 meters. The offered traffic is expressed in relation to the theoretical capacity of the system. The figures present the results for three exemplary values of population density which correspond to three cities in Poland with 3282, 208 and 94 persons per square kilometer.[1] Based on the data provided by one of Polish mobile telecom operator, we adopted the obtained data to 35% saturation of users. Finally, we obtained the average number of the mobile users in the group of seven cells in these cities: 5194, 329 and 149 users. Having the knowledge of the percentage distribution of mobile users per service (Table 15.1) in the mobile telecom operator, we can determine the numbers of users of each service. We assume the following distribution of users: 75% for voice users, 5% for video users, 10% for 144 kbps data users and for 384 kbps data users (Table 15.1). Thus, we can subsequently determine the following numbers of users in these cities:

- $n_1 = 3896, n_2 = 260, n_3 = 519$ and $n_4 = 519$ (Warszawa),
- $n_1 = 247, n_2 = 16, n_3 = 33$ and $n_4 = 33$ (Świnoujście),
- $n_1 = 112, n_2 = 7, n_3 = 15$ and $n_4 = 15$ (Sulmierzyce).

where n_i is the number of users of service i in the cell.

Figures 15.4–15.6 present the results for two values of parameter L'_j. The results for $L'_j = 0.3L_j$ are presented in Figures 15.4a, 15.5a and 15.6a, and the results for $L'_j = 0.5L_j$ are shown in Figures 15.4b, 15.5b and 15.6b. The comparison of Figures 15.4–15.6 led us to conclude that the increase in L'_j on

---

[1] The selected cities were Warszawa, Świnoujście and Sulmierzyce, respectively, and the relevant data come from '"The Report of the Main Statistical Office for the year 2006, Poland".

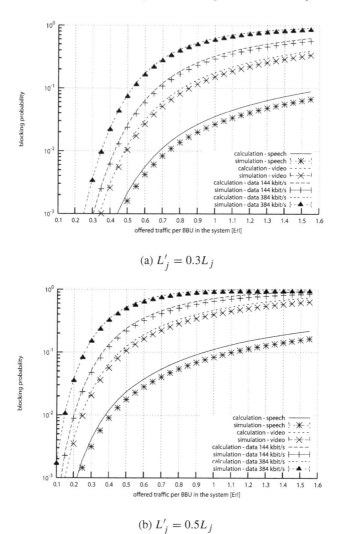

(a) $L'_j = 0.3L_j$

(b) $L'_j = 0.5L_j$

Figure 15.4 Blocking probability in the group of cells for 5194 users ($n_1 = 3896$, $n_2 = 260$, $n_3 = 519$ and $n_4 = 519$).

20% of L_j causes the increase in blocking probabilities for calls of particular traffic classes even several times.

The results of the simulations are shown in the charts (Figures 15.4–15.6) in the form of marks with 95% confidence intervals that were calculated

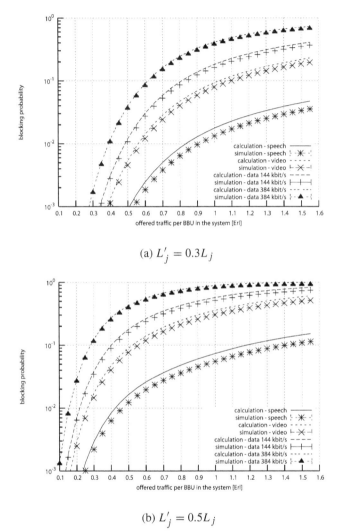

(a) $L'_j = 0.3L_j$

(b) $L'_j = 0.5L_j$

Figure 15.5 Blocking probability in the group of cells for 329 users ($n_1 = 247$, $n_2 = 16$, $n_3 = 33$ and $n_4 = 33$).

after the t-Student distribution. 95% confidence intervals of the simulation are almost included within the marks plotted in the figures.

On the basis of a number of researches carried out recently and the results shown in the figures, one can make tenable assumption that the accuracy of the proposed method does not depend on the number of the serviced traffic

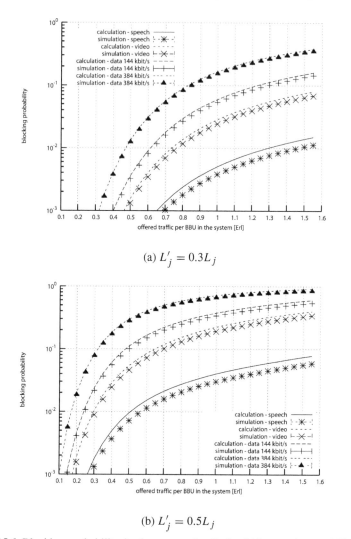

(a) $L'_j = 0.3L_j$

(b) $L'_j = 0.5L_j$

Figure 15.6 Blocking probability in the group of cells for 149 users ($n_1 = 112$, $n_2 = 7$, $n_3 = 15$ and $n_4 = 15$).

classes and on the loads introduced by them [16–18]. All the presented results (Figures 15.4–15.6) show the robustness of the proposed method of blocking probability calculation. In each case, regardless of the offered traffic load and the ratio of the number of traffic sources, the results of the calculation are characterised by fair accuracy.

15.5 Conclusions

The admission control in wireless network with WCDMA radio interface admits or blocks new calls depending on the current load situation both in access cell and in neighbouring cells. A new call is rejected if the predicted load exceeds particular thresholds set by the radio network planning. In this paper, we present uplink blocking probability calculation method with finite number of sources, which employs analytical model of the node in multiservice network. In our model, we use load factor L_j to estimate whether a new call can be allowed or blocked. In the presented model, it is assumed that sources of the voice and video traffic as well as internet traffic will be described by Poisson/Engset distribution. However, in the case of the later, with appropriately long mean time. The proposed method allows us to model cellular system more accurately due to considering limited number of customers (traffic sources). The obtained results show that taking into account decreasing number of active sources in the system yields lower blocking probability than in models assuming infinite number of sources. This can be essential for services, which demand higher capacity. The calculations are validated by a simulation. The proposed method can be easily applied to 3G network capacity calculations during planning stages of its design.

References

[1] K. Wesołowski, *Mobile Communication Systems*, 1st edition, John Wiley and Sons, 2002.

[2] S. Faruque, *Cellular Mobile Systems Engineering*, Artech House, London, 1997.

[3] H. Holma and A. Toskala, *WCDMA for UMTS. Radio Access for Third Generation Mobile Communications*, John Wiley and Sons, 2000.

[4] J. Laiho, A. Wacker and T. Novosad, *Radio Network Planning and Optimization for UMTS*, John Wiley and Sons, 2006.

[5] I. Koo and K. Kim, *CDMA Systems Capacity Engineering*, Mobile Communications. Artech House, Boston/London, 2005.

[6] P. Zwierzykowski M. Stasiak and A. Wiśniewski, Analytical method of calculating blocking probability in 3G networks with spreading spectrum and a finite number of traffic sources, *Theoretical and Applied Informatics*, vol. 19, no. 4, pp. 297–314, 2007.

[7] M. Stasiak, A. Wiśniewski and P. Zwierzykowski, Blocking probability calculation in the uplink direction for cellular systems with WCDMA radio interface, in *Proceedings of 3rd Polish-German Teletraffic Symposium*, P. Buchholtz, R. Lehnert and M. Pióro (Eds.), Dresden, Germany, September 2004, VDE Verlag, Berlin/Offenbach, pp. 65–74, 2004.

[8] M. Stasiak and P. Zwierzykowski, Analytical model of ATM node with multicast switching, in *Proceedings of 9th Mediterranean Electrotechnical Conference (MEleCon 1998)*, Tel-Aviv, Israel, May 1998, vol. 2, pp. 683–687, 1998.

[9] J.W. Roberts, V. Mocci and I. Virtamo (Eds.), *Broadband Network Teletraffic, Final Report of Action COST 242*, Commission of the European Communities, Springer, Berlin, 1996.

[10] D. Staehle and A. Mäder, An analytic approximation of the uplink capacity in a UMTS network with heterogeneous traffic, in *Proceedings of 18th International Teletraffic Congress (ITC18)*, Berlin, pp. 81–91, 2003.

[11] J. S. Kaufman, Blocking in a shared resource environment, *IEEE Transactions on Communications*, vol. 29, no. 10, pp. 1474–1481, 1981.

[12] J. W. Roberts, A service system with heterogeneous user requirements – Application to multi-service telecommunications systems, in *Proceedings of Performance of Data Communications Systems and Their Applications*, G. Pujolle (Ed.), North Holland, Amsterdam, pp. 423–431, 1981.

[13] M. Stasiak and M. Głąbowski, A simple approximation of the link model with reservation by a one-dimensional Markov chain, *Journal of Performance Evaluation*, vol. 41, nos. 2–3, pp. 195–208, July 2000.

[14] M. Głąbowski and M. Stasiak, An approximate model of the full-availability group with multi-rate traffic and a finite source population, in *Proceedings of 3rd Polish-German Teletraffic Symposium*, P. Buchholtz, R. Lehnert, and M. Pióro (Eds.), Dresden, Germany, September 2004, VDE Verlag, Berlin/Offenbach, pp. 15–204, 2004.

[15] F. P. Kelly, Loss networks, *The Annals of Applied Probability*, vol. 1, no. 3, pp. 319–378, 1991.

[16] M. Głąbowski, M. Stasiak, A. Wiśniewski and P. Zwierzykowski, Uplink blocking probability calculation for cellular systems with WCDMA radio interface and finite source population, in *Proceedings of 2nd International Working Conference on Performance Modelling and Evaluation of Heterogeneous Networks (HET-NETs)*, D. Kouvatsos (Ed.), Ilkley, June 2004, Networks UK, pp. 80/1–80/10, 2004.

[17] M. Stasiak, A. Wiśniewski and P. Zwierzykowski, Uplink blocking probability for a cell with WCDMA radio interface and differently loaded neighbouring cells, in *Service Assurance with Partial and Intermittent Resources Conference (SAPIR)*, Lisbon, June 2005, pp. 402–407, 2005.

[18] M. Głąbowski, M. Stasiak, A. Wiśniewski and P. Zwierzykowski, Uplink blocking probability calculation for cellular systems with WCDMA radio interface, finite source population and differently loaded neighbouring cells, in *Proceedings of Asia-Pacific Conference on Communications*, Perth, Australia (Best Paper Award), pp. 138–142, 2005.

[19] J. Tyszer, *Object-Oriented Computer Simulation of Discrete-Event Systems*, Kluwer Academic Publishers, 1999.

16

Teletraffic Analysis of the Uplink-WCDMA Networks with Soft Capacity

David García-Roger[1] and Vicente Casares-Giner[2]

[1]*Grupo NETCOM, Departamento de Telemática, Universidad Carlos III de Madrid, Av. Universidad 30, 28911 Leganés, Madrid, Spain;*
e-mail: dgroger@it.uc3m.es
[2]*Grupo GIRBA, Instituto ITACA, Universidad Politécnica de Valencia, Camino de vera s/n, 46022 Valencia, Spain; e-mail: vcasares@dcom.upv.es*

Abstract

In this work we formulate a new approximation method to estimate the uplink capacity in WCDMA systems. The radio link capacity is mapped into several *unit of resource* that allows to see the radio link model as a circuit switched network. Hence, we compute several parameters such as time congestion, call congestion, traffic congestion and carried traffic in a multiservice scenario. Analytical results has been obtained when varying the granularity of the *unit of resource* and the conclusion is that our approach is quite insensitive to the granularity used in the *unit of resource* definition.

Keywords: CDMA, multiservice networks, uplink capacity, soft blocking, Markovian models, noise rise, Delbrouck's algorithm, traffic congestion, call congestion, time congestion.

16.1 Introduction

Universal Mobile Telecommunications System (UMTS) is a new wireless tecnology adopted by the 3rd Generation Partnership Project (3GPP) for mobile telephony. It envisages a wide range of services: interactive, streaming,

D. D. Kouvatsos (ed.), Performance Modelling and Analysis of Heterogeneous Networks, 319–335.

background and conversational mainly. The radio access protocol used in the radio interface is *Wide-Band Code Division Multiple Access* (WCDMA). WCDMA allows the simultaneous transmission in the same band of frequencies. A wide range of rates, from a few kbps up to 2 Mbps or even more, can be transmitted thanks to the use of orthogonal codes that simultaneously occupy the same frequency band [1]. But due to several imperfections in implemented systems the perfect orthogonality between codes becomes hard to achieve. The fact is that other communications are experienced as interference that is added to the background (or thermal) noise. Hence, the more users there are in active state, the more interference is present in the air interface, so the capacity is gradually reduced when the number of on-going sessions increases.

In our work, we have considered a single UMTS cell and the *uplink* connection – from mobile terminal to base station – because it is more limitant than the *downlink* connection – from base station to mobile terminal [2]. The influence of other neighboring cells has been taken into account be means of a parameter f that express the relationship between the interference produced by the mentioned neighboring cells and the interference coming from our own cell.

The usual way to analyze the system capacity in WCDMA networks is by considering the physical layer. However, from the teletraffic point of view, it seems more suitable to see the system capacity in terms of a virtual *unit of resource* (ur) that could allow to see the radio link model as a classical circuit switched network. This fact permits the use of Markov process algorithms to compute blocking probabilities and other interesting teletraffic parameters. To that purpose, as in [3] is done, we have mapped the physical parameters into virtual units of resources that makes possible to see the number of resources demanded by a service class as an integer multiple of the basic and virtual ur. The ur is expressed in *chips per second* and we believe that a proper choice of that virtual ur should be insensitive to the granularity that is choosen.

Therefore, we have modelled the multi-service WCDMA uplink access network with a multi-dimensional Markov process with state dependent transition probabilities that are interference dependent, as in [3] is done. However, the difference between [3] and our approach is in the way the mentioned transition probabilities are chosen. As we will see our approach is significative less sensitive to the granularity of the virtual *unit of resource*.

The paper is structured as follows. Section 16.2 overviews the basic WCDMA background enough to introduce the concept of virtual *unit of resource*. Section 16.3 shortly describes the Delbrouck's algorithm and its use with the *Binomial-Poisson–Pascal* teletraffic model and also our proposal to be used in the presence of interference from other neighbor cells. Section 16.4 summarizes the use of the log-normal as probability density function to model the neighbor cell interference. Section 16.5 provides some illustrative examples that validates our proposal. Finally the paper ends with some conclusions in Section 16.6.

16.2 Some WCDMA Background

WCDMA systems are interference limited. In case of perfect power control in the reverse link, all reverse link signals (mobile terminal-to-cell) are received at the same power level [4]. Assuming N users, the signal to the total interference ratio, SIR, can be expressed as

$$\text{SIR} = \frac{S}{(N-1)S + N_o}$$

where S is the power of a target user and N_o is the background (or thermal) noise.

Of greater importance is the bit-energy, S/R, to interference-density, $[(N-1)S + N_o)]/W$, ratio. It is denoted by (E_b/I_o) and can be expressed as

$$(E_b/I_o) = \frac{S/R}{[(N-1)S + N_o)]/W} = \frac{W/R}{N-1 + N_o/S} \tag{16.1}$$

In (16.1) W is the so called "chip rate",[1] R the source bit rate, W/R is generally referred to as "the processing gain", G, and E_b/I_o is the minimum value needed to achieve a minimum QoS, with typical values of 3–5 dB or even more. This ratio depends on the bit rate R, the fast fading channel – usualy modeled according to Rayleigh distribution – the diversity of the antennas, and the speed of the mobile terminal, mainly.

In case of multiservice scenario, where signal from user k will be received with a power denoted by S_k, with R_k as the bit rate for user k and v_k as the

---

[1] The chip rate or sampling bit rate, for UMTS is equal to $W = 3.84$ Mcps *Megachips per second*. One *chip* (or *sub-bit sample*) is one bit from the direct sequence spread spectrum code.

activity factor.[2] For user k, equation (16.1) can be transformed into:

$$(E_b/I_o)_k = \frac{S_k/(v_k R_k)}{[I_{\text{total}} - S_k]/W} = \frac{W/(v_k R_k)}{[I_{\text{total}} - S_k)]/S_k} \qquad (16.2)$$

where $G_k = W/(v_k R_k)$ is the processing gain of user k and I_{total} is the total received wideband power including the thermal (or background) noise power N_o, in the base station under study. Then, solving (16.2) for S_k we get

$$S_k = \frac{1}{1 + \frac{G_k}{(E_b/I_o)_k}} I_{\text{total}} = L_k I_{\text{total}} \qquad (16.3)$$

In (16.3) we have defined the load factor L_k, for connection or session k, with $L_k < 1$. Clearly, L_k indicates the fraction that session k contributes to the total interference I_{total}. On the other hand, the total received interference, excluding the thermal noise N_o, can be written as follows:

$$I_{\text{total}} - N_o = \sum_{j=1}^{N} S_j = \sum_{j=1}^{N} L_j I_{\text{total}} = \eta_{UL} I_{\text{total}}$$

16.2.1 Noise Rise

WCDMA call admission control performs on the basis of the measured *noise rise*. It is defined as the ratio between the total interference density and the thermal noise density:

$$\text{Noise rise} = \frac{I_{\text{total}}}{N_o} = \frac{1}{1 - \eta_{UL}} \qquad (16.4)$$

Although the *noise rise* is the parameter used by the base station in the admission control, it does not reflect the real load in the system. A more suitable way to express *noise rise* is in terms of the *uplink* utility factor, η_{UL}, (see expression 8.9 from [5]). The parameter η_{UL} represents the capacity fraction in use. Hence, from (16.4) we can see that for value of the *noise rise* about 3.0 dB corresponds with a utility factor of 50%, while for a *noise rise* of 6.0 dB we get a utility factor of 75%. In real or practical scenarios typical values

---

[2] The *Activity factor*, v_k, is a physical medium parameter, typically 0.5 for voice. It increases to 0.67 when considering the signaling *overhead* via the uplink *Dedicated Physical Control Channel* (DPCCH), during a discontinuous transmission (DTX) when variable coding bit rates are used in *Adaptive Multi-Rate* (AMR) *compression* schemes.

of η_{UL} are 0.7–0.8. When η_{UL} approaches to 1, the corresponding noise rise approaches to ∞ and the system reaches its pole capacity, called *capacity pole*. This limit is given by $(I/S)_{max}$, e.g., the maximum ratio between the interference permitted and the useful WCDMA carrier signal. $(I/S)_{max}$ is equivalent to the *spreading gain* divided by the required E_b/N_0.

16.2.2 Neighbor Cell Interference Ratio f

I_{total}, the total received wideband power, includes the interference from the own cell I_{own}, the interference from other neigboring cells I_{other}, and the thermal noise power N_o. It can be expressed as

$$I_{total} = I_{own} + I_{others} + N_o$$

From (16.4) we can then write

$$\eta_{UL} = \frac{I_{own} + I_{other}}{I_{own} + I_{other} + N_o} \tag{16.5}$$

Other cell interference can be taken into account by the ratio of other cell to own cell interference, f:

$$f = \frac{\text{other cell interference}}{\text{own cell interference}} = \frac{I_{other}}{I_{own}} \tag{16.6}$$

Then, for the uplink load factor we can write

$$\eta_{UL} = (1+f) \sum_{j=1}^{N} L_j = (1+f) \sum_{j=1}^{N} \frac{1}{1 + \frac{G_j}{(E_b/I_o)_j}} \tag{16.7}$$

The maximum number of connections or sessions of service j in one cell is then given by

$$\eta_{UL} = (1+f) \frac{1}{1 + \frac{W}{\kappa_j}} N_{max,j} \tag{16.8}$$

where κ_j is defined as

$$\kappa_j = (E_b/I_o)_j \, v_j \, R_j$$

In other words, the maximum number of sessions of class j that can be carried in a cell is given by

$$N_{max,j} = \frac{\eta_{UL}}{1+f} \left(1 + \frac{W}{\kappa_j}\right) \tag{16.9}$$

Table 16.1 Service classes considered in [3] for the capacity of a WCDMA cell. Assumptions: $\eta_{UL} = 0.5$, $W = 3840$ kcps.

j	Class	R_j (in kbps)	ν_j	$\left(\frac{E_b}{N_0}\right)$ (in dB)	κ_j	$N_{\max, j}$ ($f = 0$)	$R_{s,j}$ (in kcps, $f = 0$)
1	Voice	7.95	0.67	4	13.379	144.002	26.666
2	Voice/moving	7.95	0.67	7	26.695	72.421	53.022
3	Voice	12.2	0.67	4	20.532	94.011	40.845
4	Data	16	1	3	31.924	60.642	63.321
5	Data	32	1	3	63.848	30.571	125.608
6	Data	64	1	2	101.433	19.428	197.645
7	Data	144	1	1.5	203.405	9.939	386.346

Remark. Typically $G_j/(E_b/I_o)_j \gg 1$, therefore the following approach is used quite often:

$$\frac{1}{1 + \frac{G_j}{(E_b/I_o)_j}} \simeq \frac{(E_b/I_o)_j}{G_j} = \frac{\kappa_j}{W}$$

16.2.3 Definition of a Unit of Resource

In order to frame the problem in the standard teletraffic engineering terminology, it is very convenient to convert all radio transmission parameters into teletraffic channels. Therefore let us denote by $R_{s,j}$ the spread bit rate of service j, that is, the proportion of W used by one connection of service j:

$$R_{s,j} = \frac{W}{N_{\max, j}} = \frac{1 + f}{\eta_{UL}} \frac{\kappa_j W}{(\kappa_j + W)} \tag{16.10}$$

Table 16.1 shows an illustrative example with the service classes borrowed from [3].

Then we define a *unit of resource* capacity, to be expressed in *chips per second* (cps). For our understanding, the unit of resource (ur) also called *unit channel*, is the minimum amount of chip rate used by a WCDMA service. The idea is that $R_{s,j}$ be a multiple of the basic ur, that is $ur \cdot d_j = R_{s,j}$ for all i.

The value of ur will depend of the desired precision we want to work. For instance, if we set $ur = 1$ cps, then we will work with maximum precision.

Table 16.2 Four definitions of unit of resource for seven classes of traffic with their equivalent bandwidth, for $f = 0$.

$f = 0$	1 ur = 26.666 kcps $C = 144$	1 ur = 13.333 kcps $C = 288$	1 ur = 1.333 kcps $C = 2880$	1 ur = 1 cps $C = 3.84 \cdot 10^6$
j	d_j	d_j	d_j	d_j
1	1	2	20	$26.666 \cdot 10^3$
2	2	4	40	$53.022 \cdot 10^3$
2	2	4	31	$40.845 \cdot 10^3$
4	3	5	48	$63.321 \cdot 10^3$
5	5	10	95	$125.608 \cdot 10^3$
6	8	15	149	$197.645 \cdot 10^3$
7	15	29	290	$386.346 \cdot 10^3$

Therefore taking into account the chip transmission rate W and the definition of the ur, the total number of resources C in the cell under study are given by

$$C = W/ur \qquad (16.11)$$

From equation (16.10) we can determine the number of unit of resources ($d_j = R_{s,j}/ur$) required by a single communication of class j, from the total amount of resources, C. It is worth noting that the number of resources W/ur and the number of resources required by a session of class j, $R_{s,j}/ur$, are in general, real numbers, that sometimes would be convenient to approximate by integer numbers. The approach we use in this work is conservative: the total number of resources C is taken as $C = \lfloor W/ur \rfloor$ and the number of resources required by a session of class j is taken as $d_j = \lceil R_{s,j}/ur \rceil$. $\lceil a/b \rceil$ ($\lfloor a/b \rfloor$) denotes the minimum (maximum) integer number greater (not greater) than a/b.

We have defined four units of resource capacity that have been used to analyze our proposal. From higher to lower precision we have:

- *Definition 1*: For $f = 0$; 1 ur = 26.666 kcps $\Rightarrow C = 144$ ur. Each class demands a set of resources given by column 2 of Table 16.2.
- *Definition 2*: For $f = 0$; 1 ur = 13.333 kcps $\Rightarrow C = 288$ ur. Each class demands a set of resources given by column 3 of Table 16.2.

- *Definition 3*: For $f = 0$; 1 ur = 1.333 kcps $\Rightarrow C = 2880$ ur. Each class demands a set of resources given by column 4 of Table 16.2.
- *Definition 4*: For $f = 0$; 1 ur = 1 cps $\Rightarrow C = 3.84 \cdot 10^6$ ur. Each class demands a set of resources given by column 5 of Table 16.2.

16.3 The Markovian Models

16.3.1 Delbrouck's Algorithm

We assume that calls from different service classes follow the *Binomial-Poisson–Pascal* (BPP) model [6]. The WCDMA cell is modeled according to a R-dimensional Markov process, where R is the number of services clases. In the absence of interference, the blocking is produced when no resources are available. So for a *lost-call–cleared* system and (*complete sharing*) as call admission policy we can evaluate the blocking probabilities by means of Delbrouck's algorithm.[3] Starting with an initial value of the state probability $q(x)$, $q(0) = 1$, it allows the calculation of $q(x)$, where x is the number of resources occupied:

$$q(0) = 1$$

$$q(x) = \frac{1}{x} \sum_{i=1}^{R} \left\{ \frac{d_i A_i}{Z_i} \right\} \sum_{j=1}^{\lfloor x/d_i \rfloor} \left\{ \frac{Z_i - 1}{Z_i} \right\}^{j-1} q(x - jd_i) \qquad (16.12)$$

$$x = 1, 2, \ldots, C.$$

In the above expression, R is the number of service classes, C the number of total resources available, A_i is the offered trafic from class i (expressed in number of calls), d_i is the number of resources needed to carry a call of classe i, Z_i is the *peakedness factor* associated to the arrival process i. For $Z < 1, Z = 1, Z > 1$ we have the Binomial, the Poisson and the Pascal cases, e.g. scenarios with smooth, random and bursty traffic, respectively.

In a second step, the probabilities $q(x)$ are normalized i.e.:

$$Q_C = \sum_{i=0}^{C} q(i)$$

---

[3] Delbrouck's algorithm, published in 1983, is a generalization of Kaufman and Robert's algorithm, re-discovered independenlty by these authors and published in 1981. Kaufman and Robert's algorithm was originally published in 1964 by Fortet and Grandjean.

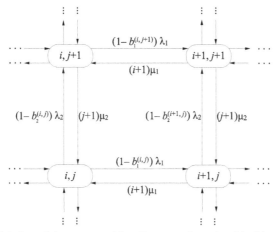

Figure 16.1 Partial view of the state transition diagram – dependent blocking probabilities – of a two-dimensional Markov process that models the soft-blocking phenomenon in a WCDMA system. The example shows two classes of services with Poisson processes as arrival processes. State (i, j) indicates the number of calls or sessions in progress of class i and j respectively.

$$p(x) = \frac{q(x)}{Q_n} \quad x = 0, 1, \ldots, C$$

Remark. There exist some procedures (for instance, the one described in [3]) that rewrites Delbrouck's algorithm such that numerical problems are avoided.

16.3.2 Markovian Model with Interferences

WCDMA systems are interference limited. The more calls are in progress the more interference exists in the uplink path. So the admission of a new call depends on the state of the system. For illustrative purposes, let us consider a system with two classes of traffic. Figure 16.1 shows a partial view of the state transition diagram when modeling the system with two multirate services, therefore as a 2-D Markov process. State (i, j) indicates the number of calls or sessions in progress of class i and j respectively. For $m = 1, 2$, let $a_m^{(i,j)} = 1 - b_m^{(i,j)}$ be the passage (or admission) factor as in [3] is named. If the state-dependent blocking probability $b_m^{(i,j)}$ is zero, then the state transition diagram is reversible and has a product form. Then, Delbrouck's algorithm [6] can be used to calculate blocking probabilities and other interesting traffic parameters.

However, when $b_m^{(i,j)}$ is greater than zero, and whenever be possible, a careful choice for $b_m^{(i,j)}$ should be done so that the state transition diagram keeps still reversible. Let d_m be the number of resources required by a call of class m ($m = 1, 2$). In [3] the following choice, that fulfills the reversibility property of the Markov process, was made:

$$a_m^{(i,j)} = 1 - b_m^{(i,j)} = \prod_{k=1}^{d_m}[1 - B(id_1 + jd_2 + k - 1)] \quad \text{for} \quad m = 1, 2 \quad (16.13)$$

In (16.13), $B(id_1 + jd_2)$ is the state dependent blocking probability when a single resource is requested and $id_1 + jd_2$ resources are occupied by i and j calls in progress of type 1 and type 2, respectively. Following other sources such as [7], $B(.)$ can be modelled as a log-normal distribution. This aspect will be treated in the next section.

The option of (16.13) corresponds to a sequential hunting of d_m single resources. But the real fact is that all d_m single resources are hunted at the same time, i.e. in a parallel way. Therefore we believe that a more realistic expression for $1 - b_m^{(i,j)}$ is:

$$a_m^{(i,j)} = 1 - b_m^{(i,j)} = 1 - B(id_1 + jd_2 + d_m - 1) \quad \text{for} \quad m = 1, 2 \quad (16.14)$$

But the choice of (16.14) does not fulfill the reversibility property and as a consequence, nice recursive schemes such as Delbrouck's algorithm can not be used. In this case the price to be paid is to solve a large set of linear equations to find the state probabilities or to use simulation techniques. However, as we will see in the illustrative examples, in contrast to [3] our approach is rather insensitive to the granularity of the *ur* definition.

16.4 Model for the Interference

The effect of the interference is a reduction of the number of resources available in the cell under study, as we have seen in section 16.2. This interference limitation has been translated into a Markov process with state dependent probabilities, as we have seen in section 16.3. The model we follow in the next lines is the identification of the blocking probabilities $b_m^{(i,j)} = b_m^H$ that reflect the neighbor cells interference. This interference is modelled as a log-normal distribution [7]. Its probability density function is given by

$$f(x; M, S) = \frac{1}{xS\sqrt{2\pi}}e^{-\frac{(\ln x - M)^2}{2S^2}}$$

with mean and standard deviation:

$$\mu = e^{(M+S^2/2)}; \quad \sigma^2 = e^{(2M+S^2)}(e^{S^2} - 1)$$

Note that M and S are the parameters that correspond to the normal distribution. They can be obtained from μ and σ as

$$M = \ln\left(\frac{\mu^2}{\sqrt{\mu^2 + \sigma^2}}\right); \quad S^2 = \ln\left(\frac{\mu^2 + \sigma^2}{\mu^2}\right)$$

For our study, in the same way as in [3], we have followed the identification of Staehle and Mäder [8], that is:

$$\mu = \frac{f}{f+1}C; \quad \sigma = \mu \quad \Rightarrow \quad M = \frac{\mu}{\sqrt{2}}; \quad S^2 = \ln 2$$

where C is the number of resources in the cell under study and f the so called interference factor.

Then, when the system is in state $H \equiv (i, j, \ldots, m, \ldots, n)$ $(h = id_1 + jd_2 + \cdots + nd_R)$, the blocking probability $B(h + d_k - 1)$ for a request of class k is

$$
\begin{aligned}
B(h + d_k - 1) &= P(x+ > C - h - d_k + 1) \\
&= 1 - P(x \le C - h - d_k + 1) \\
&= 1 - D(C - h - d_k + 1)
\end{aligned}
$$

$D(x)$ is de cumulative distribution function of the log-normal distribution:

$$D(x) = \frac{1}{2}\left(1 + \mathrm{erf}\left(\frac{\ln(x) - M}{S\sqrt{2}}\right)\right).$$

16.5 Illustrative Examples

For illustrative purposes, from Table 16.1 we have selected four classes of services with the teletrafic parameters shown in Table 16.3. The offered traffic is expressed in number of calls, and the number of resource requested by the class k, d_k in *unit of resources* which is dependent of the definition of the unit of resources. The analysis has been done considering the *complete sharing* policy with the following performance parameters as main target:

Table 16.3 Parameters for the scenario under consideration. A is the offered traffic, Z the peakdness factor.

Class	A	Z	d	Traffic Model
1	30.00	2.00	2.00	Pascal
2	15.00	1.00	4.00	Poisson
5	6.00	0.40	10.00	Binomial
6	3.75	0.25	16.00	Binomial

Table 16.4 Time, traffic and call congestions and carried traffic for a scenario free of interferences, $f = 0$, when $\diamond$: $ur = 26{,}666$ Kcps, $\triangle$: $ur = 13{,}666$ Kcps, $\square$: $ur = 1.333$ Kcps and $\bigcirc$: $ur = 1$ cps.

Class	Time c. (E)	Traffic c. (C)	Call c. (B)	Carried traffic
1 $\diamond$	0.00821	0.01815	0.00918	29.45541
2 $\diamond$	0.01727	0.01731	0.01727	14.74021
5 $\diamond$	0.04998	0.01469	0.03575	5.911838
6 $\diamond$	0.09172	0.01228	0.04729	3.703932
1 $\triangle$	0.00667	0.01510	0.00753	29.54699
2 $\triangle$	0.01410	0.01404	0.01410	14.78931
5 $\triangle$	0.04142	0.01168	0.02882	5.92989
6 $\triangle$	0.06982	0.00946	0.03674	3.71449
1 $\square$	0.00542	0.01221	0.00619	29.63342
2 $\square$	0.01152	0.01156	0.01152	14.82651
5 $\square$	0.03203	0.00892	0.02213	5.94642
6 $\square$	0.05818	0.00722	0.02828	3.72292
1 $\bigcirc$	0.00501	0.01132	0.00573	29.66031
2 $\bigcirc$	0.01064	0.01064	0.01064	14.84026
5 $\bigcirc$	0.02959	0.00818	0.02024	5.95088
6 $\bigcirc$	0.05452	0.00664	0.02610	3.72509

time congestion (E), traffic congestion (C), call congestion (B) and carried traffic. See [9] for the corresponding definitions.

16.5.1 No Interference from Other Cells, $f = 0$

In this situation, we do not have state-dependent blocking probabilities. Therefore Delbroucks's algorithm has been used. Table 16.4 shows the performances when the unit of resource ur, equals to $ur = 26.666$ Kcps, $ur = 13.333$ Kcps, $ur = 1.333$ Kcps and $ur = 1$ cps, respectively. Theoretically we could have expected an insensitive property of the blocking probabilities with respect to the ur definition. But we notice a slight difference when comparing the results. Because of the finer granularity in the ur definition, the lower is the ur in cps, the lower are the congestion probabilities we have. However we believe that for practical cases, for teletraffic engineering, they are not significative and this fact could validate the methodology.

16.5.2 With Interference from Other Cells, $f > 0$

In the same way as in previous works [3, 8], we set $f = 0.55$. In this case we have:

- *Definition 1*: 1 ur = 26.666 kcps, $C = 144$ ur, $\mu = \sigma = [f/(1+f)]C = 51.096$, $M = 3.587$, $S = 0.832$.
- *Definition 2*: 1 ur = 13.333 kcps, $C = 288$ ur, $\mu = \sigma = [[f/(1+f)]C = 102.193$, $M = 4.280$, $S = 0.832$.
- *Definition 3*: 1 ur = 1.333 kcps, $C = 2880$ ur, $\mu = \sigma = [f/(1+f)]C = 1021.935$, $M = 6.582$, $S = 0.832$.
- *Definition 4*: 1 ur = 1 cps, $C = 3.84 \cdot 10^6$ ur, $\mu = \sigma = [f/(1+f)]C = 1.362 \cdot 10^6$, $M = 13.778$, $S = 0.832$.

We have analyzed the sensitivity of the performance parameters with the choice of ur. To that end we have re-programmed the proposal in [3], based on Delbroucks' recursion, that considers a state-dependent blocking probability according to (16.13) and compared with our analytical proposal with state-dependent blocking probability according to (16.14), named as [ours]; see Tables 16.5–16.8. Assuming *complete sharing* policy, the results show the traffic congestion C, the call congestion (B) and the carried traffic. Our results have been obtained by solving the set of linear equation (Gauss-Seidel iteration) and by simulation, since definition (16.14) do not fulfill the conditions to apply Delbrouck's recursion.

Table 16.5 Traffic and call congestion and carried traffic
when $u = 26.666$ Kcps, with interferences, $f = 0.55$.
[3]: obtained with (16.12) according to definition (16.13).
[ours]: obtained by simulation according to definition
(16.14).

Class	Traffic c. (C)	Call c. (B)	Carried traffic
1 [3]	0.33107	0.19837	20.06783
2 [3]	0.35724	0.35724	9.64137
5 [3]	0.44703	0.66899	3.31778
6 [3]	0.55450	0.83274	1.67060
1 [ours]	0.48073	0.31642	15.57804
2 [ours]	0.32335	0.32335	10.14973
5 [ours]	0.16821	0.33580	4.99070
6 [ours]	0.11535	0.34280	3.31740

Table 16.6 Traffic and call congestion and carried traffic
when $u = 13.333$ Kcps, with interferences, $f = 0.55$.
[3]: obtained with (16.12) according to definition (16.13).
[ours]: obtained by simulation according to definition
(16.14).

Class	Traffic c. (C)	Call c. (B)	Carried traffic
1 [3]	0.41678	0.26324	17.49654
2 [3]	0.45757	0.45757	8.13632
5 [3]	0.59254	0.78427	2.44475
6 [3]	0.69784	0.90232	1.13309
1 [ours]	0.47204	0.30893	15.83871
2 [ours]	0.31569	0.31569	10.26458
5 [ours]	0.16303	0.32750	5.02177
6 [ours]	0.11050	0.33197	3.33558

Table 16.7 Traffic and call congestion and carried traffic when $u = 1.333$ Kcps, with interferences, $f = 0.55$. [3]: obtained with (16.12) according to definition (16.13). [ours]: obtained by simulation according to definition (16.14).

Class	*Traffic c. (C)*	*Call c. (B)*	Carried traffic
1 [3]	0.81567	0.68872	5.52967
2 [3]	0.90226	0.90226	1.46603
5 [3]	0.99116	0.99644	0.05303
6 [3]	0.99953	0.99988	0.00172
1 [ours]	0.46484	0.30280	16.05451
2 [ours]	0.30960	0.30960	10.35599
5 [ours]	0.15795	0.31924	5.052286
6 [ours]	0.10717	0.32439	3.348102

Table 16.8 Traffic and call congestion and carried traffic when $u = 1$ cps, with interferences, $f = 0.55$. [3]: obtained with (16.12) according to definition (16.13). [ours]: obtained by simulation according to definition (16.14).

Class	*Traffic c. (C)*	*Call c. (B)*	Carried traffic
1 [3]	0.99999	0.99999	0.00000
2 [3]	0.99999	0.99999	0.00000
5 [3]	0.99999	0.99999	0.00000
6 [3]	0.99999	0.99999	0.00000
1 [ours]	0.46260	0.30090	16.12173
2 [ours]	0.30729	0.30729	10.39062
5 [ours]	0.15653	0.31691	5.060807
6 [ours]	0.10630	0.32240	3.351354

At first glance, we can realize the sensitivity of the performance parameters with the definition of ur. What we could expect is that congestion probabilites and carried traffics be insensitive to the definition of ur. Then, taking a look to the referred tables, it is clear that our proposal, [ours] has a very low sensitivity compared with proposal from [3]. In fact, our proposal is quite insensitive to the definition of ur as it might be expected. Therefore, we can say that the results of Tables 16.5–16.8, validate our proposal based on definition (16.14). However of our proposal can not take advantage from Delbrouck's recursion because, as it was discussed in Section 16.3, it does not fulfil the suitable conditions.

16.6 Conclusions and Further Work

In this paper we have proposed an alternative approach to evaluate WCDMA systems. After a short introduction to the WCDMA physical layer, we have identified the *unit of resource* (ur) in terms of *chips per second* and the number of ur required by a given service, as a function of the spread bit rate of such a service. It allows us to work in the standard teletraffic engineering terminology. The interference from other cells has been modelled by a lognormal distribution and a state-dependent multi-dimensional Markov process has been used to model the situation. We have seen that this approach works pretty well, at least for practical purposes – design of WCDMA system capacity. Further work should be done to investigate a proper definition of $a_m^{(i,j))}$ of (16.13) and (16.14) that fulfills the reversibilily property of the Markov prporces in order to apply nice recursions schemes such as Delbrouck's algorithms.

Acknowledgements

This work has been supported by the Spanish Government and by the European Union (FEDER program) under project number TSI2007-66869-C02-02 and by TELEFÓNICA.

References

[1]   A. J. Viterbi, *CDMA*, Addison-Wesley, 1995.
[2]   D. Kim and D. G. Jeong, Capacity unbalance between uplink and downlink in spectrally overlaid narrow-band and wide-band CDMA mobile systems, *IEEE Transactions on Vehicular Technology*, vol. 49, no. 4, pp. 1086–1096, 2000.

[3] V. B. Iversen, V. Benetis, N. T. Ha and S. Stepanov, Evaluation of multi-service CDMA networks with soft blocking, in *Proceedings of the 16th ITC-SS (International Teletraffic Congress – Special Seminar)*, Antwerpen, Belgium, September 2004, pp. 212–216, 2004.

[4] K. S. Gilhousen, I. M. Jacobs, R. Padovani, A. J. Viterbi, L. A. Weaver and C. E. Wheatley, On the capacity of a cellular CDMA system, *IEEE Trans on Vehicular Technology* vol. 40, no. 2, pp. 303–312, 1991.

[5] H. Holma and A. Toskala, *WCDMA for UMTS: Radio Access for Third Generation Mobile Communications*, Wiley, 2002.

[6] L. E. N. Delbrouck, On the steady-state distribution in a service facility carrying mixtures of traffic with different peakedness factors and capacity requirements, *IEEE Transactions on Communications*, vol. 31, no. 11, pp. 1209–1211, 1983.

[7] A. M. Viterbi and A. J. Viterbi, Erlang capacity of a power controlled CDMA system, *IEEE Journal on Selected Areas on Communications*, vol. 1, no. 6, pp. 892–900, 1993.

[8] D. Staehle and A. Mäder, An analytic approximation of the uplink capacity in a UMTS network with heterogeneous traffic, in *Proceedings of the 18th International Teletraffic Congress*, Berlin, Germany, September 2003, pp. 81–91, 2003.

[9] V. B. Iversen, *Teletraffic Engineering Handbook*, ITU-D SG 2 and ITC, 2002, URL: http://www.tele.dtu.dk/teletraffic/

PART FIVE
ALL-OPTICAL NETWORKS

17

Circuit Emulation Service Technologies and Modified Packet Bursting in Metropolitan Optical Networks

Viêt Hùng Nguyên and Tülin Atmaca

TELECOM & Management SudParis, 9 Rue Charles Fourier, 91011 Evry, France; e-mail: {viet_hung.nguyen, tulin.atmaca}@it-sudparis.eu

Abstract

With the growth in data traffic, modern metropolitan area networks are moving away from traditional circuit-switched networking technologies towards high-speed packet-switched networking technologies, supporting best data traffic, offering high flexibility, cost-effective and easy-installation solution. Modified packet bursting (MPB) is introduced in packet-switched networks to improve protocol transmission efficiency and, accordingly, network performance. Circuit emulation service (CES) is introduced in packet-switched networks in order to support traditional TDM (circuit-switched) traffic. This paper presents the performance analysis of these two technologies on an all-optical Ethernet ring, demonstrating CES performance improvement thanks to MPB.

Keywords: Circuit Emulation Service (CES), TDM, Quality of Service (QoS), Modified Packet Bursting (MPB), Bursting Timer (BT), Metropolitan Optical Ethernet Ring, performance evaluation, network simulation.

Abbreviations

AAD	Average Access Delay
APE	Average Protocol Efficiency

D. D. Kouvatsos (ed.), Performance Modelling and Analysis of Heterogeneous Networks, 339–370.

ATM	Asynchronous Transfer Mode
BMT	Burst Mode Transceiver
BT	Bursting Timer
CBR	Constant Bit Rate
CES	Circuit Emulation Service
CSMA/CA	Carrier Sense Multiple Access with Collision Avoidance
DBORN	Dual Bus Optical Ring Network
EL	Effective Load
FD	Frame Delay
FDL	Fiber Delay Line
FJ	Frame Jitter
FL	Frame Loss
IETF	Internet Engineering Task Force
IPG	Inter-Packet Gap
IPP	Interrupted Poisson Process
IWF	Inter-Working Function
MAC	Media Access Control
MAN	Metropolitan Area Network
MEF	Metro Ethernet Forum
MPB	Modified Packet Bursting
MPLS	Multi-Protocol Label Switching
MTU	Maximum Transmission Unit
NSP	Native Service Processor
OECF	Optical Ethernet Concatenated Frame
OEF	Optical Ethernet Frame
PDH	Plesiochronous Digital Hierarchy
PDU	Protocol Data Unit
PLR	Packet Loss Rate
PWE3	Pseudo-Wire Emulation Edge to Edge
QoS	Quality of Service
RPR	Resilient Packet Ring
SDH	Synchronous Digital Hierarchy
SLA	Service Level Agreement
SONET	Synchronous Optical NETwork
TCARD	Traffic Control Architecture using Remote Descriptors
TDM	Time Division Multiplexing
UL	Useful Load

17.1 Introduction

Today's Metropolitan Area Networks (MANs) are principally TDM (circuit-switched) based networking technologies, including SONET/SDH rings. These networks were primarily designed to transport voice traffic, offering high reliability, survivability and superior quality to voice communications. However, they were not able to handle data traffic efficiently and cannot scale cost-effectively to accommodate the rapid growth in data traffic. The advent of high bandwidth applications such as music/video sharing, video conferencing, video on demand, etc, has increased the volume of data traffic in MAN and, as a result, the bandwidth required to carry that traffic. Therefore modern MANs are moving away from traditional circuit-switched networking technologies towards high-speed packet-switched networking technologies, supporting best data traffic, offering high flexibility, cost-effective and ease of installation solution.

In order to cope with this new trend, an experimental optical ring architecture (DBORN [1]) has been designed as a flexible and cost-efficient solution for short-term metropolitan networks. This architecture uses mature Ethernet technology at the electronic level and passive technology at the optical level. Thus, it is more efficient than circuit switching or wavelength switching thanks to the Ethernet based packet-switching granularity. It also offers better cost perspective than currently proposed Resilient Packet Ring (RPR [2]) solution, thanks to optical passive technology used at ring nodes.

This paper presents two novel technologies applied to this network. The first one is Modified Packet Bursting (MPB) technology that aims to improve the classical Media Access Control (MAC) protocol based on Optical Carrier Sense Multiple Access with Collision Avoidance (Optical CSMA/CA). MPB allows reducing optical overhead volume used for electronic packet transmission, hence improves network resource utilisation efficiency and, as a result, network performance. It employs the principle of packet bursting (PB) technology introduced by Gigabit Ethernet network [3].

The second technology presented in this paper is Circuit Emulation Service (CES) which allows the transport of TDM service (e.g., voice and video) over a packet-switched network. Recall that TDM service, which represents the major revenues of service providers, is traditionally transported over circuit-switched networks. Thus, an important challenge for MAN service providers while developing packet-switched networks is the ability to converge both TDM and data services on the same network infrastructure, providing a cost-effective and high service integration solution. A technology such as CES should provide the same reliability and quality of transporting TDM service as that provided by existing circuit-based networks. This work analyses the feasibility and performance impact of realising CES on DBORN. It extends our previous works in [4] while providing deeper performance analysis. Concretely we specially investigate the performance analysis of the interaction between MPB and CES technologies, and their influence on the global network performance.

The rest of this paper is organised as follows. Section 17.2 describes the optical ring architecture and its main features. Section 17.3 focuses on MPB technology with its principle and process. It also introduces a bursting timer mechanism, which is associated with MPB mechanism, enhancing the MPB performance. Some definitions on performance parameters used to evaluate MPB are also given in this section. Section 17.4 presents CES technology with the studied CES model, CES performance parameters and QoS requirements. Sections 17.5 and 17.6 respectively provide simulation results on MPB and CES performance. Finally, Section 17.7 finishes with some conclusions and discussions on future work.

17.2 Optical Ring Architecture and Main Features

17.2.1 Optical Ring Architecture

The network considered in this work is a metropolitan optical Ethernet ring network, the so-called DBORN for Dual Bus Optical Ring Network. We briefly describe this network architecture in this section. For more details the lectors are invited to refer to [1]. DBORN uses a ring topology with two separated buses (Figure 17.1) which are spectrally disjoint. Indeed, this architecture can be seen as a unidirectional optical fibre split into downstream and upstream channels using different wavelengths. The downstream bus, starting at the hub node, is a shared media for reading purpose (i.e., all ring nodes receive packets sent by the hub node on this bus). The upstream bus, initialising at ring nodes, is a multiple access writing media (i.e., all ring nodes send packets to the hub node on this bus). The hub node plays the role of a concentrator, all communications are performed between ring nodes and the hub node, and there is no direct communication between two ring nodes. The hub node also assures the communication of ring nodes with the core network.

The spectral separation allows the use of a simple passive structure for the optical part of ring nodes. This means that optical packets travel along the

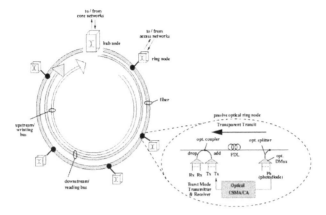

Figure 17.1 Metropolitan all-optical ring architecture.

ring without any opto-electronic conversion at intermediate nodes. Indeed, at the optical level, ring nodes use an optical splitter to separate an incoming signal into two identical signals: the main transit signal and its copy used for the control. A Fibre Delay Line (FDL) creates a fixed delay on the transit path between the control and the add/drop function. Thus, neither active optical device nor opto-electronic conversion is employed to handle transit optical packets. Once the optical packet is transmitted on the ring, it cannot be dropped before arriving to the hub.

Before continuing to describe some features of the network, we notice that in this work we only interested in the analysis and improvement of the performance of ring nodes on the upstream bus, because the later is a multiple access sharing resource (multipoint-to-point communication) that raises many problems of resource access control, sharing fairness, etc. These problems do not come up for the communication on the downstream bus since it simply is point-to-multipoint communication.

17.2.2 Optical Ethernet Frame Format

To have a simple and flexible architecture, Ethernet is used in DBORN as the convergence layer for the data plane. The structure of an optical Ethernet frame (OEF) is shown in Figure 17.2. The PDU Ethernet frame size is kept unchanged (from 64 bytes to 1518 bytes). The existing standard extensions (IEEE 802.1Q/802.1p /802.3ad) are still applicable. As DBORN uses burst mode transceiver (BMT) at the hub node for the packet-by-packet detection on the upstream bus, an optical preamble field of 16 bytes is required for

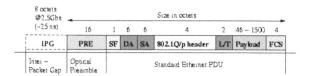

Figure 17.2 Optical Ethernet Frame (OEF) structure.

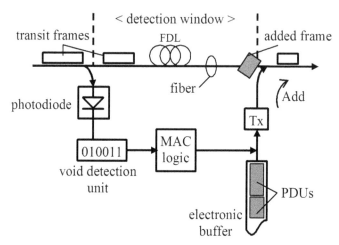

Figure 17.3 Schema of CSMA/CA in DBORN.

optical frame. Finally, the optical frame includes an inter-packet gap (IPG) of 25 ns (i.e., approximately 8 bytes at 2.5 Gbs) that allows an asynchronous insertion of frames on the media.

17.2.3 Media Access Control Protocol

The main issue in terms of logical performance when using Ethernet is the collision-free packet insertion on a multiple access writing bus (i.e., the upstream bus). In order to avoid collision with transit packets during local packets transmission, a media access control (MAC) protocol is required at ring nodes. DBORN has adopted an optical Carrier Sense Multiple Access with Collision Avoidance (CSMA/CA) based on void detection to ensure packet insertion on the upstream bus. Figure 17.3 shows the schema of optical CSMA/CA used in DBORN.

As stated earlier, incoming transit packets pass transparently through the local FDL of ring nodes. The local FDL size is slightly larger than the

Maximum Transmission Unit (MTU) of the used communication protocol, to provide the MAC logic unit with sufficient time to listen and to measure the media occupancy. It also creates a sliding detection window for the MAC logic unit. Once a free state (i.e., a *void*) of the media is detected thanks to a signal detection photodiode, the MAC unit measures the size of the progressing void. The ring node will begin inserting a local packet to fill a void only if the void is large enough. Otherwise, local packet is buffered in the electronic memory of the ring node until a large enough void is detected. Notice that with this mechanism, a detected void size is not necessarily limited by the size of the FDL, since the sliding observation window allows a progressing void measurement. Thus, a ring node can observe a void whose size is larger than the FDL size ($\sim$ MTU).

17.2.4 Traffic Control Mechanism

In a ring topology such as DBORN, fairness issues for bandwidth access are likely to raise between ring nodes sharing a common transmission channel (wavelength). Indeed, considering the upstream bus, ring nodes that start the bus (i.e., upstream nodes) can grab all available bandwidth, and, as a result, block the transmission of ring nodes close to the hub node (i.e., downstream nodes). This leads to performance degradation of downstream nodes. Therefore, in order to face the unfairness issues on the upstream bus, a traffic control mechanism called TCARD (Traffic Control Architecture using Remote Descriptors) [5] has been designed in the framework of DBORN.

TCARD is a preventive mechanism that statistically guarantees the access to the resource for downstream nodes. With TCARD, free bandwidth is preserved by a ring node according to traffic requirements of downstream nodes (e.g., based on Service Level Agreements – SLA). This is completed by the generation of anti-tokens (ATOK), whose size is around the Ethernet MTU ($\sim$ 1500 bytes), that forbid the node emission, and hence preserving voids for downstream nodes. Indeed, a flow of ATOK is generated with constant rate at each ring node. Each time an ATOK is generated, the ring node's transmission is forbidden until a void of size equal to ATOK size has been reserved by the node. Since the ATOK size is around the Ethernet MTU, this reserved void is sufficient for the insertion of any packet size at downstream nodes. In this mechanism, the ATOK rate is easily computed from the average required bandwidth of all downstream nodes. For example, consider an upstream bus that is shared by three ring nodes, each node requires 100 Mbs. Then the ATOK rate at the first ring node is equal to 200 Mbs/(1500 $\times$ 8 bits) $\sim$ 16667

ATOK/s, and that at the second ring node should be 100 Mbs/(1500 × 8 bits) ~ 8333 ATOK/s. Obviously the last ring node does not need to reserve bandwidth for any node, therefore the ATOK rate at this node is zero.

More details about the DBORN architecture and the implementation of TCARD mechanism can be found in [1, 5]. Note that depending on the volume/demand of client traffic, TCARD mechanism can be combined (enabled) or not (disabled) with the MAC protocol in order to ensure the functioning of the network.

17.3 Modified Packet Bursting Technology

This section discusses a technology aiming to improve the performance of the optical CSMA/CA protocol previously described.

17.3.1 Packet Bursting in Gigabit Ethernet

The recent development of the Gigabit Ethernet standard [3] introduced a new technique, called Packet Bursting (PB), in the MAC layer. In Gigabit Ethernet, the "carrier extension" was introduced in order to guarantee the interoperability of Gigabit Ethernet with existing 802.3 Ethernet networks. The slot time was increased from 64 bytes to 512 bytes. This is required for the collision detection in the half-duplex CSMA/CD (CSMA with Collision Detection) transmission mode. Thus, if a packet size is smaller than 512 bytes, the extension field is filled with extension symbols (i.e., padding pattern) to bring the minimum length of the transmission frame up to 512 bytes. This technique highly decreases the protocol efficiency for transmission of small size packets (e.g., for a packet of 64 bytes, the padding is up to 448 bytes).

Packet bursting mechanism was introduced to counteract the disadvantage of "carrier extension" used in Gigabit Ethernet technology. It allows a given transmission node to keep control of the media without relinquishing it between two consecutive frames. When a node has several packets to be transmitted, the first packet is transmitted with carrier extension if necessary. Subsequent packets are transmitted back to back with the minimum Interpacket gap (IPG), until a burst timer (of 1500 bytes) expires. More details on packet bursting process and performances can be found in [3].

17.3.2 MPB Basic Principle

The *modified packet bursting* (MPB) mechanism presented in this paper is based on the same principle as native packet bursting (i.e., Gigabit Ethernet packet bursting), with some enhancement for better adapting to the peculiar architecture of DBORN. The minimum time slot in DBORN is kept to 64 bytes, therefore there is no consideration of the carrier extension introduced in Gigabit Ethernet. However, the issue of protocol efficiency still arises for small size packets, due to the proportion of optical preamble and IPG. (From here we use the terms *optical overhead* as the reference to both optical preamble and IPG). For instance, the transmission at 2.5 Gbs of an electronic packet of 64 bytes needs an optical overhead of 24 bytes. The protocol efficiency is 46/70 = 72% in this case, which is poorly efficient.

As stated above, ring nodes and hub node in DBORN uses BMT (burst mode transceiver) for an asynchronous packet transmission and reception. The BMT at ring nodes (transmission side) requires the IPG field in the optical frame for an asynchronous packet insertion on the media. The BMT at the hub node (reception side) requires an optical preamble field (PRE) for a packet-by-packet detection. However, considering optical frames transmitted by the same BMT of a ring node, once the IPG has been preserved for the burst mode transmitter and the burst mode receiver has been synchronised thanks to the optical preamble, there is no more a need to repeat those fields for consecutive frames. This observation led us to consider the MPB mechanism, which consists in minimising the use of optical overheads for transmission of electronic packets (therefore maximising protocol efficiency).

Figure 17.4 illustrates the principle of MPB. If one node has several packets to transmit, the first packet is emitted with an optical overhead. Other packets are transmitted back to back without optical overhead and without any gap. This process creates on the transmission media a sequence of concatenated electronic frames led by one optical overhead. We call that an Optical Ethernet Concatenated Frame (OECF).

An important modification on MPB, compared to the native packet bursting in Gigabit Ethernet, is that no limit on OECF length is imposed. OECF length can be dynamically sized according to the distribution of free bandwidth (voids) on the media. It is limited only by the volume of traffic to be transmitted, and by the availability of free bandwidth. Concretely, MPB allows one node, which is transmitting on the detected progressing void, to keep control of the media to emit its electronic packets, until the end of the current void occurs or there are no more electronic packets to transmit. Thus

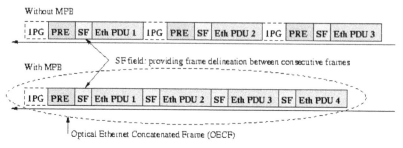

Figure 17.4 Sequence of frames with and without MPB.

the length of the OECF formed by those electronic packets is not limited to 1500 bytes as in native packet bursting. However, note that even if the length of an OECF can be in theory infinite, the size of each electronic packet composing this OECF is always limited to the MTU of the used transmission protocol (i.e., ~ 1500 bytes in Ethernet case).

17.3.3 Bursting Timer Principle

The *protocol efficiency* is defined as the proportion of optical overhead used to transport electronic packets. Thus the OECF length is an important factor that impacts the protocol efficiency of MPB. Indeed, as the optical overhead length is fixed (e.g., 24 bytes at transmission rate of 2.5 Gbs) for each optical frame transmission, the increase of OECF length leads to the increase of MPB protocol efficiency. Unfortunately, the basic concept of MPB improves the protocol efficiency of a node only if there exist, at a given moment, at least two electronic packets waiting for transmission at the node. This means that in the case where there is only one packet waiting for transmission in the electronic buffer of the node, the MPB protocol efficiency remains the same as in case of native optical CSMA/CA protocol.

From this observation, we proposed a new mechanism, called *bursting timer (BT)*, in order to improve the MPB protocol efficiency. Bursting timer mechanism tries to impose (if possible) a minimum length for each OECF. As a result, it tries to impose an inferior limit for MPB protocol efficiency. More specifically, BT unit forbids MPB unit at each ring node from emitting OECF until a bursting timer expires. During this time, many electronic packets are (probably) accumulated in the local electronic buffer waiting for transmission. Once the bursting timer expires, MPB unit can build up a big OECF thanks to gathered electronic packets, hence increases its protocol efficiency. However, introducing BT mechanism in MPB process can lead to an increase of access delay for electronic packets (i.e., additional waiting time during bursting timer period). Thus a compromise between the protocol efficiency and the access delay should be considered in the parameterisation of BT. Also, BT must be designed in such a way that it should not introduce excessive delay for high priority service, notably for delay-sensitive service such as voice.

17.3.4 MPB and Bursting Timer Processes

The BT and MPB processes implemented at an access node are respectively described in Figures 17.5 and 17.6. Each time the MAC unit detects a void on the media, it informs BT and MPB units. BT unit first looks at its local electronic buffer, allowing OECF transmission (i.e., BT is off) if the current buffer length is sufficient to build a big OECF (i.e., buffer length must at least equal to a threshold defined by BT size parameter). In order not to introduce excessive delay for electronic packets, notably for high priority (e.g., voice) service, BT unit also permits OECF transmission when the first packet on the top of local buffer has high priority, or has waiting time exceeding a threshold given by BT time parameter. BT unit naturally does not block OECF transmission if an OECF is currently under construction. In all others cases, OECF transmission of the node is blocked (i.e., BT is on).

Informed by MAC unit about the detection of a void, MPB unit first verifies whether it has permission of transmission given by BT unit. In case where BT is off, MPB unit looks at its local electronic buffer for the first electronic packet. If an OECF is under construction, the electronic packet is padded at the end of the current OECF if its size matches the detected void size. In case where no OECF is under construction, if the total size of the electronic packet plus an optical overhead (e.g., 24 bytes at 2.5 Gbs) matches the void size, the optical overhead is added to the electronic packet and a new optical frame is transmitted within the void. As we can notice, MPB process does not impose any limit on the length of OECF. But with the introduction of BT mechanism, the minimum length of OECF should be increased thanks to a number of electronic packets accumulated in local buffer during the on-time (i.e., transmission forbidden) of BT, hence the MPB protocol efficiency should be increased.

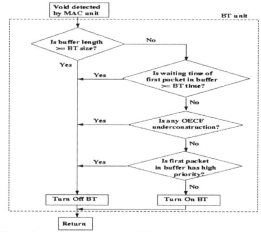

- *BT size*: size in bytes, a parameter of BT
- *BT time*: duration in seconds, a parameter of BT

Figure 17.5 Bursting timer process.

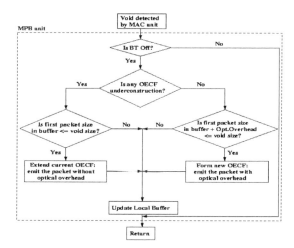

Figure 17.6 MPB process.

17.3.5 MPB Performance Parameters

We focused on the main performance parameters below: packet loss rate, average protocol efficiency, average access delay, network effective and useful loads.

Packet Loss Rate (PLR) at a ring node is the number of lost electronic packets divided by the total number of electronic packets received by the node during a simulation time. It is expected to be lowest possible.

Average protocol efficiency (APE) for optical frames (OEF/OECFs) is the average proportion of electronic data that they contain. The APE is expected to be highest possible. If we call n the number of OEF/OECF emitted by an access node, and m_i the number of electronic packets in each $OEF_i/OECF_i$, then we have the formula for computing the APE for an access node below:

$$\text{APE} = \frac{1}{n} \sum_{i=1}^{n} \frac{\sum_{j=1}^{m_i} \text{EthernetPDU}_j}{\text{Overhead}_i + \sum_{j=1}^{m_i} \text{EthernetPDU}_j} \tag{17.1}$$

Average access delay (AAD) is the average waiting time from the moment when a electronic packet is inserted in the electronic local buffer of a node, until it is transmitted successfully on the media. AAD is expected to be lowest possible.

Effective load (EL) is the number of bits transmitted through the network in a time unit divided by the bit rate of the network. The EL includes electronic packets as well as optical overhead used to transport them. *Useful load* (UL) is the size in bits of all electronic packets transmitted through the network in a time unit divided by the bit rate of the network. The UL is not aware of optical overhead transmitted in the network. EL is expected to be lowest possible, but UL is expected to be highest possible.

17.4 Circuit Emulation Service Technology

17.4.1 Introduction

Traditionally, voice has been carried over Time Division Multiplexed (TDM) based networks. TDM-based networks such as SONET/SDH offer high reliability and survivability for connections and predictable delays for voice samples, thus providing a superior quality. Nevertheless, the widespread development and deployment of packet-switched networks due to data service growth led network provider to a new challenge: transport of TDM (not only voice but also video) and data services over the same packet-switched network architecture. This convergence of TDM and data service in an existing packet-switched architecture could save considerable equipment and installation cost. The problem to be resolved for network providers is to find out a technology that provides the same reliability and quality of transporting TDM service as in existing TDM-based networks.

Circuit Emulation Service (the so-called CES) is a technology allowing the transport of TDM service such as PDH (E1/T1/E3/T3) as well as SONET/SDH circuit over a packet-switched network. Circuit emulation originally comes from Asynchronous Transfer Mode (ATM) world [6]. The idea has been taken up in the packet-switched world by a number of organisms, including the Internet Engineering Task Force (IETF), the Metro Ethernet Forum (MEF) and the Multi-Protocol Label Switching (MPLS) forum. The Pseudo-Wire Emulation Edge to Edge (PWE3) working group in the IETF [7] is setting the main CES standards. This group is chartered to develop methods to carry Layer-1 and Layer-2 services across a packet-switched network (principally IP or MPLS). Hence the group is looking at TDM circuit emulation, and also carriage of Layer-2 service such as ATM, Frame Relay and Ethernet across a packet-switched network. The Metro Ethernet Forum [8] is looking to extend the work of the PWE3 group to make it applicable to a metropolitan Ethernet context. Similarly, the MPLS Forum [9] is also looking at the same standards from the point of view of an MPLS network.

DBORN that was designed as a metropolitan packet-switched network should naturally support the transport of TDM service. Our previous work in [4] has described the studied model of CES on DBORN network, and also provided the first CES performance analysis. We briefly summarize our previous work on model, quality of service (QoS) requirements and performance parameters of CES in the following subsections.

17.4.2 CES Model and QoS Requirements

17.4.2.1 Reference Model of CES

The reference model for CES on DBORN [4] is based on the global model for circuit emulation described in PWE3 draft [7]. Figure 17.7 presents the general model of CES on DBORN. We have two TDM customers' edges (CE) communicating via DBORN. One CE is connected to a ring node (ingress CE), the other CE is connected to the hub node (egress CE). TDM service generated by ingress CE is transported by DBORN to egress CE. The emulated TDM service between two CEs is managed by two inter-working functions (IWF) implemented at appropriate nodes.

CES has two main modes of operation. In the first one, called "unstructured" emulation mode, the entire TDM service bandwidth is emulated transparently. The frame structure of TDM service is ignored. The ingress bit stream is encapsulated into an emulated TDM flow (called also CES flow) and is reproduced at the egress side. The second mode, called "structured"

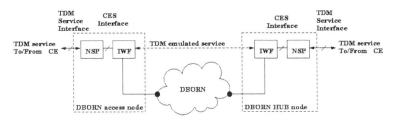

Figure 17.7 CES reference model.

emulation, requires the knowledge of TDM frame structure being emulated. Individual TDM frames are visible and are byte aligned in order to preserve the frame structure. "Structured" mode allows frame-by-frame treatment, permitting overhead stripping, flow multiplexing/demultiplexing. In the reference model of CES, the Native Service Processor (NSP) block performs some necessary operations (in TDM domain) on native TDM service such as overhead treatment or flow multiplexing/demultiplexing, terminating the native TDM service coming/going from/to CE.

The main functions of an Inter-Working Function are to encapsulate TDM service in transport packets (i.e., Ethernet packets in our case), to perform TDM service synchronisation, sequencing, signalling, and also to monitor performance parameters of emulated TDM service. Each TDM emulated service requires a pair of IWF installed respectively at the ingress and egress provider edges (PE). The aim of our study is to evaluate the logical performance of CES on DBORN. Hence, we ignore some operations and functional blocks in the CES reference model which are outside the scope of this work, such as the synchronisation and signalling aspect of the IWF block, as well as the native TDM operation of the NSP block.

17.4.2.2 Logical Model of CES Implemented at a Ring Node

The logical model of a DBORN access node supporting CES is described in Figure 17.8. Each access node (CES ingress side) is composed of an electronic part and an optical part. The incoming TDM service (treated as "structured" or "unstructured") is mapped into Ethernet packets thanks to the Inter-Working Function (IWF) block. A static segmentation mechanism, which fragments TDM frames into smaller segments according to a predefined threshold, is applied on large TDM frames in order to fit them to Ethernet packet. Ethernet packets transporting data service and Ethernet packets transporting TDM service are aggregated into local electronic buffers.

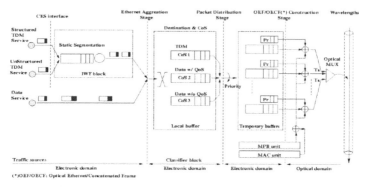

Figure 17.8 Logical DBORN access node architecture.

Here all packets are classified, according to their destinations and classes of service (CoS), into three separated buffers corresponding to three CoS. A scheduler, taking into account the CoS priority, distributes all packets from local buffer to temporary sending electronic buffers. OEFs are built by adding an optical preamble (Pr) to each electronic packet, and then are sent on appropriate wavelength. In case where MPB mechanism is enabled, MPB unit builds OECFs and sends them over the ring.

At the egress side (the HUB node), the same architecture with some modifications concerning IWF block is used. A jitter buffer (or playout buffer) is introduced in IWF block in order to accommodate the expected TDM frame jitter. The main function of the egress IWF at the hub node is to measure the CES flow performance (e.g., delay, jitter, and loss) based on delivered Ethernet packets transporting TDM service. The other aspects of egress IWF, such as jitter buffer dimensioning and reconstruction of native TDM frames, are not considered in this work.

17.4.2.3 CES Ethernet Packet Format

We adopted the CES packet format proposed in the PWE3 draft [7] (Figure 17.9). A CES control word is added to each TDM payload. The main functions of the CES control word are to differentiate the network outage and the emulated service outage, to signal problems detected at the IWF egress to the IWF ingress, to save bandwidth by not transferring invalid data, and to perform packet sequencing if Real Time Protocol (RTP) header is not used. An RTP header may be added to the resulting packet for the synchronisation and sequencing. The new resulting packet is encapsulated in the CES Ethernet

Figure 17.9 CES Ethernet packet format.

packet by adding Ethernet and multiplexing headers. All details about the structure of CES control word and RTP header are described in [7].

17.4.2.4 Segmentation Mechanism for TDM Frames

As we explained above, in order to perform CES on DBORN, TDM frames are encapsulated into Ethernet packets. A TDM frame would ideally be relayed across the emulated TDM service as a single unit. However, when the combined size of TDM frame and its associated headers exceeds the MTU size of DBORN, a segmentation and re-assembly process should be performed in order to deliver TDM service over DBORN.

We proposed two segmentation mechanisms. The first one, called *dynamic segmentation*, fragments a TDM frame into smaller segments according to void size detected on the media (wavelength) by the MAC unit. This approach promises a good use of wavelength bandwidth, but it is technically complex to implement. The second one, called *static segmentation*, segments the TDM packet according to a pre-defined threshold. This technique is simple to implement, and it provides resulting TDM segments with predictable size. Thus current TDM monitoring methods could be reused, simplifying the management of CES. In the framework of this study, we used the static segmentation method to evaluate the performance of CES on DBORN.

Segmentation threshold is a parameter that we have to determine during this work. In [7] authors recommended some rules to determine the segmentation threshold. First, the segmentation threshold should be either an integer multiple or an integer divisor of the TDM payload size. For example, for all unstructured SONET/SDH services, the segmentation threshold could be an integer multiple of STS-1 (or STM-0) frame of 810 bytes. Second, for unstructured E1 and DS1 services, the segmentation threshold for E1 could be 256 bytes (i.e., multiplexing of 8 native E1 frames), and for DS1 could be 193 bytes (i.e., multiplexing of 8 native DS1 frames).

Table 17.1 QoS definition.

CoS type	Service	QoS			
		Priority	PLR	Delay	Jitter
CoS1	TDM	High	10^{-9}	Strictly limited	Strictly limited
CoS2	Data with QoS guarantee	Media	10^{-6}	Limited	Limited
CoS3	Best-Effort (BE)	Low	N/A	N/A	N/A

17.4.2.5 QoS Definition

To be able to support circuit emulation in DBORN, we defined three CoS for electronic packets as given in Table 17.1. Ethernet packets transporting TDM service require very high quality of service, therefore they are given the highest priority. The media class (data with guarantee of QoS) can be considered as pseudo real-time traffic (e.g., video streaming). The CoS3 or Best-Effort (BE) class is sporadic Internet data, which has no guarantee of QoS.

17.4.3 CES Performance Parameters

We focused on three main parameters for CES performance evaluation: CES Ethernet Frame Loss (FL), CES Ethernet end-to-end Frame Delay (FD) and CES Ethernet Frame Jitter (FJ).

The CES Ethernet Frame Loss is defined as the ratio of lost Ethernet frames carrying TDM service among total sent Ethernet frames carrying TDM service. The CES Ethernet end-to-end Frame Delay (FD) is the maximum delay measured for a percentile P (superior to 95%) of successfully delivered Ethernet frames carrying TDM service over a measured interval T. The CES Ethernet Frame Jitter (FJ) is derived from the FD measured over the same measurement interval T and percentile P. FJ is obtained by subtraction of the lowest frame delay from FD. FJ is typically used to size the Jitter buffer at the egress side. These parameters must meet the MEN requirements for CES given in [8]. More specifically, FL and FD shall be kept to a minimum, and FJ shall not exceed *10 ms*.

17.5 Simulation Parameters and Results on MPB Performance

We use simulations with Network Simulator tool (version NS-2.1b8) to evaluate the performance of MPB. In our works each mean value is measured with 95% confidence intervals.

We now consider an upstream bus that consists of eight access nodes and one hub node. All ring nodes share one wavelength running at 2.5 Gbs for sending packets to the hub node. The fixed propagation delay between consecutive ring nodes is assumed to be 1 millisecond. Ring nodes are numbered from 1 to 8 according to their position (i.e., rank) on the ring: the shorter the distance from the node to the hub, the higher the rank number of the node. Therefore the bus starts at the first rank node and ends at the hub node.

For the performance study of the MPB mechanism, we suppose that the total offered traffic is sporadic data traffic that can be interpreted as the worst case for the network performance. (Notice that for the study of CES performance in Section 17.6, we will introduce other types of traffic such as TDM constant bit rate traffic that will obviously reduce the burstiness level of the total offered traffic). Each node is equipped with only one buffer, and electronic packets in the buffer are served in a First-Come-First-Serve (FCFS) order. The choice of the buffer size could be completely arbitrary (note that if it is too small, it may evoke packet loss, if it is too big, it may evoke high access delay for electronic packets). However in the context of a high bit rate network such as DBORN, client traffic usually does not need very big electronic buffers. Hence we choose for our simulations a buffer size equal to 250 Kbytes (relatively small).

To model the sporadic packets arrival process at each access node, we use the Interrupted Poisson Process (IPP) traffic source. The packet size for IPP source is assumed to be 50 bytes (small), 500 bytes (media) or 1500 bytes (big). The packet length repartition is as follows: 10% of small size packets, 40% of media size packets and 50% of big size packets. This distribution of packet length is inspired from the Internet traffic statistic published in [10]. In terms of packets number this distribution becomes: 64% of small size packets, 26% of media size packets and 10% of big size packets. Each IPP source is configured to have a burstiness level of 10 (i.e., the maximum source rate is 10 times higher than the average source rate). We also suppose that the offered traffic is repartitioned uniformly (i.e., each access node is fed with the same arrival processes, and the average arrival rate at each access node is identical). The bandwidth reservation TCARD mechanism is not used in this experimentation.

In the following paragraphs we will compare the network performance when MPB is used (*MPB case*) with the case where only optical CSMA/CA MAC protocol is used (*MAC case*). We set the average arrival rate at each access node to around 250 Mbs, thus the average offered ring load is about 80%. For the choice of BT parameters, we first notice that the protocol efficiency for the transmission at 2.5 Gbs of an optical frame, whose size is equal to the MTU size of DBORN, is high (1500/1524 ≈ 98.43%). This value can be considered as a good protocol efficiency. Therefore we believe that the BT size value should be superior or equal to the DBORN MTU size in order to effectively improve the MPB protocol efficiency. Note that if we set the BT size to zero, we will obtain the basic MPB mechanism without bursting timer. The BT time actually is the transmission duration at a given bit rate (e.g., 2.5 Gbs) of the number of bits defined by BT size. As the traffic is uniformly distributed on the ring, the BT size and BT time values are set identically for all ring nodes.

We first look at the packet loss event in this simulation. Figure 17.10 plots the PLR measured at each ring node, comparing the MAC case to the MPB case with different values of BT size. We observe that the MAC case evokes highest PLR at several downstream nodes, followed by the MPB case with BT size equal to zero, to MTU, to 4 times MTU and so on. The very high PLR (almost 100% at the 8th node) measured in MAC case shows the inefficiency of the optical CSMA/CA protocol in sharing the wavelength bandwidth, notably at heavy ring load. Indeed, in this architecture, upstream nodes (i.e., low rank nodes) always have higher priority to access bandwidth than downstream nodes (i.e., high rank nodes). Therefore, when the network is heavily loaded (e.g., in this case, the offered network load is 80%), upstream nodes can grab all the shared bandwidth. They consequently block the transmission of downstream nodes. This phenomenon is often known as the "positional priority" or "Ethernet capture" problem. However we can see in Figure 17.10 that the use of MPB mechanism considerably reduces the PLR at downstream nodes, notably when the BT size is big (e.g., no loss any more when BT size exceeds 10 times MTU).

The behaviour of MPB in function of BT size parameter can be better understood by examining the Figure 17.11 which depicts the average protocol efficiency for each value of BT size at all ring nodes. As expected we notice the lowest APE in MAC case and the higher APE in MPB case at all ring nodes. (The behaviour of the curve with BT size equal to zero will be investigated later in Figure 17.12.) Moreover, higher BT size values provide higher APE for all ring nodes (e.g., APE becomes higher than 90% when

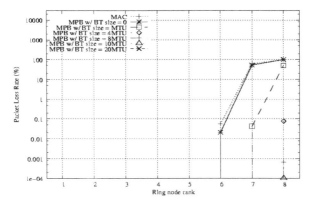

Figure 17.10 PLR versus ring node rank.

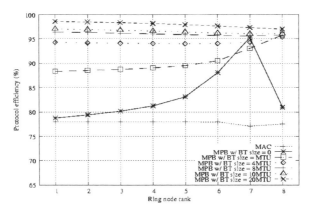

Figure 17.11 APE versus ring node rank.

the BT size exceeds 4 times MTU, and it is almost 99% when the BT size reaches 20 times MTU). The gain in terms of protocol efficiency of MPB case compared to MAC case varies between 10% with BT size equal to MTU and around 20% with BT size equal to 20 times MTU. Here we can see that MPB efficiently improves the protocol efficiency, notably when BT size is high, hence outstandingly reduces the bandwidth used for electronic packets transmission. It allows upstream nodes to use less bandwidth for transmitting their traffic, therefore they preserve more free bandwidth for downstream nodes. This also explains how the PLR at downstream nodes are decreased in MPB case in Figure 17.12.

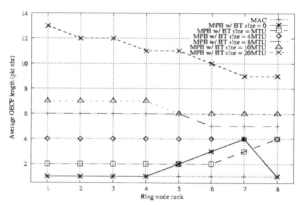

Figure 17.12 Average OECF length versus ring node rank.

Figure 17.12 shows the curves representing the average number of electronic packets per OECF measured at each ring node. Those curves clarify the trend of the APE curves in Figure 17.11. Indeed, the APE value, by definition, depends on the length of OECF. In other words, it is proportional to the number of electronic packets within an OECF (i.e., the higher the OECF length, the higher the APE value). Look at Figure 17.12, we observe that the behaviour of OECF length in all cases is very similar to the behaviour of APE curves in Figure 17.11. For instance, the OECF length is shortest (equal to 1) at all ring nodes in MAC case, but it becomes higher than 1 in MPB cases. Higher BT size value provides higher OECF length and higher APE as well at all ring nodes. For the same value of BT size, the OECF length at upstream nodes is shorter than that at downstream nodes when BT size is under 4 times MTU, and the OECF length decreases at downstream nodes when BT size exceeds 4 times MTU. Especially for the case of BT size equal to zero we observe that the OECF length increases at some downstream nodes and suddenly drops at the last ring node. This is simply due to the fact that almost 100% of electronic packets are dropped at the last ring node (see Figure 17.10), which leads to an inexact mean value computation of OECF length in this case. This also explains the "strange" behaviour of the curve with BT size equal to zero in Figure 17.11.

Now we explain the behaviour of the curves plotted in Figure 17.12. The explanation for the MAC case (OECF length equal to 1) is obvious, since each ring node emits packet by packet with an optical overhead. With MPB, ring nodes concatenate and transmit many electronic packets with only one optical overhead, hence increasing OECF length to superior to 1. The number of accumulated electronic packets in the local buffer, during BT time period, naturally grows with the increase of BT size, leading to the higher probability of building longer OECF. For the case where BT size is set to zero, we observe that even if the emission of ring nodes is not constrained by the BT time, the OECF length at some downstream nodes is still higher than that with positive BT size. This is mainly due to the above stated positional priority problem: the emission of downstream nodes is dominated by upstream nodes traffic, thus it produces a similar phenomenon as using positive BT size at downstream nodes. Also because of the positional priority that limits the availability of free bandwidth at downstream nodes, they can not always build long length OECF as upstream nodes can do, even if there exist many accumulated electronic packets in their local buffers. This is the explanation for the decrease of OECF length at downstream nodes when the BT size is superior to 4 times MTU in Figure 17.12. Therefore a global remark is that when BT size is big enough, the APE values at all ring nodes becomes quite stable (e.g., around 98% with BT size equal to 20 times MTU) even if there is a difference of OECF length at ring nodes, promising an APE-fair solution for a ring topology such as DBORN.

Let us now analyse the average access delay curves in Figure 17.13. We observe that the MAC case provides lowest access delay at upstream nodes, but highest access delay at downstream nodes (except at the last node the computation of average access delay is inexact due to very high (99%) electronic packets loss rate as shown in Figure 17.10). On the contrary the MPB cases give higher access delay for electronic packets at upstream nodes but offer very low access delay (notably when BT size is big) compared to MAC case at downstream nodes. The explanation for this behaviour is very similar to that given above. Indeed, the behaviour of the curve for MAC case is due to the positional priority. In MPB case with BT size equal to zero, thanks to the reduction of optical overhead volume by including many electronic packets in one OECF, ring nodes can: on one hand reduce the waiting time for other electronic packets in the local buffer; on the other hand preserve more free bandwidth for other downstream nodes. This leads to a slight decrease of AAD at downstream nodes. When BT size becomes higher, because of the time constraint introduced by MPB at upstream nodes, the AAD of electronic packets at upstream nodes increases. Higher BT time value obviously produces higher AAD at upstream nodes. However, during the period where upstream nodes are forbidden from transmission by MPB mechanism, downstream nodes benefit from free bandwidth left by upstream nodes to transmit their traffic and, as a result, they reduce access delay for

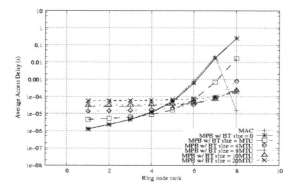

Figure 17.13 AAD versus ring node rank.

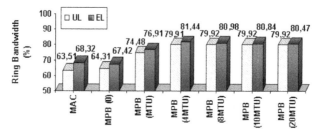

Figure 17.14 EL & UL for different configurations.

their electronic packets. Thus MPB balances access delay at ring nodes on the bus. Furthermore, as BT mechanism is easily customisable with its BT size and BT time parameters, we can completely expect a scenario where all access nodes have the same AAD, promising an access delay fair protocol for a shared media architecture.

Finally we investigate other interesting parameters: the effective and useful ring loads. Figure 17.14 shows the comparison of EL and UL between different configurations. Recall that in this experimentation the offered network load is around 80% (exactly 79.92%). We notice that the MAC case provides lowest UL (only 63.51% that is largely lower than the offered ring load), due to the important packet loss rate at downstream nodes. And it needs an effective bandwidth of 68.72%, meaning that almost 5.21% of bandwidth are occupied by optical overheads volume. In MPB cases, we observe that the UL increases with the growth of the BT size value and it reached its maximum value of 79.92% when BT size becomes higher than 4 times MTU. Beside, the EL remains always slightly higher than the UL for the same value of BT size, but this difference decreases when BT size increases (e.g., from 3.11% with BT size equal to zero to 0.55% with BT size equal 20 times MTU). From these observations, we can conclude that MPB considerably reduces the volume of optical overheads needed for the transmission of electronic packets, notably with big BT size, thus effectively improves the network transport capacity. We have already proved in [11] that with a good choice of BT size, we can improve the acceptable ring load to almost 90%, with low packet loss ratio at the last ring node only and small access delay at all ring nodes.

17.6 Simulation Parameters and Results on CES Performance

We use the same simulation model as in the study of MPB performance, except some modifications below. Each ring node is now equipped with three CoS buffers as shown in Figure 17.8. At each access node, the size of the CoS3 buffer which stocks sporadic BE data traffic is set to 250 Kbytes according to above stated reason. The CoS2 buffer is used to stock QoS-guarantee data traffic. In reality the volume of data traffic requiring QoS is usually smaller than the volume of best-effort data traffic, thus we suppose that the CoS2 traffic requires smaller buffer than the best-effort data traffic. Then we set CoS2 buffer size to 100 Kbytes. We also set the CoS1 buffer capacity to 100 Kbytes since it only serves constant TDM traffic that requires generally small buffer.

To model the packet arrival process at each access node, we used three types of traffic source. The Constant Bit Rate (CBR) source is used to model TDM traffic. The CoS2 traffic is modelled by an exponential source for the reason of simplicity. It generates packets with size of 250 bytes. The same source type (i.e., IPP) and parameters studied in Section 17.5 are used to model the sporadic BE data traffic.

The offered traffic is also repartitioned uniformly on the bus. This time, TCARD mechanism is enabled at each access ring node, statistically allocating for downstream nodes a mean bandwidth equal to their mean bit rate.

17.6.1 Preliminary Performance Analysis

In [4] we have proved that a segmentation threshold of 810 bytes, which is the STS-1 frame size, is appropriate to emulate unstructured SONET/SDH circuits. Other unstructured TDM services whose frame size is less than 810 bytes (e.g., E1/T1/E3/T3 PDH services) could be emulated entirely (without frame segmentation) in DBORN. It has been shown in [4] that with a simple asynchronous reservation mechanism (i.e., TCARD), DBORN can support CES with satisfied performance (i.e., no loss for Ethernet packets transporting TDM service, low access delay and FJ lower than 10 ms). However, we observed that CES performance at downstream nodes may not satisfy QoS requirements when the volume of BE traffic inserted on the bus by upstream nodes is high (e.g., exceeding 70% of the total offered traffic). The reason is that BE traffic inserted by upstream nodes disturbs and consumes free bandwidth for premium service (TDM service) at downstream nodes. Moreover, with optical CSMA/CA protocol and simple reservation mechanism used in DBORN, free bandwidth left by upstream nodes for the emission of downstream nodes can be *fragmented* into small voids or unusable voids, which probably are not sufficient for packet insertion at downstream nodes. By consequent it degrades the performance of CES and other services, notably at heavy network load. The work presented in this paper aims to improve CES performance on DBORN by introducing MPB mechanism in the MAC layer, improving bandwidth utilisation efficiency. Therefore this could counteract the negative impact of sporadic BE traffic and of inefficient bandwidth fragmentation on the CES performance at downstream nodes.

17.6.2 CES in Combination with MPB

At high volume of BE traffic. In [4], we have shown that the important volume of BE traffic inserted on the ring by upstream nodes can degrade CES performance at downstream nodes. We focus in this work on CES performance analysis when the volume of BE traffic is higher than 70% of the total offered traffic. For TDM traffic parameters, each access node is fed with 6 unstructured E-1 TDM flows, each flow corresponds to frame of size 32 bytes generated each 125 microseconds. In order to fit E-1 frames to Ethernet packets, we group 8 native E-1 frames ($8 \times 32 = 256$ bytes) in one Ethernet packet according to PWE3 draft [7] recommendation. The volume distribution of the offered traffic at each access node is set as following: BE traffic volume increases to 77%, TDM traffic occupies 5% and the 18% remaining is for CoS2 traffic. We set the average offered load of the upstream bus to 80%. In the following paragraphs we will analyse the performance of CES in three cases: MAC case, BT size equal to zero and BT size equal to 3 times MTU (which gives approximately 14.4 microseconds to BT time at 2.5 Gbs).

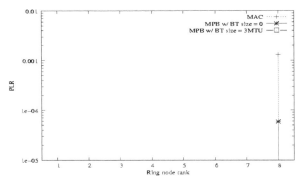

Figure 17.15 PLR versus ring node rank.

For this experimentation, we only observe loss for BE traffic (Figure 17.15). There is no loss for CoS2 traffic and TDM traffic, meaning the FL for TDM traffic is zero, satisfying one of the MAN requirements on CES performance [8]. We notice that even if offered ring load is acceptable (80% according to the precedent work [1]), the high volume of BE traffic evokes loss at the last node in MAC case (highest PLR) and in MPB case with BT size equal to zero. However, the MPB case with BT size equal to 3 times MTU provides no loss at all nodes, proving its efficiency in terms of bandwidth utilisation. Lectors can refer to Section 17.5 for more explanation about the behaviour of PLR curves in function of BT size value.

Since the propagation delay between consecutive ring nodes is fixed, the total end-to-end delay of electronic packets is mainly influenced by their access delay. Therefore we only investigate the average access delay for TDM traffic in this work. Figure 17.16 shows the AAD for TDM traffic at each node. We observe that in all cases the AAD is under 300 μs. And the lowest delay is provided by the MPB case with BT size equal to 3 times MTU, followed by the MPB case with BT size equal to zero and the MAC case. The same explanation as in the precedent section can clarify this tendency.

Figure 17.17 plots the curves representing the FJ for TDM traffic measured at each ring node. We notice that in MAC case, the FJ at the last access node exceeds 10 ms, unsatisfying QoS requirements on CES performance. This is because the high volume of BE traffic inserted on the ring by upstream nodes disturbs the transmission of TDM traffic at downstream nodes. Nevertheless, the MPB cases decrease FJ for TDM traffic to highly lower than 10 ms, thanks to optical overhead reduction and better bandwidth utilisation. We can see that MPB, notably when associated with BT mechanism, can coun-

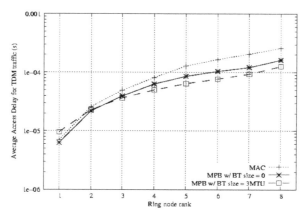

Figure 17.16 AAD for TDM traffic versus ring node rank.

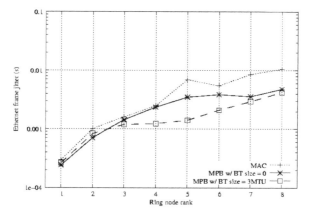

Figure 17.17 FJ versus ring node rank.

teract negative effect of sporadic BE traffic on the QoS of premium traffic (TDM) in a shared media network. It consequently improves the availability and reliability of the realising CES on DBORN.

At high total offered load.   We now focus on CES performance at high offered bus load in order to show the impact of MPB on network performance, notably on the performance of high QoS requirement traffic such as TDM. In this experimentation, the volume distribution of the offered traffic at each access node is set as follows: 10% of TDM traffic, 19% of CoS2 traffic and

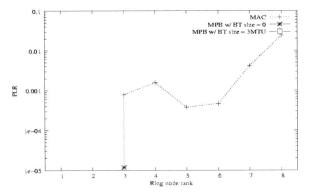

Figure 17.18 PLR versus ring node rank.

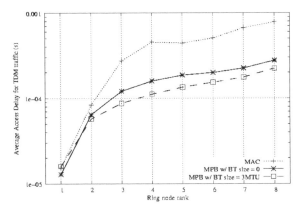

Figure 17.19 AAD versus ring node rank: load = 0.87.

71% of BE traffic. We set the average offered load of the upstream bus to approximately 87%. The same BT parameters are kept.

We observe in this simulation work that MAC case causes loss (for BE traffic only) at several access nodes, while almost no loss is observed with MPB cases (Figure 17.18). This demonstrates that MPB considerably improves the network performance and capacity.

Figure 17.19 depicts the AAD for TDM traffic versus ring node rank. We notice the same trend of AAD at access nodes as discussed in the precedent sections. MPB case with BT size equal to 3 times MTU always provides lowest access delay at downstream access nodes. Moreover this AAD remains acceptable (lower than 1 ms) even if the offered ring load is very high (87%).

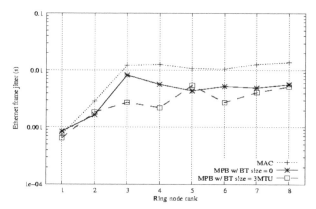

Figure 17.20 FJ versus ring node rank: load = 0.87.

Observing Figure 17.20, which plots the FJ versus ring node rank, we note that at high offered ring load, the FJ in MAC case exceeds 10 ms at many nodes, unsatisfying CES requirements. On the contrary the FJ in MPB cases still remains under 10 ms, again insuring CES performance requirements.

17.7 Conclusion

In this paper we have analysed two novel technologies for a metropolitan all-optical Ethernet ring network (DBORN): Modified Packet Bursting associated with Bursting Timer mechanism, and Circuit Emulation Service. The performance of MPB with BT mechanism has been carefully studied via simulation works. The paper also presents extended performance analysis on CES technology, which is a complementary contribution for the work in [4]. We observed that MPB is useful and always offers higher network performance than optical CSMA/CA protocol, under any traffic condition and network configuration. This is a simple mechanism but very efficient in improving network capacity and stability, notably in combination with Bursting Timer mechanism. Indeed, MPB with BT is easy to parameterize (only two parameters to control: BT size and BT time). The higher the value of BT size (or BT time), the higher the protocol efficiency, the lower the packet loss ratio, the higher the useful network load and the lower the effective network load. However we noticed that high value of BT size leads to the increase of access delay for electronic packets at upstream ring nodes as well.

The CES technology, with the help of MPB and BT mechanisms, can be realised with high reliability and offer superior quality to TDM service in DBORN, even under worse network configurations (e.g., high BE traffic volume, high network offered load, etc). For instance, with big enough value of BT time (or BT size) for MPB, we can increase the acceptable offered bus load to superior to 80% (e.g., 87%), while still satisfying QoS requirements for high priority TDM (voice, video) traffic.

The main purpose of this paper is not to perform an exhaustive study of the above technologies, but to demonstrate their performance, their feasibility and the interoperability between them. Therefore the present simulation works are restricted on some simple assumptions on traffic profiles: exponential source model for CoS2 traffic, IPP source model for BE traffic, uniform traffic partition on the bus, etc. We hope to present in our future work more pragmatic scenarios that could confirm the availability and reliability of these technologies. For instance, a next work could be to identify BT parameter values which provide the best network performance when the ring traffic repartition is not uniform (i.e., each node may have different values of BT parameters).

Acknowledgments

The authors would like to express their gratitude to the collaborators from TELECOM SudParis and Alcatel-Lucent for their many valuable technical discussions on the subject.

References

[1] N. Le Sauze, E. Dotaro, A. Dupas et al., DBORN: A Shared WDM Ethernet Bus Architecture for Optical Packet Metropolitan Network, Photonic in Switching, July 2002.
[2] IEEE Draft P802.17, draft 3.0, Resilient Packet Ring, December 2003.
[3] H. Frazier and H. Johnson, Gigabit Ethernet: From 100 to 1,000 Mbps, *IEEE Internet Computing*, vol. 3, no. 1, pp. 24–31, February 1999.
[4] V. H. Nguyên, M. Ben Mamoun, T. Atmaca et al., Performance evaluation of Circuit Emulation Service in a metropolitan optical ring architecture, in *Telecommunications and Networking – ICT 2004*, Lecture Notes in Computer Science, vol. 3124, Springer, Berlin/Heidelberg, pp. 1173–1182, 2004.
[5] N. Bouabdallah, N. Le Sauze, E. Dotaro and L. Ciavaglia, *TCARD*, OPTICOM 2003, Dallas, TX, October 2003.
[6] ATM Forum, Circuit Emulation Service Interoperability Specification 2.0, http://www.atmforum.com, January 1997.

[7] PWE3 – IETF working document, TDM Circuit Emulation over Packet-switched Network (CESoPSN), http://www.ietf.org/internet-drafts/draft-vainshtein-cesopsn-06.txt, June 2003.

[8] Metro Ethernet Forum (MEF), Requirements for circuit emulation services in metro ethernet networks, MEF working document, revision 3.0, April 2003.

[9] MPLS forum, http://www.mplsforum.org

[10] IP packet length distribution, http://www.caida.org/analysis/AIX/plein_hist, June 2002.

[11] V. H. Nguyên and T. Atmaca, Performance analysis of the Modified Packet Bursting mechanism applied to a metropolitan WDM ring architecture, in *Proceedings of IFIP Open Conference on Metropolitan Area Networks: Architecture, Protocols, Control and Management*, Technical Proceedings, HCMC Viet Nam, 11–13 April 2005, pp. 199–215, 2005.

18

A Generalized Markovian Queue and Its Applications to Performance Analysis in Telecommunications Networks

Ram Chakka[1], Tien Van Do[2] and Zsolt Pandi[2]

[1]*Meerut Institute of Engineering and Technology, Meerut 250005, India;
e-mail: ramchakka@yahoo.com*
[2]*Department of Telecommunications, Budapest University of Technology and
Economics, H-1117 Budapest, Hungary; e-mail: do@hit.bme.hu*

Abstract

In this paper the MM $\sum_{k=1}^{K}$CPP$_k$/GE/c/L G-queue (also termed as the *Sigma queue*) is introduced and proposed as a *generalized Markovian* queueing model for the performance analysis of telecommunications networks. An exact and computationally efficient solution is obtained for the steady-state probabilities and performance measures. Issues concerning the computational effort are also discussed.

In order to demonstrate the versability of the proposed queue, the application of the queue to two problems (the performance analysis of optical burst switching nodes and High Speed Downlink Packet Access) in telecommunication networks is demonstrated. The new queueing model is quite promising as a viable performance predictor.

Keywords: Sigma queue, generalized Markovian queueing model, optical burst switching, HSDPA, spectral expansion method.

*D. D. Kouvatsos (ed.), Performance Modelling and Analysis of Heterogeneous
Networks,* 371–387.

18.1 Introduction

A number of models are being developed for the traffic, service and queuing in the emerging telecommunication networks. The task is rather difficult because the traffic is bursty and correlated. These models include the compound Poisson process (CPP) [9], the Markov modulated Poisson process (MMPP) and self-similar traffic models such as the Fractional Brownian Motion (FBM) [14]. The CPP and the $\sum_{k=1}^{K} CPP_k$ traffic models often give a good representation of the burstiness (batch size distribution) of the traffic from one or more sources [7], but not the auto-correlations of the inter-arrival time observed in real traffic. Conversely, the MMPP models can capture the auto-correlations but not the burstiness [10, 15]. Self-similar traffic models such as the FBM can represent both burstiness and auto-correlations, but they are analytically intractable in a queuing context. The traffic arriving to a node is often the superposition of traffic from a number of sources (homogeneous or heterogeneous), which further complicates the analysis of the system.

In order to model such systems in the Markovian framework, a new traffic/queuing model, the Markov modulated $\sum_{k=1}^{K} CPP_k/GE/c/L$ G-queue is introduced in this paper. This is a homogeneous multi-server queue with c servers, GE (generalized exponential) service times. The arrivals are the superposition of K independent positive customers (referred simply as customers) and an independent negative customer streams.[1] Each of these arrival streams is a CPP, i.e. a Poisson point process with batch arrivals of geometrically distributed batch size. Furthermore, all the $K + 2$ independent GE distributions (K positive and one negative customer inter-arrival times and also the service time) are jointly modulated by a continuous time Markov phase process (CTMP), also termed as the modulating process. This MM $\sum_{k=1}^{K} CPP_k/GE/c/L$ G-queue and its extensions can capture a large class of traffic and queuing models applicable to telecommunication networks, in the Markovian framework.

We propose the MM $\sum_{k=1}^{K} CPP_k/GE/c/L$ G-Queue (the Sigma queue) and present a methodology for obtaining the steady state probabilities of the new queue. The methodology includes a series of non-trivial transformations of the balance equations in order to reach a computable form, that of the QBD-M (Quasi Simultaneous-Bounded-Multiple Births and simultaneous-Bounded-Multiple Deaths) type. As a consequence, existing methods (either

---

[1] Negative customers remove (positive) customers from the queue and have been used to model random neural networks, task termination in speculative parallelism, faulty components in manufacturing systems and server breakdowns [6].

the spectral expansion method [11] or an appropriate extension of Naoumov's method [12] for QBD processes based on the matrix-geometric solution approach [13]) can be employed to obtain the steady state probabilities.

Moreover, we illustrate the utility of the new queue by two applications, viz. the performance analysis of an Optical Burst Switching (OBS) and the High Speed Downlink Packet Access (HSDPA) protocol in UMTS networks.

The rest of the paper is organized as follows. The MM $\sum_{k=1}^{K} CPP_k/GE/c/L$ G-queue is described in Section 18.2. In Section 18.3, a solution technique is presented for obtaining the exact steady state probability distribution of the Markov process that represents the system. The overview of the application is carried out in Section 18.4, and the paper concludes in Section 18.5.

18.2 The MM $\sum_{k=1}^{K} CPP_k/GE/c/L$ G-Queue

18.2.1 The Modulating Process

The arrival and service processes are modulated by the same continuous time, irreducible Markov phase process with N states. Let Q be the generator matrix of this process, given by

$$
Q = \begin{bmatrix}
-q_1 & q_{1,2} & \cdots & q_{1,N} \\
q_{2,1} & -q_2 & \cdots & q_{2,N} \\
\vdots & \vdots & \ddots & \vdots \\
q_{N,1} & q_{N,2} & \cdots & -q_N
\end{bmatrix},
$$

where $q_{i,k}$ ($i \neq k$) is the instantaneous transition rate from phase i to phase k, and $q_i = \sum_{j=1}^{N} q_{i,j}$ ($i = 1, \ldots, N$). Let $\mathbf{r} = (r_1, r_2, \ldots, r_N)$ be the vector of equilibrium probabilities of the modulating phases. Then, $\mathbf{r}$ is uniquely determined by the equations $\mathbf{r}Q = 0$ and $\mathbf{r}\mathbf{e}_N = 1$, where $\mathbf{e}_N$ stands for the column vector with N elements, each of which is unity.

18.2.2 The Customer Arrival Process

The arrival process, in any given modulating phase, is the superposition of K independent CPP arrival streams of customers (or packets, in packet-switched networks) and an independent CPP of negative customers. Customers of different arrival streams are not distinguishable. The parameters of the GE inter-arrival time distribution of the kth ($1 \leq k \leq K$) customer arrival stream in phase i are $(\sigma_{i,k}, \theta_{i,k})$, and (ρ_i, δ_i) are those of the negative customers. That is, the inter-arrival time probability distribution function is $1-(1-\theta_{i,k})e^{-\sigma_{i,k}t}$, in phase i, for the kth stream of customers and $1-(1-\delta_i)e^{-\rho_i t}$ for the negative customers. Thus, all the $K+1$ arrival *point*-processes are batch-Poissons, with batches arriving at each point having geometric size distribution. Specifically, in phase i, the probability that a batch is of size s is $(1-\theta_{i,k})\theta_{i,k}^{s-1}$ for the kth stream of customers, and $(1-\delta_i)\delta_i^{s-1}$ for the negative customers.

Let $\sigma_{i,.}, \overline{\sigma_{i,.}}$ be the average arrival rate of customer batches and customers in phase i respectively. Let $\sigma, \overline{\sigma}$ be the overall average arrival rate of batches and customers respectively. Then,

$$\sigma_{i,.} = \sum_{k=1}^{K}\sigma_{i,k}, \quad \overline{\sigma_{i,.}} = \sum_{k=1}^{K}\frac{\sigma_{i,k}}{(1-\theta_{i,k})}, \quad \sigma = \sum_{i=1}^{N}\sigma_{i,.}r_i, \quad \overline{\sigma} = \sum_{i=1}^{N}\overline{\sigma_{i,.}}r_i.$$

$$(18.1)$$

18.2.2.1 Arrival Batch Size Distribution of Positive Customers

The overall arrival stream of customers is thus an MM $\sum_{k=1}^{K}$CPP$_k$. Because of the superposition of many CPPs, the overall arrivals in phase i can be considered as bulk-Poisson ($M^{[x]}$) with arrival rate $\sigma_{i,.}$ and with a batch size distribution $\{\pi_{i,l}\}$, that is more general than mere geometric. The probability that the batch size is l ($\pi_{i,l}$) and the overall batch size distribution ($\pi_{.,l}$), can be given by

$$\pi_{i,l} = \sum_{k=1}^{K}\frac{\sigma_{i,k}}{\sigma_{i,.}}(1-\theta_{i,k})\theta_{i,k}^{l-1}, \quad \pi_{.,l} = \sum_{i=1}^{N}r_i\pi_{i,l}. \qquad (18.2)$$

Clearly, by choosing K and other parameters appropriately it may be possible to approximate $\{\pi_{i,l}\}$ or $\{\pi_{.,l}\}$ to suit certain given classes of batch size distribution, however this is only an objective of further research.

18.2.3 The GE Multi-Server

The service facility has c homogeneous servers in parallel, each with GE-distributed service times with parameters (μ_i, ϕ_i) in phase i. The service discipline is FCFS and each server serves at most one positive customer at any given time. Negative customers neither wait in the queue, nor are served. The operation of the GE server is similar to that described for the CPP arrival processes above. L denotes the queuing capacity, in all phases, including the customers in service. L can be finite or infinite. We assume, when the number of customers is j and the arriving batch size of positive customers is greater than $L - j$ (assuming finite L), then only $L - j$ customers are taken in and the rest are rejected.

However, the batch size associated with a service completion is bounded by one more than the number of customers waiting to commence service at the departure instant. For queues of length $c \leq j < L + 1$ (including any customers in service), the maximum batch size at a departure instant is $j - c + 1$, only one server being able to complete a service period at any one instant under the assumption of exponentially distributed batch-service times. Thus, the probability that a departing batch has size s is $(1 - \phi_i)\phi_i^{s-1}$ for $1 \leq s \leq j - c$ and ϕ_i^{j-c} for $s = j - c + 1$. In particular, when $j = c$, the departing batch has size 1 with probability one, and this is also the case for all $1 \leq j \leq c$ since each customer is already engaged by a server and there are then no customers waiting to commence service.

It is assumed that the first positive customer in a batch arriving at an instant when the queue length is less than c (so that at least one server is free) *never* skips service, i.e. always has an exponentially distributed service time. However, even without this assumption the methodology described in this paper is still applicable.

18.2.4 Negative Customer Semantics

A negative customer removes a positive customer in the queue, according to a specified *killing discipline*. We consider here a variant of the RCE killing discipline (removal of customers from the end of the queue), where the most recent positive arrival is removed, but which does *not* allow a customer actually in service to be removed: a negative customer that arrives when there are no positive customers waiting to start service has no effect. We may say that customers in service are immune to negative customers or that the service itself is *immune servicing*. Such a killing discipline is suitable for modeling e.g. load balancing, where work is transferred from overloaded queues but never work, that is, actually in progress.

When a batch of negative customers of size l ($1 \leq l < j - c$) arrives, l positive customers are removed from the end of the queue leaving the remaining $j - l$ positive customers in the system. If $l \geq j - c \geq 1$, then $j - c$ positive customers are removed, leaving none waiting to commence service (queue length equal to c). If $j \leq c$, the negative arrivals have no effect.

$\overline{\rho_i}$, the average arrival rate of negative customers in phase i and $\overline{\rho}$, the overall average arrival rate of negative customers are given by

$$\overline{\rho_i} = \frac{\rho_i}{1 - \delta_i}, \quad \overline{\rho} = \sum_{i=1}^{N} r_i \overline{\rho_i}. \tag{18.3}$$

18.2.5 Condition for Stability

When L is finite, the system is ergodic since the representing Markov process is irreducible. Otherwise, i.e. when $L = \infty$, the overall average departure rate increases with the queue length, and its maximum (the overall average departure rate when the queue length tends to ∞) can be determined as

$$\overline{\mu} = c \sum_{i=1}^{N} \frac{r_i \mu_i}{1 - \phi_i}. \tag{18.4}$$

Hence, we conjecture that the necessary and sufficient condition for the existence of steady state probabilities is

$$\overline{\sigma} < \overline{\rho} + \overline{\mu}. \tag{18.5}$$

The above condition is obvious and intuitively appealing. However, we have neither a rigorous proof of the same nor a suitable reference to such a proof.

18.2.6 The Steady State Balance Equations

The state of the system at any time t can be specified completely by two integer-valued random variables, $I(t)$ and $J(t)$. $I(t)$ varies from 1 to N, representing the phase of the modulating Markov chain, and $0 \leq J(t) < L + 1$ represents the number of positive customers in the system at time t, including any in service. The system is now modeled by a continuous time discrete state Markov process, $\overline{Y}$ (Y if L is infinite), on a rectangular lattice strip. Let $I(t)$, the phase, vary in the horizontal direction and $J(t)$, the queue length or *level*, in the vertical direction. We denote the steady state probabilities by $\{p_{i,j}\}$, where $p_{i,j} = \lim_{t \to \infty} \text{Prob}(I(t) = i, J(t) = j)$, and let $\mathbf{v}_j = (p_{1,j}, \ldots, p_{N,j})$.

The process Y evolves due to the following instantaneous transition rates:

(a) $q_{i,k}$ – purely lateral transition rate – from state (i, j) to state (k, j), for all $j \geq 0$ and $1 \leq i, k \leq N$ $(i \neq k)$, caused by a phase transition in the Markov chain governing the arrival phase process $(q_{i,i} = 0)$;

(b) $B_{i,j,j+s}$ – s-step upward transition rate – from state (i, j) to state $(i, j+s)$, for all phases i, caused by a new batch arrival of size s positive customers. For a given j, s can be seen as bounded when L is finite and unbounded when L is infinite;

(c) $C_{i,j,j-s}$ – s-step downward transition rate – from state (i, j) to state $(i, j - s)$, $(j - s \geq c + 1)$ for all phases i, caused by either a batch service completion of size s or a batch arrival of negative customers of size s;

(d) $C_{i,c+s,c}$ – s-step downward transition rate – from state $(i, c + s)$ to state (i, c), for all phases i, caused by a batch arrival of negative customers of size $\geq s$ or a batch service completion of size s $(1 \leq s \leq L - c)$;

(e) $C_{i,c-1+s,c-1}$ – s-step downward transition rate, from state $(i, c - 1 + s)$ to state $(i, c - 1)$, for all phases i, caused by a batch departure of size s $(1 \leq s \leq L - c + 1)$;

(f) $C_{i,j+1,j}$ – 1-step downward transition rate, from state $(i, j + 1)$ to state (i, j), $(c \geq 2; 0 \leq j \leq c - 2)$, for all phases i, caused by a single departure;

where

$$B_{i,j-s,j} = \sum_{k=1}^{K} (1 - \theta_{i,k})\theta_{i,k}^{s-1}\sigma_{i,k} \quad (\forall i; 0 \leq j - s \leq L - 2; j - s < j < L);$$

$$B_{i,j,L} = \sum_{k=1}^{K} \sum_{s=L-j}^{\infty} (1 - \theta_{i,k})\theta_{i,k}^{s-1}\sigma_{i,k} = \sum_{k=1}^{K} \theta_{i,k}^{L-j-1}\sigma_{i,k} \quad (\forall i; \ j \leq L - 1);$$

$$C_{i,j+s,j} = (1 - \phi_i)\phi_i^{s-1}c\mu_i + (1 - \delta_i)\delta_i^{s-1}\rho_i$$
$$(\forall i; \ c + 1 \leq j \leq L - 1; \ 1 \leq s \leq L - j);$$
$$= (1 - \phi_i)\phi_i^{s-1}c\mu_i + \delta_i^{s-1}\rho_i \quad (\forall i; \ j = c; \ 1 \leq s \leq L - c);$$
$$= \phi_i^{s-1}c\mu_i \quad (\forall i; \ j = c - 1; \ 1 \leq s \leq L - c + 1);$$
$$= 0 \quad (\forall i; \ c \geq 2; \ 0 \leq j \leq c - 2; \ s \geq 2);$$
$$= (j + 1)\mu_i \quad (\forall i; \ c \geq 2; \ 0 \leq j \leq c - 2; \ s = 1).$$

Define

$$B_{j-s,j} = \text{Diag}\left[B_{1,j-s,j}, B_{2,j-s,j}, \ldots, B_{N,j-s,j}\right] \quad (j-s < j \le L);$$

$$B_s = B_{j-s,j} \quad (j < L)$$

$$= \text{Diag}\left[\ldots, \sum_{k=1}^{K} \sigma_{i,k}(1-\theta_{i,k})\theta_{i,k}^{s-1}, \ldots\right];$$

$$\Sigma_k = \text{Diag}\left[\sigma_{1,k}, \sigma_{2,k}, \ldots, \sigma_{N,k}\right] \quad (k = 1, 2, \ldots, K);$$

$$\Theta_k = \text{Diag}\left[\theta_{1,k}, \theta_{2,k}, \ldots, \theta_{N,k}\right] \quad (k = 1, 2, \ldots, K);$$

$$\Sigma = \sum_{k=1}^{K} \Sigma_k; \quad R = \text{Diag}\left[\rho_1, \rho_2, \ldots, \rho_N\right];$$

$$\Delta = \text{Diag}\left[\delta_1, \delta_2, \ldots, \delta_N\right];$$

$$M = \text{Diag}\left[\mu_1, \mu_2, \ldots, \mu_N\right]; \quad \Phi = \text{Diag}\left[\phi_1, \phi_2, \ldots, \phi_N\right];$$

$$C_j = jM \quad (0 \le j \le c);$$

$$= cM = C \quad (j \ge c);$$

$$C_{j+s,j} = \text{Diag}\left[C_{1,j+s,j}, C_{2,j+s,j}, \ldots, C_{N,j+s,j}\right]; \quad E = \text{Diag}(\mathbf{e}'_N).$$

Then, we get

$$B_s = \sum_{k=1}^{K} \Theta_k^{s-1}(E-\Theta_k)\Sigma_k; \quad B_1 = B = \sum_{k=1}^{K}(E-\Theta_k)\Sigma_k;$$

$$B_{L-s,L} = \sum_{k=1}^{K} \Theta_k^{s-1}\Sigma_k;$$

$$C_{j+s,j} = C(E-\Phi)\Phi^{s-1} + R(E-\Delta)\Delta^{s-1}$$
$$(c+1 \le j \le L-1; \ s = 1, 2, \ldots, L-j);$$

$$= C(E-\Phi)\Phi^{s-1} + R\Delta^{s-1} \quad (j = c; \ s = 1, 2, \ldots, L-c);$$

$$= C\Phi^{s-1} \quad (j = c-1; \ s = 1, 2, \ldots, L-c+1);$$

$$= 0 \quad (c \ge 2; \ 0 \le j \le c-2; \ s \ge 2);$$

$$= C_{j+1} \quad (c \ge 2; \ 0 \le j \le c-2; \ s = 1).$$

The steady state balance equations are

$$\sum_{s=1}^{L} \mathbf{v}_{L-s} B_{L-s,L} + \mathbf{v}_L \left[Q - C - R \right] = 0;$$ (18.6)

$$\sum_{s=1}^{j} \mathbf{v}_{j-s} B_s + \mathbf{v}_j \left[Q - \Sigma - C_j - R I_{j>c} \right] + \sum_{s=1}^{L-j} \mathbf{v}_{j+s} C_{j+s,j} = 0;$$ (18.7)

$$(0 \le j \le L - 1)$$

$$\sum_{j=0}^{L} \mathbf{v}_j \mathbf{e}_N = 1,$$ (18.8)

where $I_{j>c} = 1$ if $j > c$ else 0, and $\mathbf{e}_N$ is a column vector of size N with all ones.

18.3 Solution Methodology and Technique

18.3.1 Transforming the Balance Equations

When $L < c + K + 4$, then the number of the states of the Markov process $\overline{Y}$ is not large and can be solved by traditional methods [16]. However, when L is large, computationally more efficient other methods are available.

In this section the balance equations are essentially transformed into a computable form of the QBD-M type. The necessary mathematical transformations are based on profound theoretical analysis and proofs [3]. Moreover, they are convenient for software implementation.

Let the balance equations for level j be denoted by $\langle j \rangle$. Hence, $\langle 0 \rangle$, $\langle 1 \rangle$, ..., $\langle j \rangle$, ..., $\langle L \rangle$ are the balance equations for the levels $0, 1, \ldots, j, \ldots, L$ respectively. Substituting $B_{L-s,L} = \sum_{k=1}^{K} \Theta_k^{s-1} \Sigma_k$ and $B_s = \sum_{k=1}^{K} \Theta_{.k}^{s-1} (E - \Theta_k) \Sigma_k$ in (18.6, 18.7), we get the balance equations for level L and for all the other levels as:

$\langle L \rangle$:

$$\sum_{s=1}^{L} \sum_{k=1}^{K} \mathbf{v}_{L-s} \Theta_k^{s-1} \Sigma_k + \mathbf{v}_L \left[Q - C - R \right] = 0$$ (18.9)

$$\vdots$$

$\langle j \rangle$:

$$\sum_{s=1}^{j} \sum_{k=1}^{K} \mathbf{v}_{j-s} \Theta_k^{s-1} (E - \Theta_k) \Sigma_k$$

$$+ \mathbf{v}_j \left[Q - \Sigma - C_j - RI_{j>c} \right] + \sum_{s=1}^{L-j} \mathbf{v}_{j+s} C_{j+s,j} = 0$$

$$(j = L - 1, \ldots, 0). \tag{18.10}$$

Define the functions, $F_{k,l}$ $(k = 1, 2, \ldots, K; l = 1, 2, \ldots, k)$, recursively as

$$F_{k,0} = E, \quad F_{k,k} = \prod_{n=1}^{k} \Theta_n \quad (k = 1, 2, \ldots, K);$$

$$F_{k,l} = 0 \quad (k = 1, 2, \ldots, K; \ l < 0);$$

$$F_{k,l} = 0 \quad (k = 1, 2, \ldots, K; \ l > k);$$

$$F_{1,0} = E; \quad F_{1,1} = \Theta_1;$$

$$F_{k,l} = F_{k-1,l} + \Theta_k F_{k-1,l-1} \quad (2 \le k \le K, \ 1 \le l \le k - 1). \tag{18.11}$$

Transformation 1. *Modify simultaneously the balance equations for levels j $(L - 1 \ge j \ge K)$, by the transformation:*

$$\langle j \rangle^{(1)} \longleftarrow \langle j \rangle + \sum_{l=1}^{K} (-1)^l \langle j - l \rangle F_{K,l} \quad (K \le j \le L - 1);$$

$$\langle j \rangle^{(1)} \longleftarrow \langle j \rangle \quad (j = L \ or \ j < K).$$

Apply the following transformation to the resulting equations:

Transformation 2. *Modify simultaneously the balance equations for levels j $(L - 3 \ge j \ge c + K + 1)$, by the transformation:*

$$\langle j \rangle^{(2)} \longleftarrow \langle j \rangle^{(1)} - \langle j + 1 \rangle^{(1)} (\Phi + \Delta) + \langle j + 2 \rangle^{(1)} \Phi \Delta$$

$$(c + K + 1 \le j \le L - 3);$$

$$\langle j \rangle^{(2)} \longleftarrow \langle j \rangle^{(1)} \quad (j > L - 3 \ or \ j < c + K + 1).$$

After these two transformations, the transformed balance ($\langle j \rangle^{(2)}$) equations for the rows ($c + K + 1 \leq j \leq L - 3$), will be of the form:

$$\mathbf{v}_{j-K} Q_0 + \mathbf{v}_{j-K+1} Q_1 + \ldots + \mathbf{v}_{j+2} Q_{K+2} = 0$$
$$(j = L - 3, L - 4, \ldots, c + K + 1), \tag{18.12}$$

where $Q_0, Q_1, \ldots, Q_{K+2}$ are $K + 3$ number of j-*independent* matrices and can be easily derived and computed.

Thus, the resulting equations (18.12) corresponding to the rows from $c + K + 1$ to $L - 3$, are of the same form as those of the QBD-M processes and hence have an efficient solution by several alternative methods such as the spectral expansion method, Bini-Meini's method or the matrix-geometric method with folding or block size enlargement [8].

18.3.2 Spectral Expansion Solution of the Balance Equations

The set of equations (18.12) concerning the levels $c + K + 1$ to $L - 3$ have the coefficient matrices $Q_0, Q_1, \ldots, Q_{K+2}$ that are independent of j and hence have an efficient solution by the spectral expansion method [11]. These Q_l's ($l = 0, 1, \ldots, K + 2$) can be obtained quite easily following the computational procedure.

Define the matrix polynomials $Z(\lambda)$ and $\overline{Z}(\xi)$ as

$$Z(\lambda) = Q_0 + Q_1\lambda + Q_2\lambda^2 + \cdots + Q_{K+2}\lambda^{K+2}, \tag{18.13}$$

$$\overline{Z}(\xi) = Q_{K+2} + Q_{K+1}\xi + Q_K\xi^2 + \cdots + Q_0\xi^{K+2}. \tag{18.14}$$

Then the spectral expansion solution for $\mathbf{v}_j$ ($c + 1 \leq j \leq L - 1$) is given by

$$\mathbf{v}_j = \sum_{l=1}^{KN} a_l \boldsymbol{\psi}_l \lambda_l^{j-c-1} + \sum_{l=1}^{2N} b_l \boldsymbol{\gamma}_l \xi_l^{L-1-j} \quad (c + 1 \leq j \leq L - 1), \tag{18.15}$$

where λ_l ($l = 1, 2, \ldots, KN$) are the KN eigenvalues of least absolute value of the matrix polynomial $Z(\lambda)$ and ξ_l ($l = 1, 2, \ldots, 2N$) are the $2N$ eigenvalues of least absolute value of the matrix polynomial $\overline{Z}(\xi)$. $\boldsymbol{\psi}_l$ and $\boldsymbol{\gamma}_l$ are the left-eigenvectors of $Z(\lambda)$ and $\overline{Z}(\xi)$ respectively, corresponding to the eigenvalues λ_l and ξ_l respectively. a_l's and b_l's are arbitrary constants to be determined later.

The matrix $\sum_{l=0}^{K+2} Q_l$ can be proved to be singular, so that $\lambda = 1$ is an eigenvalue on the unit-circle for both $Z(\lambda)$ and $\overline{Z}(\xi)$. If (18.5) is satisfied, the number of eigenvalues of $Z(\lambda)$ that are strictly within the unit circle is KN. If (18.5) is not satisfied, this number is $KN - 1$. These and also certain other properties of these eigenvalues, Eigenvectors, also the relevant spectral analysis are dealt with (some of them are proved, others explained in detail) in [2,11]. Some of them are summarized below. Let the rank of Q_0 be $N - n_0$ and that of Q_{K+2} be $N - n_{K+2}$. Then,

(a) $Z(\lambda)$ would have $d = (K + 2)N - n_{K+2}$ eigenvalues of which n_0 are zero eigenvalues (also referred to as null eigenvalues), whereas $\overline{Z}(\xi)$ would have n_{K+2} zero eigenvalues and $(K + 2)N - n_0 - n_{K+2}$ non-zero eigenvalues.

(b) If $(\lambda \neq 0, \psi)$ is a non-zero eigenvalue-eigenvector pair of $Z(\lambda)$, then there exists a corresponding non-zero eigenvalue-eigenvector pair, $(\xi = 1/\lambda, \gamma = \psi)$ for $\overline{Z}(\xi)$. Thus, the non-zero eigenvalues of these two matrix polynomials are mutually reciprocal.

(c) The KN eigenvalues of least absolute value of $Z(\lambda)$ and the $2N$ eigenvalues of least absolute value of $\overline{Z}(\xi)$ lie either strictly inside, or on, their respective unit-circles, but not outside.

(d) There is no problem posed by multiple eigenvalues, i.e. independent eigenvectors having coincident eigenvalues, since each pair (λ, ψ) is *distinct*.

If the unknowns a_l's and b_l's are determined in such a way that all the balance equations are satisfied, then the vectors v_j $(c+1 \leq j \leq L-1)$ can be computed from the steady state solution (18.15). Hence, the unknowns are the scalars, $a_1, a_2, \ldots, a_{K \cdot N}, b_1, b_2, \ldots, b_{2N}$, and the vectors $v_0, v_1, \ldots, v_c, v_L$. These are totally, $(c + 2)N + K \cdot N + 2N = (c + K + 4)N$ scalar unknowns. In order to solve for them, we still have the transformed balance equations concerning the levels $0, 1, \ldots, c+K, L-2, L-1, L$ and also equation (18.8). These are $(c + K + 4)N + 1$ linear simultaneous equations in the above $(c + K + 4)N$ scalar unknowns. Of these equations only $(c + K + 4)N$ equations (including equation (18.8)) are independent. Hence, all these unknowns can be solved for. The number of linear simultaneous equations can be reduced further by following a similar simplification as in [4]. The computational time of our algorithm does not depend on L, is of the order $O(K^3 \cdot N^3)$. We believe Naomov's extended matrix-geometric algorithm would also be of the same order. The spectral expansion method was applied because its implementation was readily available, and the comparison of these two standard methods is out of the scope of the present paper. For some detailed comparison with respect to other modeling problems one may look into [2, 8, 17].

Required performance measures such as the average number of customers in the system, customer-loss probability, loss caused by negative customer, mean response times can be derived and computed quite easily from $\mathbf{v}_j$'s (see [3]). These are exact computations and we have verified that fact by simulation.

18.3.3 System with Infinite Queuing Capacity

Similar analysis when the queuing capacity is *unbounded*, gives

$$\mathbf{v}_j = \sum_{l=1}^{KN} a_l \boldsymbol{\psi}_l \lambda_l^{j-c} \quad (j = c+1, c+2, \ldots), \qquad (18.16)$$

which also tallies with equation (18.15). Obviously, the required computational effort is even less in this case.

18.4 Applications

In this section, we overview the applications of the proposed queue to the performance analysis of two problems in telecommunication networks. The first problem is the performance analysis of an OBS multiplexer, the second problem is the performance evaluation of the High Speed Downlink Packet Access protocol in wireless networks.

18.4.1 Optical Burst Switching

Optical Burst Switching [18] is an attempt to unify the advantages of circuit and packet switching paradigms and to exploit the capabilities of state-of-the-art optical networking technology. In an OBS network there are two types of nodes: edge nodes that interface with other networks and core nodes that can only be connected to other edge or core nodes. Edge nodes receive and buffer data coming from outside the OBS network and construct data bursts that contain data to be forwarded to other edge nodes. When a data burst is ready for transmission a Burst Header Cell (BHC) is injected to the OBS network that informs the core nodes of the arrival of the burst so that they could make the necessary resource allocations. After a certain amount of time indicated in the BHC the data burst is also injected into the network. Optimally, it simply cuts through the network and arrives at the destination edge node, where it is decomposed into data units that are forwarded to the interfacing network.

Optical switching nodes, like core OBS nodes, typically have incoming and outgoing fiber links, each of which has a number of wavelengths for carrying data bursts and a distinguished wavelength for carrying control information, such as BHC's. In order to avoid resource allocation conflicts of data bursts, which is often termed as contention in the optical networking context, and to facilitate more effective resource usage, core nodes may use buffers for temporary storage of data bursts. These buffers, or burst storage locations are most often implemented by Fiber Delay Loops (FDLs). An FDL is a span of optical fiber of certain length wound into a coil. An FDL causes a fixed amount of delay to the signal transmitted over it, and thus an appropriate combination of FDLs might be used as a buffer with limited capabilities. It is worth emphasizing that due to the technology limitations of optical buffering, bursts cannot wait for an indefinite amount of time in an FDL buffer in contrast to, e.g., electronic buffers in ATM switches where cells can wait almost as long as the control function allows. Thus, the packets in the FDLs may get lost after certain period of waiting for service. We have used negative customers in our queuing model, in an innovative way, to reflect and account for the loss of packets due to this phenomenon. The detailed study of this problem is presented in [3].

18.4.2 High Speed Downlink Packet Access

High Speed Downlink Packet Access (HSDPA) was introduced by the 3rd Generation Partnership Project (3GPP) to satisfy the demands for high speed data transfer in the downlink direction in UMTS networks. It can offer peak data rates of upto 10 Mbps, which is achieved essentially by the use of Adaptive Modulation and Coding (AMC), extensive multicode operation and a retransmission strategy [1].

In the implementation of HSDPA, several channels are introduced. The transport channel carrying the user data, in HSDPA operation, is called the High-Speed Downlink Shared Channel (HS-DSCH). The High-Speed Shared Control Channel (HS-SCCH), used as the downlink (DL) signaling channel, carries key physical layer control information to support the demodulation of the data on the HS-DSCH.

The uplink (UL) signaling channel, called the High-Speed Dedicated Physical Control Channel (HS-DPCCH), conveys the necessary control data in the UL to Node B (Node B is responsible for the transmission and reception of data across the radio interface). User Equipment sends feedback information about the received signal[2] quality on HS-DPCCH. That is, the UE calculates the DL Channel Quality Indicator (CQI) based on the received signal quality measured at the UE. Then, it sends the CQI on the HS-DPCCH channel to indicate which estimated transport block size, modulation type and number of parallel codes (i.e., physical channels) could be received correctly with reasonable block error rate in the DL. The CQI is integer valued, with a range between 0 and 30. The higher the CQI is, the better the condition of the channel and the more information can be transmitted.

To enable a large dynamic range of the HSDPA link adaptations and to maintain a good spectral efficiency, a user may simultaneously until up to 15 codes (physical channels) in parallel. The available code resources are primarily shared in the time domain but it is possible to share the code resources using code multiplexing too. The model and numerical evaluation of the adaptive channel coding versus the channel quality and the allocation of the physical channels are presented in [5].

18.4.3 Computational Effort

To compare the computational effort of the proposed solution approach, the steady state balance equations (18.6), (18.7) and the normalization equation (18.8) are also solved as a general system of linear equations using LU factorization [16].

Figure 18.1 demonstrates the trade-off in computation time between the two solution methods. Both implementations are run on a SUN Ultra 60 Workstation for different values of L and K, while c is always 3. The displayed times are averaged over five runs. Our solution outperforms the traditional method over a broad range of parameters, and its computation time does not depend on the L, where as the computation time of the latter does increase with L.

The computational effort of our model can be reduced further by reducing the number of linear simultaneous equations to be solved which can be done by following a similar simplification as in [17].

---

[2] In wireless communications, the quality of a received signal depends on a number of factors – the distance between the target and interfering base stations, the path-loss exponent, shadowing, channel-fading and noise.

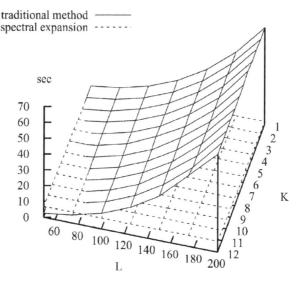

Figure 18.1 Runtimes of different solutionmethods for different input parameters.

18.5 Conclusions

One of the aims of research in the performance evaluation of telecommunications networks is to find analytically tractable queuing models with the capability of capturing the burstiness and correlation of the traffic. In this context, our first contribution is the introduction of the *Sigma* queue, its exact and efficient solution. It is explained how this queuing model can capture the burstiness and correlations of the traffic. It can also accommodate environmentally sensitive service times.

On the practical side, we show the application of the new queue to two problems in telecommunication networks. Our model is found to be accurate, when comparison with simulation is done using traces from the Internet Traffic Archive.

We believe our model has potential to evolve into a good performance evaluation tool in telecommunications networks.

References

[1] 3GPP Technical Report 25.848, version 4.0.0, Physical layer aspects of UTRA High Speed Downlink Packet Access, March 2001.

[2] R. Chakka, Performance and reliability modelling of computing systems using spectral expansion, PhD Thesis, University of Newcastle upon Tyne, 1995.

[3] R. Chakka, T. V. Do and Z. Pandi, Generalized Markovian queues and applications in performance analysis in telecommunication networks, in *Proceedings of the First International Working Conference on Performance Modelling and Evaluation of Heterogeneous Networks (HET-NETs 03)*, Ilkley, UK, July 21–23, 2003, pp. 60/1–10, 2003.

[4] R. Chakka and P. G. Harrison, A Markov modulated multi-server queue with negative customers – The MM CPP/GE/c/L G-queue, *Acta Informatica*, vol. 37, pp. 881–919, 2001.

[5] T. V. Do, R. Chakka and P. G. Harrison, An integrated analytical model for computation and comparison of the throughputs of the UMTS/HSDPA user equipment categories, in *MSWiM'07: Proceedings of the 10th ACM Symposium on Modeling, Analysis, and Simulation of Wireless and Mobile Systems*, ACM, New York, pp. 45–51, 2007.

[6] J. M. Fourneau, E. Gelenbe and R. Suros, G-networks with multiple classes of positive and negative customers, *Theoretical Computer Science*, vol. 155, pp. 141–156, 1996.

[7] R. J. Fretwell and D. D. Kouvatsos, ATM traffic burst lengths are geometrically bounded, in *Proceedings of the 7th IFIP Workshop on Performance Modelling and Evaluation of ATM & IP Networks*, Antwerp, Belgium, pp. 1/12–12/12, Chapman and Hall, 1999.

[8] B. Haverkort and A. Ost, Steady state analyses of infinite stochastic Petri nets: A comparison between the spectral expansion and the matrix geometric methods , in *Proceedings of the 7th International Workshop on Petri Nets and Performance Models*, Saint Malo, France, pp. 335–346, 1997.

[9] D. Kouvatsos, Entropy maximisation and queueing network models, *Annals of Operations Research*, vol. 48, pp. 63–126, 1994.

[10] K. S. Meier-Hellstern, The analysis of a queue arising in overflow models, *IEEE Transactions on Communications*, vol. 37, pp. 367–372, 1989.

[11] I. Mitrani and R. Chakka, Spectral expansion solution for a class of Markov models: Application and comparison with the matrix-geometric method, *Performance Evaluation*, vol. 23, pp. 241–260, 1995.

[12] V. Naoumov, U. Krieger and D. Wagner, Analysis of a multi-server delay-loss system with a general Markovian arrival process, in *Matrix-Analytical Methods in Stochastic Models*, S. R. Chakravarthy and A. S. Alfa (Eds.), Marcel Dekker, pp. 43–66, 1997.

[13] M. Neuts, *Matrix-Geometric Solutions in Stochastic Models: An Algorithmic Approach*, Dover Publications, 1995.

[14] I. Norros, A storage model with self-similar input, *Queueing Systems and Their Applications*, vol. 16, pp. 387–396, 1994.

[15] S. Shah-Heydari and T. Le-Ngoc, Mmpp models for multimedia traffic. *Telecommunication Systems*, vol. 15, nos. 3–4, pp. 273–293, 2000.

[16] W. J. Stewart, *Introduction to Numerical Solution of Markov Chains*. Princeton University Press, 1994.

[17] H. T. Tran and T. V. Do, Computational aspects for steady state analysis of QBD processes, *Periodica Polytechnica, Ser. El. Eng.*, vol. 44, no. 2, pp. 179–200, 2000.

[18] J. S. Turner, Terabit burst switching, *Journal of High Speed Networks*, vol. 8, pp. 3–16, 1999.

19

A Novel Issue for the Design of Access Interfaces in All Optical Slotted Networks

Mohamad Chaitou, Gérard Hébuterne and Hind Castel

Institut National des Télécommunications, 9 Rue Charles Fourrier, 91011 Evry, France; {mohamad.chaitou, gerard.hebuterne, hind.castel}@int-evry.fr

Abstract

In this contribution, we consider a novel approach for efficiently supporting IP packets directly into a slotted optical wavelength-division-multiplexing (WDM) layer with several quality of service (QoS) requirements. The approach is based on aggregating IP packets, regardless of their final destinations, at fixed time intervals before the optical conversion phase. A QoS support access mechanism based on the strict priority discipline is evaluated by means of an analytical model. In this case, IP packet aggregation is performed in a loop manner by always beginning the aggregation cycle with the highest priority class. The aggregation cycle ends if the aggregate packet cannot accommodate more IP packets, or if the lowest priority class is reached. In order to overcome the drawbacks of the strict priority discipline, an algorithm depicting a probabilistic priority discipline is presented. The aggregation technique leads to increasing the filling ratio of an optical packet, and consequently, the bandwidth efficiency of an optical network, due to the multicast nature of the aggregation technique. In this context, the support of multicast in metropolitan area networks (MANs), with ring architectures, is presented. Furthermore, a new architecture is proposed in order to support the multicast aggregation technique in wide area networks (WANs).

Keywords: Optical networking, packet aggregation, bandwidth efficiency, Quality of Service (QoS), traffic engineering.

D. D. Kouvatsos (ed.), Performance Modelling and Analysis of Heterogeneous Networks, 389–415.

19.1 Introduction

Recently, the explosive growth of the Internet protocol (IP)-based traffic has accelerated the development of high-speed transmission systems and the emergence of wavelength division multiplexing (WDM) technology [1] that, in the near future, will support hundreds of wavelengths of several gigabits per second each. In a long-term scenario, the optical packet switching (OPS), based on fixed-length packets and synchronous node operation, can provide a simple transport platform based on a direct IP over WDM structure which can offer high bandwidth efficiency, flexibility, and fine granularity [2]. An example of optical fixed packet size engineering can be found in [3]. Two major challenges face the application of packet switching in an optical domain. First, the adaptation of IP traffic, which mainly consists of asynchronous and variable length packets, with the considered synchronous OPS network. Second, the handling of QoS requirements in the context of a multi-service packet network. To cope with the first problem, IP packet aggregation at the interface of the optical network (e.g., [4–6]) presents an efficient solution among few other proposals in literature (e.g., [7,8]). This is because in the current OPS technology, a typical guard time of 50 ns must be inserted between optical packets [9]. This requires that optical packets must be long enough to overcome the resulting link efficiency problems and hence, a possible issue is the aggregation of several IP packets into a single electronic macro-packet with fixed size, called an aggregate packet. Furthermore, it is necessary to perform the aggregation process regardless of the destinations of IP packets. This is due to the permanent increase in the number of IP networks, and consequently, in the number of destinations, which leads to a poor filling ratio of the optical packet if the aggregation process is performed by destination (i.e., IP packets with same destinations are aggregated together).

The QoS problem is treated by adopting a class-based scheme in the edge nodes, which simplifies the core of the optical network by pushing the complexity towards the edge nodes.

According to the above discussion, this paper proposes an aggregation technique where, variable-length and multi-CoS IP packets are aggregated into the payload of an optical packet, regardless of their final destinations. The idea consists of the separation of IP traffic into $\{J \geq 2\}$ prioritized queues according to the desired CoS. At each fixed interval of time (τ), an aggregation cycle begins and an aggregate packet is constructed from several IP packets belonging to the different queues. In addition, we apply an aggregation priority mechanism by collecting IP packets, at the beginning of each aggregation cycle, from higher priority queues before those of lower priority. However, in order to relieve the drawbacks of such strict priority discipline, an hybrid version of the probabilistic priority algorithm presented in [10, 11] may be used. Since the size of an IP packet at the head of a queue i, $\{1 \leq i \leq J\}$ may be greater than the gap remaining into the aggregate packet, the IP packet can be segmented in this case.

The proposed aggregation technique exhibits two particular characteristics. First, an aggregate packet is generated at regular time intervals, which may vary dynamically in order to sustain a prefixed amount of the filling ratio. Second, a multicast optical packet is constructed. We present possible applications of such aggregation method in MANs and WANs respectively. As candidate applications in MANs, we mention the family of slotted ring networks deploying destination stripping such as the network studied in [12]. In addition, a broadcast network, which is a slotted version of the multi-channel packet-switched network called DBORN (Dual Bus Optical Ring Network) [13], represents an important application. This is because DBORN matches very well the multicast nature of the generated optical packets without the addition of any complexity in the node architecture. Furthermore, DBORN, coupled with the aggregation technique, can be adapted to use an access scheme based on TDMA, but avoids the lack of efficiency exhibited by the latter in the case of unbalanced traffic.

The present paper is organized as follows. The aggregation technique is analyzed in Section 19.2. In Section 19.3 we show how to extend the analytical model to J, $\{J > 2\}$, CoS. Section 19.4 describes the algorithm of the probabilistic priority discipline. In Section 19.5, some numerical examples are presented and commented. Sections 19.6 and 19.7 explain how to apply the aggregation technique to MANs and WANs, respectively. Finally, Section 19.8 concludes the paper.

19.2 Mathematical Model

The aggregation mechanism operates as follows. Let there be J classes of packets (throughout the paper the term "packet" stands for "IP packet"), where packets with a smaller class number have a higher priority than packets with a larger class number. Each class of packets has its own queue and the buffer of the queue is infinite. Packets in the same queue are served in FCFS fashion. Each packet is modeled by a batch of blocks having a fixed size of b bytes. Let X be the batch size random variable with probability generating function (PGF) $X(z)$, and probability mass function (pmf)

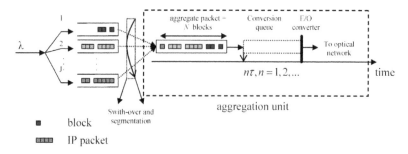

Figure 19.1 The aggregation mechanism at the optical interface.

$\{x_n = P(X = n), n \geq 1\}$. The size of the aggregate packet is fixed to N blocks ($N > \max(X)$), and a timer with a time-out value τ is implemented as shown in Figure 19.1. At each timer expiration (i.e at instants $\{n\tau, n = 0, 1, 2, \ldots\}$) the aggregation unit takes min(N, the whole queue 1 length) blocks to attempt filling the aggregate packet. If a gap still remains, the aggregation unit attempts to fill it from queue 2, then from queue 3, $\ldots$, until the aggregate packet becomes full or until queue J is reached. If the whole of the IP packet cannot be inserted (e.g., the packet at the head of queue J in Figure 19.1), only a part of it, needed to fill the aggregate packet, is transferred to the aggregation unit (a segmentation interface performs packets segmentation just before the aggregation unit). The aggregate packet is then sent to a queue (called conversion queue) preceding the stage of the electronic to optical conversion (E/O), and a new aggregation cycle is performed by polling queue 1 again. Note that in the discipline described above, it may take several times for the aggregation unit to switch from one queue to the other to actually perform packets aggregation in each aggregation cycle. Throughout the rest of this paper, we assume that the switch-over time can be much small as compared to the sojourn time and will not be taken into account in the analysis. It is worthwhile mentioning that the discrimination between successive IP packets, contained in an aggregate packet, is performed by using a delineation protocol such as the protocol proposed in [14].

In the analytical study, we first begin by considering only two classes of packets ($J = 2$). However, the analysis can be extended easily to a multi-class system, as it will be shown later. The following assumptions and notations are used throughout this paper. We assume that packets arrive at the corresponding queues according to independent Poisson processes with rates λ_1 and λ_2, i.e. the total arrival rate is $\lambda_0 = \lambda_1 + \lambda_2$. We define $\{A_t^c, c = 0, 1, 2\}$ as the number of blocks arriving at queue c, (for $c = 0$ the queue corresponds to the combination of queues 1 and 2), during an interval of time t, and we design by $\{A_i^c(z) = e^{\lambda_c t(X(z)-1)}\}$ its PGF and $\{a_i^c(n), n \geq 0\}$ its pmf.

19.2.1 Blocks Number Pre-Departure Probabilities and Filling Ratio

We define $\{Y^c(t), c = 0, 1, 2\}$ by the number of blocks in queue c at time t, and we suppose that $Y_n^c = Y^c(t_n^-)$. We choose a set of embedded Markov points as those points in time which are just before timer expirations. Let $t_0, t_1, \ldots, t_n, \ldots$, be the epochs of timer expirations. Since the whole system and queue 1 behave in a similar way, the steady state distribution for $\{Y_n^c, n = 0, 1, 2, \ldots\}$ is obtained by the same manner for $\{c = 0, 1\}$:

$$y_k^c = \lim_{n \to \infty} P(Y_n^c = k), \qquad k \geq 0$$

The following state equation holds:

$$Y_{n+1}^c = |Y_n^c - N|^+ + A_\tau^c \tag{19.1}$$

where $|c|^+$ denotes $\max(0, c)$. The equilibrium queue length distribution (in blocks number) at an arbitrary time epoch is then described by the probability generating function $Y^c(z)$, which can be derived from (19.1) in a straightforward and well-known fashion. It is given by (see Appendix A)

$$Y^c(z) = \frac{A_\tau^c(z)(z-1)(N - E[A_\tau^c])}{z^N - A_\tau^c(z)} \prod_{k=1}^{N-1} \frac{z - z_k}{1 - z_k} \tag{19.2}$$

where $z_1, z_2, \ldots, z_{N-1}$ are the $N - 1$ zeros of $z^N - A_\tau^c(z)$ inside the unit circle of the complex plane, and $E[\ldots]$ is the expectation value of the expression between square brackets. Using the inverse fast fourier transform (ifft) of MATLAB, y_n^c can be derived from (19.2) in few seconds. Equation (19.2) allows us to obtain the pmf of the aggregate packet filling value (i.e the number of blocks in the aggregate packet). Let F be the filling value random variable, and define the filling ratio random variable by $F_r = F/N$ (which is equivalent to the bandwidth efficiency). If we denote by $\{f_n = P(F = n), 0 \leq n \leq N\}$ the pmf of F, we obtain:

$$f_n = \begin{cases} y_n^0, & 0 \leq n < N - 1 \\ 1 - \sum_{i=0}^{N-1} y_i^0, & n = N \end{cases} \tag{19.3}$$

To obtain the steady state distribution for $\{Y_n^2, n = 0, 1, 2, \ldots\}$, the state equation can be written as

$$Y_{n+1}^2 = |Y_n^2 - G|^+ + A_\tau^2 \qquad (19.4)$$

where G represents the gap random variable. It is given by $G = N - F$, and hence, its pmf defined by $\{g_n, n = 0, 1, 2, \ldots, N\}$ can be obtained easily from (19.3). The PGF of Y^2 is then given by (see Appendix A):

$$Y^2(z) = \frac{A_\tau^2(z)(z - 1)(N - E[U])}{z^N - U(z)} \prod_{k=1}^{N-1} \frac{z - z_k}{1 - z_k} \qquad (19.5)$$

where U is the random variable defined by $U = N + A_\tau^2 - G$, and $z_1, z_2, \ldots, z_{N-1}$ are the $N - 1$ zeros of $z^N - U(z)$ inside the unit circle.

19.2.2 Blocks Number Random Instant Probabilities

Let $\{K^c, c = 1, 2\}$ denote the number of blocks, at a random instant t, in queues 1 and 2 respectively , and let $\{q_k^c = P(K^c = k), k \geq 0\}$ be its pmf.

Lemma 1. *K^c is related to Y^c by*

$$K^c(z) = Y^c(z) \frac{1 - e^{-\lambda \tau (X(z) - 1)}}{\lambda \tau (X(z) - 1)}$$

Proof. The proof is given for $c = 2$, the case of $c = 1$ is a particular case obtained by replacing the random variable G (the gap) with the constant parameter N (the length of the aggregate packet). Let T_e be the elapsed time since the last timer expiration. By conditioning on T_e, the following state equation can be obtained:

$$K^2 \mid (T_e = t) = |Y^2 - G|^+ + A_t^2$$

Then (superscript "2" is omitted for simplicity),

$$K(z \mid T_e = t) = \sum_{k=0}^{\infty} P\left[|Y - G|^+ + A_t = k\right] z^k$$

$$= \sum_{k=0}^{\infty} \sum_{i=0}^{N} g_i P\left[|Y - i|^+ + A_t = k\right] z^k$$

$$= \sum_{k=0}^{\infty} \sum_{i=0}^{N} g_i \left\{ \sum_{j=0}^{i} y_j P\left[A_t = k\right] + \sum_{j=i+1}^{\infty} y_j P\left[A_t = k - j + i\right] \right\} z^k$$

$$= A_t(z) \left\{ \sum_{i=0}^{N} g_i \left(\sum_{j=0}^{i} y_j + z^{-i} \left(Y(z) - \sum_{j=0}^{i} y_j z^j \right) \right) \right\}$$

With (see (19.16))

$$\left(\sum_{i=0}^{N} g_i \sum_{j=0}^{i} y_j - \sum_{i=0}^{N} g_i z^{-i} \sum_{j=0}^{i} y_j z^j \right) = \frac{Y(z) \left(1 - A_\tau(z) \sum_{i=0}^{N} g_i z^{-i} \right)}{A_\tau(z)}$$

$$\Rightarrow K(z \mid T_e = t) = \frac{A_t(z)}{A_\tau(z)} Y(z)$$

To obtain $K(z)$, it is sufficient to remove the condition on T_e, thus,

$$K(z) = \int_{t=0}^{\tau} \frac{A_t(z)}{A_\tau(z)} Y(z) \left(\frac{dt}{\tau} \right)$$

and the proof is completed. □

19.2.3 Mean Delay Analysis

In this section we present a method to obtain the mean aggregation delay of an IP packet belonging to class 1 or to class 2. The aggregation delay random variable of a class c packet, $\{c = 1, 2\}$, is represented by $\{D^c, c = 1, 2\}$, and it is defined as the time period elapsed between the arrival instant of the packet to its corresponding queue, and the instant when the last block of the packet leaves the queue. The delay can be decomposed in two parts: the waiting time of the packet first block until it becomes at the head of the queue, and the delay due to the packet segmentation when the packet cannot be inserted directly into the remaining gap of the aggregate packet. The decomposition is written, in term of mean delays, as

$$E[D^c] = E[D_b^c] + D_s^c$$

where D_b^c denotes the packet first block waiting time in the queue and D_s^c stands for the packet segmentation delay. By using the Little theorem we get

$$E[D_b^c] = \frac{E[K^c]}{\lambda_c \times E[X]} \tag{19.6}$$

where $E[K^c]$ can be obtained by putting $z = 1$ in the first derivative of $K^c(z)$, and it is given by

$$E\left[K^c\right] = E[Y^c] - \frac{\lambda_c \tau E[X]}{2}$$

$$= \left\{ E[A_\tau^c] + \left[\frac{\text{VAR}[C(U, A_\tau^1)]}{2(N - E[C(U, A_\tau^1)])} \right.\right.$$ (19.7)

$$\left.\left. - \frac{E[C(U, A_\tau^1)]}{2} + \frac{1}{2} \sum_{k=1}^{N-1} \frac{1 + z_k^c}{1 - z_k^c} \right] \right\} - \frac{\lambda_c \tau E[X]}{2}$$

with

$$C(x, y) = \begin{cases} x & \text{if } c = 1 \\ y & \text{if } c = 2 \end{cases}$$

$\text{VAR}[X]$ is the variance of X, $\{z_k^1, 1 \le k \le N - 1\}$ and $\{z_k^2, 1 \le k \le N - 1\}$ are the roots inside the unit circle of the two respective following equations: $\{z^N - A_\tau^1(z)\}$ and $\{z^N - U(z)\}$, with $U(z)$ defined in Section 19.2.1. Note that the roots are found using a numerical method with a precision of 10^{-10}.

19.2.3.1 Computation of D_s^1

We consider the case where $N > \max(X)$, where X denotes the batch size random variable (i.e the size of an IP packet in term of blocks), which implies that a class 1 packet cannot be segmented more than once. If we denote by $p_s^{1,n}$ the probability that a packet is segmented n times, then the segmentation delay of a class 1 packet will be given by

$$D_s^1 = \sum_{n=1}^{\infty} p_s^{1,n} \times (n\tau) = p_s^{1,1} \times \tau$$ (19.8)

The following is a method to obtain $p_s^{1,1}$: let N_s be the random variable representing the number of blocks that enters the aggregation unit before the first block of a random packet, given that the latter (the first block of the packet) has entered the aggregation unit. The pmf of N_s, $\{P(N_s = n, n = 0, \ldots, N - 1\}$, can be obtained as follows:

$$P[N_s = n] = \sum_{k=0}^{\infty} P[K^{1,a} = kN + n]$$

$$= \sum_{k=0}^{\infty} \sum_{i=0}^{\infty} k_i^{1,a} \delta(i - kN - n)$$

$$= \sum_{i=0}^{\infty} k_i^{1,a} \sum_{k=-\infty}^{\infty} \delta(i - kN - n)$$

where $K^{1,a}$ is the number of blocks presented in queue 1 seen at the arrival of a random packet. PASTA property [15] implies that $K^{1,a} = K^1$. $\delta(n)$ is the Kronecker delta function, which equals 1 for $n = 0$ and 0 for all other n, and $\{k_i^1 = 0,$ for $i < 0\}$. Now we make use of the following identity:

$$\sum_{k=-\infty}^{\infty} \delta(i - kN - n) = \frac{1}{N} \sum_{s=0}^{N-1} a^{s(i-n)} \qquad (19.9)$$

with $a = \exp(j(2\pi/N))$. In words, the right-hand side of (19.9) equals zero unless the integer $i - n$ is a multiple of N, when it equals unity. Thus,

$$P[N_s = n, 0 \le n \le N - 1] = \sum_{i=0}^{\infty} k_i^1 \frac{1}{N} \sum_{s=0}^{N-1} a^{s(i-n)}$$

$$= \frac{1}{N} \sum_{s=0}^{N-1} a^{-sn} K^1(a^s) \qquad (19.10)$$

where $K^1(a^s)$ is $K^1(z)$ evaluated at $z = a^s$.

Obtaining the pmf of N_s by using (19.10) for each value of n leads to a very long computation time, especially when N becomes large. However, we give an equivalent matrix equation for this relation. This approach reduces the computation time considerably since it gives the pmf of N_s by using only one matrix equation. If P_{N_s} denotes the row vector representing the pmf of N_s, i.e $P_{N_s} = (P(N_s = 0) \quad P(N_s = 1) \quad \ldots \quad P(N_s = N - 1))$, and if we define R_{K^1} by $(K^1(a^0) \quad K^1(a^1) \quad \ldots \quad K^1(a^{(N-1)}))$, we will have:

$$P_{N_s} = \frac{R_{K^1}}{N} \times \begin{pmatrix} a^0 & a^0 & a^0 & \cdots & a^0 \\ a^0 & a^{-1} & a^{-2} & \cdots & a^{-(N-1)} \\ a^0 & a^{-2} & a^{-4} & \cdots & a^{-2(N-1)} \\ \cdots \cdots \cdots \cdots \cdots \cdots \cdots \cdots \\ a^0 & a^{-(N-1)} & a^{-2(N-1)} & \cdots & a^{-(N-1)^2} \end{pmatrix} \tag{19.11}$$

where the last matrix in (19.11) is an $N \times N$ matrix. Now, $p_s^{1,1}$ can be obtained easily by $p_s^{1,1} = P(X > N - N_s)$.

19.2.3.2 Computation of D_s^2

Unlike the class 1 case, where a packet cannot be segmented more than once, a class 2 packet may encounter several segmentations before its complete transmission. This is because when a class 2 packet is segmented for the first time, its remaining blocks cannot enter the aggregation unit unless queue 1 is polled again.

If we denote by $p_s^{2,n}$ the probability that a packet is segmented n times, we obtain:

$$D_s^2 = \sum_{n=1}^{\infty} p_s^{2,n} \times (n\tau) \tag{19.12}$$

However, considering only the first two terms of (19.12) is sufficient as we will see in the numerical examples. To obtain the pmf of N_s we proceed as in the previous section with the difference that here the number of blocks that enter the aggregation unit before a class 2 packet is equal to the number of blocks presented in queue 1 and queue 2 when the packet arrives (i.e, $K^{1,a} + K^{2,a}$), plus the number of class 1 blocks that arrive during the waiting time of the packet first block (represented by the random variable $N_{b,2}^1$). Now we use the PASTA property $\{K^{c,a} = K^c, c = 1, 2\}$ and we approximate $N_{b,2}^1$ by its mean ($E[N_{b,2}^1] = \lambda_1 E[X]E[D_b^2]$ thanks to the Little theorem). Then, by replacing the random variable K^1 with $\{K^1 + K^2 + E[N_{b,2}^1]\}$ in the method presented in the previous section, and by supposing that queue 1 and queue 2 are independent (i.e, K^1 and K^2 are two independent random variables), we can express the pmf of N_s by

$$P[N_s = n, 0 \leq n \leq N - 1] = \frac{1}{N} \sum_{s=0}^{N-1} a^{-sn} \left((a^s)^{E[N_{b,2}^1]} K^1(a^s) K^2(a^s) \right) \tag{19.13}$$

The last term between parentheses in (19.13) is the z-transform of $\{K^{1,a} + K^{2,a} + E[N_{b,2}^1]\}$, $(= z^{E[N_{b,2}^1]} K^1(z) K^2(z))$, evaluated at $z = a^s$.

Now the first two terms of (19.12) can be obtained as follows:

$$p_s^{2,1} = P(X > N - N_s) \quad \text{and} \quad p_s^{2,2} = P(X - (N - N_s) > G) \quad (19.14)$$

where G stands for the gap random variable.

19.3 Extension to J, $\{J > 2\}$ Classes

In this section, we explain how to extend the analysis when more than two packet classes are desired.

The filling ratio can be obtained by the same manner used in the case of two classes. However, for the delay analysis of packets belonging to queue i, $\{i = 2, \ldots, J\}$, we combine queues $\{1, 2, \ldots, i-1\}$ in a single queue and the analysis is reduced to two queues with respective arrival rates: $\lambda_1 = \sum_{k=1}^{i-1} \lambda_k$, and $\lambda_2 = \lambda_i$.

19.4 The Probabilistic Priority Discipline

In order to overcome the drawbacks of the strict priority discipline that has been used in the analytical model, one may apply a probabilistic algorithm such as the one presented in [10]. For this purpose, a parameter p_i, $1 \le i \le J$, $0 \le p_i \le 1$, is assigned to each arrival queue i, and a relative weight $r_i = p_i \prod_{j=1}^{i} (1 - p_{j-1})$, where $p_0 = 0$, is computed. Then, the following steps are applied:

1. At each aggregation cycle, the set of non-empty queues, NQ is determined, and a normalized relative weight,

$$\tilde{r}_i = \frac{r_i}{\sum_{j \in NQ} r_j}$$

   is calculated. The latter is regarded as the probability with which queue i is served among all non-empty queues in an aggregation cycle.
2. Fill the aggregate packet from the polled queue in Step 1. If the aggregate packet becomes full, send it to the conversion queue and apply Step 1 to the next aggregation cycle, elsewhere exclude the polled queue from NQ and apply Step 1 to the same aggregation cycle if NQ is not empty, or apply Step 1 to the next aggregation cycle if NQ is empty.

19.5 Numerical Examples

In this section we give some numerical examples showing the usefulness of the model. All the computations (probabilities and means) have been done in double precision. The mean delay and the mean filling ratio are obtained from their corresponding pmfs. We consider the following assumptions.

1. From experimental measurements [16], the size distribution of IP packets has three predominant values: 40 bytes, 552 bytes and 1500 bytes, with the corresponding probabilities 0.6, 0.25 and 0.15 respectively. To discretize the size distribution, we suppose that the size of a random packet is a batch of 40 bytes blocks, and hence the packet size PGF is given by $\{X(z) = 0.6z + 0.15z^{14} + 0.25z^{38}\}$, where each power of z represents the first integer greater or equal to the division quotient of the corresponding packet size by the block size (e.g., $\lceil 552/40 \rceil = 14$, $\lceil \cdot \rceil$ is the first integer greater or equal to $\cdot$).

2. Each node has two classes (real-time applications and non-real-time applications) with proportions 0.6 and 0.4, respectively. The arrival processes of the two classes are Poisson processes with rates λ_1 (packet/s) and λ_2 (packet/s) respectively; the total arrival rate is $\lambda_0 = \lambda_1 + \lambda_2$. In the examples, we suppose that the arrival rates are represented, in Mb/s, by $\{\theta_c, c = 0, 1, 2\}$, where $c = 0$ represents the case of the combination of queues 1 and 2. It is easy to verify that λ_c is related to θ_c by

$$\lambda_c = \frac{\theta_c \times 10^6}{8bE[X]}.$$

In the rest of the paper, we use θ and θ_0 interchangeably. Note that the choice of two classes in the numerical examples is adopted for comparison purposes only since the model is scalable regardless of the number of classes. Furthermore, Poisson traffic is considered because we aim at proving that the multicast aggregation technique increases the filling ratio of an optical packet. Clearly, this conclusion remains true regardless of the traffic profile.

3. The size of the aggregate packet (N blocks) is supposed to be fixed by the operator, and the maximal arrival rate $\theta_{0,\max}$ (or simply $\theta_{\max}$) is supposed to be known a priori (by effectuating measurements over several timescales).

19.5.1 Nodes Stability Condition and Stability Region

The stability condition of the node is respected if (see (19.2)): $N > E[A_\tau^0]$, with $E[A_\tau^0] = \lambda_0 \tau E[X]$.

Stability region. Let the parameter a be the following: $a = E[A_\tau^0]$ (this is the mean of blocks number that arrive during τ). For a given value of N and θ_0 (or simply θ), the parameter a must be strictly less than N to maintain the stability and then we must choose τ according to

$$\tau_\theta < \frac{N}{\lambda_0 E[X]} \tag{19.15}$$

where τ_θ is the value of the time-out when operating at θ. (19.15) defines what we call the stability region at the arrival rate θ. To obtain a desired value of a inside the stability region we choose τ_θ according to: $\tau_\theta = a/\lambda_0 E[X]$. For instance, for $N = 76$ blocks, and $\theta = 900$ Mb/s, (19.15) implies that the limit of the stability region is $\tau_{max} = 27\ \mu s$.

19.5.2 Impact of the Time-Out

We present in Figure 19.2 the impact of τ on the filling ratio. The pmf of the filling ratio is shown for two values of τ ($\tau = 12.5\ \mu s$ and $\tau = 25\ \mu s$), with $\theta = 900$ Mb/s and $N = 76$. We observe that when τ increases, the probability that the aggregate packets are sent with better filling ratio increases. This is because when τ increases, while remaining inside the stability region, the number of packets presented in queues 1 and 2 at a random epoch increases. Note that for $\tau = 25\ \mu s$ the aggregate packets are sent full with probability 0.82952. This is because τ is very close from the stability limit (27 μs).

19.5.3 Effect of the Arrival Rate

The impact of the arrival rate on both mean packet delay and mean filling ratio is given in Figures 19.3 and 19.4 respectively. N is fixed to 76 and τ to 25 μs. It can be observed that the mean delay of class 1 packets remains approximately unchanged when θ varies, while the mean delay of class 2 packets degrades when θ increases. The mean filling ratio increases also as θ increases. At $\theta = 900$ Mb/s, the mean filling ratio attains 93.506% since at this arrival rate, $\tau = 25\ \mu s$ becomes very close to the limit of the stability region. Note that when θ decreases, the enhancement in the delay is compensated by a loss in the filling ratio. This is because τ remains constant. To avoid this limitation, we can adapt the value of τ with respect to

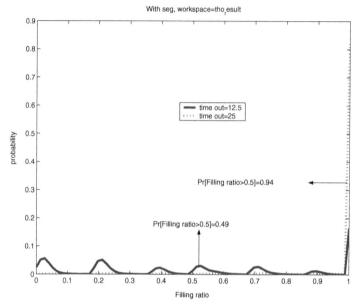

Figure 19.2 PMF of the filling ratio.

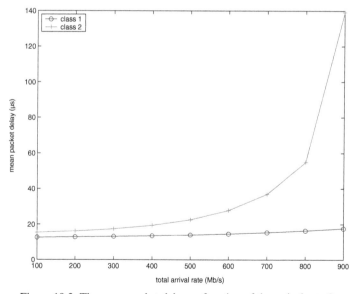

Figure 19.3 The mean packet delay as function of the arrival rate θ.

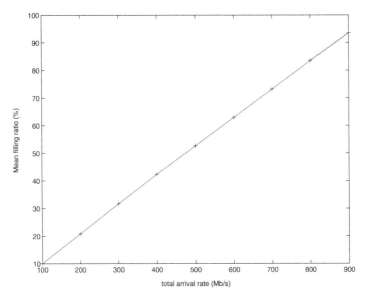

Figure 19.4 The mean filling ratio as function of the arrival rate θ.

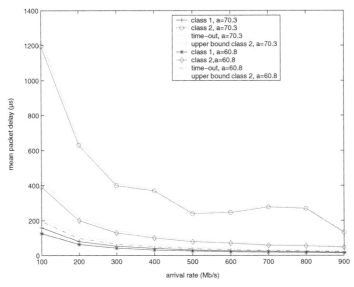

Figure 19.5 The mean packet delay in the case of the adaptive τ scenario.

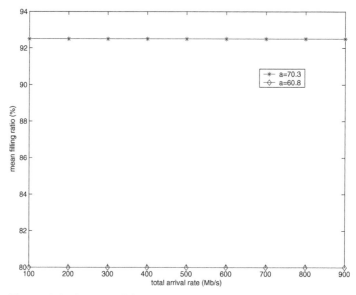

Figure 19.6 The mean filling ratio in the case of the adaptive τ scenario.

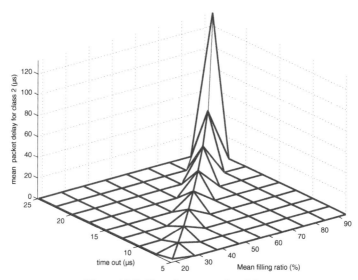

Figure 19.7 Class 2 mean packet delays.

the variation of θ in order to conserve a fixed value of the parameter a (see Section 19.5.1), which represents the mean size (in blocks) of the packets that arrive between two consecutive timer expirations. In Figure 19.5 we show the effect of a on the mean packet delay (we take $a = 70.3$ blocks and $a = 60.8$ blocks). It is clear that when a decreases the delay decreases (decreasing a means that for a given θ the time-out τ has a less value which implies less delays). The advantage of the adaptive τ scenario is to maintain the mean filling constant as shown in Figure 19.6. We can also deduce that reducing a leads to a decrease in the mean filling ratio as expected (because of the decrease in τ). Finally, Figure 19.7 represents how an industrial operator can make use of the aggregation model. If a certain value of the mean filling ratio is desired, then by using the curve we can obtain the value of class 2 mean packet delay (z-axis) and the value of the time-out to setup (y-axis). We can adapt the value of the filling ratio to obtain a desired value of class 2 mean packet delay, while class 1 mean packet delay remains below τ (as shown in Figure 19.5).

19.6 Application to MANS

19.6.1 Hub Stripping Network: DBORN

The term hub stripping refers to the case where a single node in the ring network, called the hub, drops optical packets. This is the case of DBORN, where the hub can be regarded as the single destination of ring nodes in the transmission phase. DBORN has been initially proposed for asynchronous systems [13, 17, 18]. However, we consider in this work an hybrid slotted version of the original system. DBORN is an optical metro ring architecture connecting several edge nodes, e.g., metro clients like enterprise, campus or local area networks (LAN), to a regional or core network. The ring consists of a unidirectional fiber split into downstream and upstream channels spectrally disjointed (i.e., on different wavelengths) [17]. The upstream wavelength channels are used for writing (transmitting), while downstream wavelength channels are used for reading (receiving). In a typical scenario, the metro ring has a bit-rate 2.5 Gb/s, 10 Gb/s or 40 Gb/s per wavelength. In order to keep the edge node interface cards as simple as possible, all traffic (external and intra-ring) has to pass the hub. Specifically, no edge node receives or even removes traffic on upstream channels or inserts traffic on downstream channels. Thus, both upstream and downstream channels can be modeled as shared unidirectional buses. Packets circulate around the ring without any

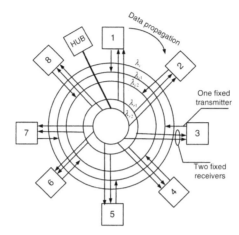

Figure 19.8 A simple DBORN topology of 8 nodes, one control channel (λ_c), two transmission wavelengths (λ_{e1} and λ_{e2}) and two reception wavelengths (λ_{r1}, and λ_{r2}).

electro-optic conversion at intermediate nodes. The hub is responsible of terminating upstream wavelengths and hence, the first node in the upstream bus receives always free slots from the hub. Moreover, the hub electronically processes the packets, which may leave to the backbone or go through the downstream bus to reach their destinations. In the latter case, the ring node must pick up a copy of the signal originating from the hub by means of a splitter in order to recover its corresponding packets by processing them electronically. In the following, we consider that each node is equipped with one fixed transmitter and as many fixed receivers as reception channels, i.e., for each node we assign only one transmission wavelength at the upstream bus. This is illustrated in Figure 19.8, where nodes 1, 3, 5 and 7 transmit on λ_{e1} and nodes 2, 4, 6 and 8 transmit on λ_{e2}. Two MAC protocols may be used. The empty slot procedure and a slot reservation mechanism.

In the case of empty slot procedure, the slot header is detected to determine the status of the slot (i.e., empty/full) and a node may transmit on every empty slot.

In the case of the reservation approach, we implement a slot reservation mechanism at the upstream bus. Indeed, a fixed number of slots are reserved for each node. We suppose that the slot assignment is performed in the hub which writes the address of the node, for which the slot is reserved, in the slot's header. Hence, when a node receives an incoming slot, it detects the header to determine the status of the slot (empty/full) and the reservation information to decide whether to transmit or not. In addition to its reserved slots, the node can use empty slots reserved to any of its upstreams. This is because all slots are emptied by the hub before being sent on the upstream bus. This interesting feature of DBORN makes it possible to avoid the inefficiency behavior of a TDMA scheme. For instance, in Figure 19.8, if node 1 does not fill a reserved slot because its local queue is empty, then any downstream node which transmits on channel λ_{e_1} can use this slot for transmission because the hub will empty the slot before it reaches node 1 again. This enables efficient use of the available bandwidth in the case of non uniform traffic. Furthermore, at overloaded conditions, i.e., when the local queues of all nodes are always non empty, the MAC protocol reduces to a TDMA scheme, since each node will consume all its reserved slots for transmission.

Note that in order to process the control information, only the control channel is converted to the electrical domain at each ring node, while the bulk of user information remains in the optical domain until it reaches the hub which is viewed as the destination of upstream data. This is in conformity with the notion of all-optical (or transparent) networks in literature (e.g., [19, 20]). The slots in the control channel have a locked timing relationship to the data slots and arrive earlier at each node by a fixed amount of time (this is achieved via an optical delay line) allowing the process of the control slot content.

At the downstream reading bus, ring nodes preserve the same behavior initially proposed in DBORN and hence, the optical signal is split and IP packets are recovered at each node. The latter drops packets which are not destined to it. This means that DBORN supports multicast without addition of any complexity in its MAC protocol and hence, it represents an interesting application of the multicast packet aggregation method presented in this work.

19.6.2 Destination Stripping Networks

Figure 19.9 shows the general WDM ring architectures for packet-switched communications with *destination stripping*. The network nodes are interconnected by a single optical fiber supporting multiple WDM channels, which are time-slotted. Logically, we thus obtain multiple rings whose available bandwidths can be equally accessed. At each node an OADM (Optical Add Drop Multiplexer) drops a prescribed wavelength from the ring and allows the addition of data at any arbitrary wavelength. Each node may be equipped with one or more fixed-tuned and/or tunable transmitters and receivers. A

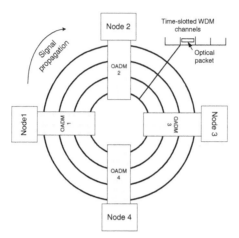

Figure 19.9 General slotted WDM ring architecture with destination stripping for $M = 4$ nodes and $C = 4$ channels.

node transmits data on the added wavelength while it receives data on the dropped wavelength. Data on the dropped wavelength are removed from the ring and optical-electronically converted. That is, an incoming full slot is emptied by the corresponding destination node. The header contains all necessary control information such as the status of the slot (empty/full) and the set of destination addresses of the multicast packet (the fanout set F_s). A self routing scheme [21], can be adopted to encode F_s into the slot's header. Thus, each node may be reserved one bit into the header. The bits corresponding to all destinations of the multicast packet are set to one. A ring node detects these bits to determine if it is within the set of destinations of the multicast packet. In this case the node resets its corresponding bit to zero and makes a copy from the slot. When the node discovers that it is the last destination node of the multicast packet (i.e. all bits, excluding its corresponding bit, are set to zero), it removes completely the multicast packet from the ring by means of an OADM. Thus the slot becomes empty and the node may reuse the same slot by inserting a multicast packet from its local queue into the slot's payload. This is called the *destination stripping with slot-reuse* scheme.

19.7 Application to WANs

The multicast aggregation technique analyzed throughout this paper can be easily adapted to the studied MANs. One one hand, the support of multicast traffic is facilitated due to the simplicity of ring architecture. On the other

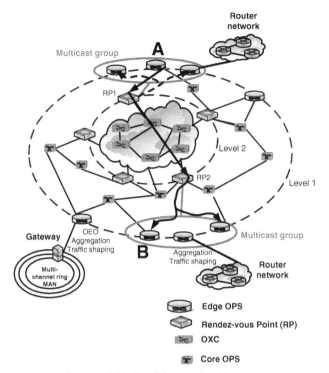

Figure 19.10 Possible WAN architecture.

hand, aggregation of IP packets belonging to different QoS classes in one aggregate packet is justified since there is no packet loss inside such ring networks. This is because in-transit traffic has always higher priority than local traffic at intermediate nodes.

The problem of multicast handling and QoS support in WAN may introduce some adjustments to the proposed aggregation techniques because wide area networks have a mesh topology in general. In Figure 19.10 we propose a possible future WAN architecture. The scenario proposed in [22] is adopted. Indeed, optical packet switching provided through optical packet switches (OPSs) and circuit switching ensured by OXCs (Optical Cross-Connects), coexist within the network. The former is deployed in areas where granularity is below the wavelength level, while the latter interconnects high-capacity points that will fully utilize the channel capacity in the core of the network. To do this, some optical channels (wavelength paths) may interconnect OXCs. Other channels might be reserved to OPSs to support optical packet transmis-

OXC

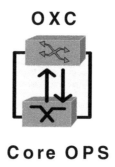

Core OPS

Figure 19.11   Architecture of the Rendez-vous point.

sion. Once the technology matures and the need for a more flexible fully IP-centric network is dominant, optical packet switches may replace the OXCs, or alternatively reduce their size and cost significantly, as the wavelength channels are more efficiently used and hence, equipment requirements are reduced. Furthermore, we propose a network with two hierarchy levels. The first level is constituted from edge and core OPSs. Edge OPSs differ from core OPSs in that they are connected to client layer networks, such as IP networks and to metropolitan area networks such as the networks studied in this work. Edge OPSs are responsible of performing IP packet aggregation in order to improve bandwidth efficiency. Indeed, optical packets received through MANs must be converted to electronic in order to join a new aggregation process. The latter must take into consideration the two fundamental questions (multicast and QoS) in the context of the WAN architecture. This is may be given as follows. In WANs the number of edge nodes is much greater than that in MANs and hence, aggregating IP packets regardless of their destinations leads to generating a big number of broadcast optical packets which may waste the network resources. Instead, we suppose that edge nodes are separated in multicast groups. Each multicast group has a designed router, called Rendez-vous point (RP).

The structure of the RP is given in Figure 19.11. The OPS is responsible of communicating with the level-1 hierarchy nodes, while the interconnection between RPs through lightpaths between OXCs makes the level-2 hierarchy. An IP-centric control plane is responsible of four major tasks:

1. Constructing the multicast groups by using the IGMP protocol [23].
2. Electing an RP for each group by using an approach similar to that of the PIM-SM multicast protocol [24].

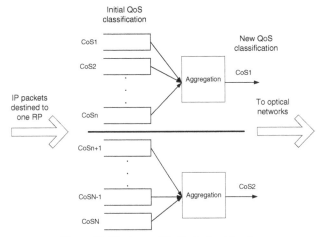

Figure 19.12 A possible QoS classification.

3. Constructing lightpaths between RPs.
4. Constructing a shared multicast tree between nodes of each group and fixed lightpaths between RPs.

GMPLS [25] is a candidate to perform these tasks after doing the necessary extensions. Now each edge node performs a separate aggregation process per RP. That is, IP packets destined to edge nodes belonging to the same multicast group are aggregated together. If the multicast optical packet is destined to the designated RP of the multicast group, the RP multicasts it on the shared multicast tree, elsewhere the RP sends the packet to the corresponding RP through a lightpath on the level-2 hierarchy and the latter multicasts it on the shared tree. For instance, in Figure 19.10, suppose that node A wishes to send two optical packets: one for its group and another one for node B and other nodes in a different multicast group. In the former case, the packet is forwarded to RP_1 which multicasts it through its multicast tree (dashed arrows). In the latter case, RP_1 sends the packet to RP_2 which multicasts it on its shared tree to get the destination nodes including node B.

The QoS support must consider that OPSs have limited optical buffering capacity, in terms of fiber delay lines[1] and hence it may be important to generate aggregate packets with several classes of service, contrary to the aggregation in MANs, in order to reduce packet loss of higher priority traffic. Typically, four QoS classes may be sufficient. However, as the number of

---

[1] Fiber delay lines are used due to the absence of random access memory in optic.

classes increases as the complexity of the RPs increases where separate control information, such as restoration of lightpaths in case of failure, must be guaranteed for each class. In order to reduce this complexity, two QoS classes may be adopted as shown in Figure 19.12.

19.8 Conclusion

We proposed and analyzed a novel approach for efficiently supporting IP packets in a slotted WDM optical layer with several QoS requirements. A simple analytical model, allowing the evaluation of IP packets aggregation delay and the bandwidth efficiency, has been presented. The results showed an immense increase in the bandwidth efficiency when using the aggregation compared to the standard approach. For instance, if no aggregation is performed and if the optical packet length (N) is equal to the maximum IP packet length ($\max(X)$) (i.e, an IP packet fits exactly into one optical slot, and hence $N = 38$), then the bandwidth efficiency (the mean filling ratio) will be equal to 25.79%.[2] However using the aggregation method, the bandwidth efficiency attains 93.42% at $\theta = 900$ Mb/s, $\tau = 25$ µs and $N = 76$ blocks. Concerning IP packets aggregation delay, a trade-off with the bandwidth efficiency results by modifying the value of the parameter a. The QoS support mechanism based on assigning higher aggregation priorities for higher QoS classes, has been evaluated by means of an analytical study. Moreover, we proposed a solution to control the level of priority between classes by adopting the probabilistic priority discipline. The application of the aggregation technique has been shown in the context of MANs and WANs, respectively.

Appendix

A Proof of (19.2) and (19.5)

We present the proof of (19.5). The proof of (19.2) is obtained as a particular case. Taking the PGF of both two sides of (19.4), and omitting the subscript "2" for simplicity, yields:

[2] $= ((1/38) \times 0.6 + (14/38) \times 0.25 + (38/38) \times 0.15) \times 100.$

$$Y(z) = \sum_{k=0}^{\infty} P(|Y - G|^+ A_\tau = k)z^k$$

$$= \sum_{i=0}^{N} g_i \sum_{k=0}^{\infty} P(|Y - i|^+ + A_\tau = k)z^k$$

$$= \sum_{i=0}^{N} g_i \sum_{k=0}^{\infty} \left\{ \sum_{j=0}^{i} y_j P(A_\tau = k) + \sum_{j=i+1}^{\infty} y_j P(j - i + A_\tau = k) \right\} z^k$$

$$= \sum_{i=0}^{N} g_i \left\{ \sum_{j=0}^{i} y_j A_\tau(z) + \sum_{j=i+1}^{\infty} y_j A_\tau(z)z^{j-i} \right\}$$

$$= \sum_{i=0}^{N} g_i A_\tau(z) \left\{ \sum_{j=0}^{i} y_j + \left(Y(z) - \sum_{j=0}^{i} y_j z^j \right) z^{-i} \right\}$$

$$= \frac{\left(\sum_{i=0}^{N} g_i \sum_{j=0}^{i} y_j - \sum_{i=0}^{N} g_i z^{-i} \sum_{j=0}^{i} y_j z^j \right) A_\tau(z)}{1 - A_\tau(z) \sum_{i=0}^{N} g_i z^{-i}} \tag{19.16}$$

$$= \frac{\sum_{i=0}^{N} g_i \left\{ \sum_{j=0}^{i-1} y_j \left(z^N - z^{N+j-i} \right) \right\}}{z^N - A_\tau(z) \sum_{i=0}^{N} g_i z^{N-i}} A_\tau(z) \tag{19.17}$$

Now let U be the random variable defined by $U = N + A_\tau - G$, then $U(z) = A_\tau(z) \sum_{i=0}^{N} g_i z^{N-i}$, and the denominator of (19.17) is $z^N - U(z)$. Rouche theorem indicates that $(z^N - U(z))$ has exactly N zeros inside the unit circle (1 being the intuitive zero) if $N > E[U]$. Under this condition, let $z_1, z_2, \ldots, z_{N-1}$ be these zeros (without the intuitive zero 1). The numerator of (19.17) must vanish at $\{1, z_1, z_2, \ldots, z_{N-1}\}$, and so it is equal to $C(z-1) \prod_{k=1}^{N-1} (z - z_k) A_\tau(z)$. Where C is a constant. To determine C we put $Y(1) = 1$ in (19.17) which leads to

$$C = \frac{N - E[U]}{\prod_{k=1}^{N-1}(1 - z_k)}$$

and then

$$Y(z) = \frac{A_\tau(z)(z-1)(N - E[U])}{z^N - U(z)} \prod_{k=1}^{N-1} \frac{z - z_k}{1 - z_k}$$

which completes the proof of (19.5). Now (19.2) is obtained by replacing the random variable G with the constant N.

References

[1] C. A. Brackett, Dense wavelength division multiplexing: Principles and applications, *IEEE J. Select. Areas Commun.*, vol. 8, no. 6, pp. 948–964, August 1990.

[2] S. Yao, B. Mukherjee and S. Dixit, Advances in photonic packet switching:an overview, *IEEE Commun. Mag.*, vol. 38, no. 2, pp. 84–94, February 2000.

[3] F. Callegati, Which packet length for a transparent optical network?, Paper presented at the SPIE Symp. Broadband Networking Technologies, Dallas, TX, 1997.

[4] M. Chaitou, G. Hébuterne and H. Castel, On aggregation in almost all optical networks, in *Proceedings of the Second IFIP International Conference on Wireless and Optical Communications Networks WOCN 2005*, Dubai, United Arab Emirates UAE, pp. 210–216, March 2005.

[5] G. Hébuterne and H. Castel, Packet aggregation in all-optical networks, in *Proceedings of the First International Conference on Optical Communication and Networks*, Singapore, pp. 114–121, 2002.

[6] D. Careglio, J. S. Pareta and S. Spadaro, Optical slot size dimensioning in IP/MPLS over OPS networks, in *Proceedings of the 7th International Conference on Telecommunications, ConTEL 2003*, Zagreb, Croatia, pp. 759–764, 2003.

[7] A. Srivatsa et al., CSMA/CA MAC protocols for IP-hornet: An IP over WDM metropolitan area ring network, in *Proceedings of GLOBECOM'00*, San Francisco, CA, vol. 2, pp. 1303Ű–1307, 2000.

[8] K. Bengi, Access protocols for an efficient and fair packet-switched IP-over-WDM metro network, *Computer Networks*, vol. 44, no. 2, pp. 247–265, February 2004.

[9] L. Dittmann et al., The European IST project DAVID: A viable approach towards optical packet switching, *IEEE J. Select. Areas Commun.*, vol. 21, no. 7, pp. 1026–1040, 2003.

[10] Y. Jiang, C. Tham and C. Ko, A probabilistic priority scheduling discipline for high speed networks, in *Proceedings of the IEEE Workshop on High Performance Switching and Routing*, pp. 1–5, May 2001.

[11] Y. Jiang, C. Tham and C. Ko, A probabilistic priority scheduling discipline for multi-service networks, *Computer Communications*, vol. 25, no. 13, pp. 1243–1254, August 2002.

[12] M. Scheutzow, P. Seeling, M. Maier and M. Reisslein, Multicast capacity of packet-switched ring WDM networks, in *Proceedings of IEEE INFOCOM*, vol. 1, pp. 706–717, March 2005.

[13] N. L. Sauze et al., A novel, low cost optical packet metropolitan ring architecture. in *Proceedings of the European Conference on Optical Communication (ECOC2001)*, vol. 3, pp. 66–67, October 2001.

[14] A. Detti, V. Eramo and M. Listanti, Performance evaluation of a new technique for IP support in a WDM optical network: Optical composite burst switching (OCBS), *J. Lightwave Technol.*, vol. 20, no. 2, pp. 154–165, February 2002.

[15] R. W. Wolff, *Stochastic Modeling and the Theory of Queues*, Prentice-Hall, 1989.

[16] K. Thompson, G. J. Miller and R. Wilder, Wide-area internet traffic patterns and characteristics, *IEEE Network*, vol. 11, no. 6, pp. 10–23, November/December 1997.

[17] N. Bouabdallah et al., Resolving the fairness issue in bus-based optical access networks, *IEEE Commun. Mag.*, vol. 42, no. 11, pp. 12–18, November 2004.

[18] N. Bouabdallah, A.-L. Beylot, E. Dotaro and G. Pujolle, Resolving the fairness issues in bus-based optical access networks, *IEEE J. Select. Areas Commun.*, vol. 23, no. 8, pp. 1444–1457, August 2005.

[19] P. Gambini et al., Transparent optical packet switching: network architecture and demonstrators in the KEOPS project, *IEEE J. Select. Areas Commun.*, vol. 17, no. 7, pp. 1245–1259, 1998.

[20] C. Guillemot, Transparent optical packet switching: the european ACTS KEOPS project approach, *J. Lightwave Technol.*, vol. 16, no. 12, pp. 2117–2134, December 1998.

[21] X. C. Yuan, V. O. K. Li, C. Y. Li and P. K. A. Wai, A novel self-routing address scheme for all-optical packet-switched networks with arbitrary topologies, *J. Lightwave Technol.*, vol. 21, pp. 329–339, February 2003.

[22] M. J. O'Mahony, The application of optical packet switching in future communication networks, *IEEE Commun. Mag.*, vol. 39, no. 3, pp. 128–135, March 2001.

[23] B. Cain et al., Internet group management protocol, version 3, *RFC 3376*. Online, available at http://www.ietf.org/rfc/rfc3376.txt?number=3376

[24] D. Estrin et al., Protocol independent multicast-sparse mode (PIM-SM): Protocol specification, *RFC 2362*. Online, available at http://www.ietf.org/rfc/rfc2362.txt?number=2362

[25] A. Banerjee, Generalized multiprotocol label switching: An overview of signaling enhancements and recovery techniques, *IEEE Commun. Mag.*, vol. 39, no. 7, pp. 144–151, July 2001.

20

Parallel Traffic Generation of a Decomposed Optical Edge Node on a GPU

Harry Mouchos and Demetres D. Kouvatsos

Networks and Performance Engineering (PERFORM) Research Unit,
School of Informatics, Department of Computing, University of Bradford,
Bradford BD7 1DP, West Yorkshire, England, UK;
e-mail: c.mouchos@bradford.ac.uk, d.d.kouvatsos@scm.brad.ac.uk

Abstract

The results of Optical Edge node simulations, employing the Graphics Processing Unit (GPU) on a commercial NVidia GeForce 8800 GTX, which was initially designed for gaming computers, are being presented. Parallel generators of optical bursts are implemented and simulated in the "Compute Unified Device Architecture" (CUDA) and favourable comparisons are made against simulations run on general-purpose CPUs. It is found that the GPU is a cost-effective platform, which can significantly speed-up calculations, making simulations of more complex and demanding networks easier to develop.

Keywords: Compute Unified Device Architecture (CUDA), parallel processing, Wavelength Division Multiplexing (WDM), Dense-wavelength Division Multiplexing (DWDM), Optical Packet Switching (OPS), Optical Burst Switching (OBS), self-similarity, Long-Range Dependence (LRD), Generalised Exponential (GE) distribution, Graphics Processing Unit (GPU).

D. D. Kouvatsos (ed.), Performance Modelling and Analysis of Heterogeneous Networks, 417–439.

20.1 Introduction

Optical networks with wavelength division multiplexing (WDM) have recently received considerable attention by the research community, due to the increasing bandwidth demand, mostly driven by Internet applications such as peer-to-peer networking and voice over IP traffic. In this context, several routing and wavelength reservation schemes, applicable to present and future optical networks, have been proposed [1–4].

More specifically, optical burst switching (OBS) [1, 2] was proposed as the next step towards the Optical Internet, since the lack of optical buffering technology makes optical packet switching (OPS) extremely difficult to implement [2]. The discovery of self-similarity and the inherent complexity of optical networks influenced the research community to employ mostly simulation towards the performance evaluation of such networks. These simulations had usually huge durations and were extremely intensive in terms of processing power.

In this paper, a novel approach in simulating OBS networks is presented using a graphics processing unit (GPU), which was originally designed for gaming. An investigation on its potentials is carried out and a comparison with general-purpose CPUs is conducted.

The paper is organized as follows: Section 20.2 carries out a brief review on wavelength division multiplexing (WDM) and dense-wavelength division multiplexing (DWDM). Section 20.3 discusses various bursty and correlated traffic models for wide area networks (WANs) such as the generalized exponential (GE) distribution, self-similarity and long-range dependence (LRD), the Pareto distribution as well as self-similar traffic generators based on *on/off sources* and the $M/G/\infty$ delay system. Section 20.4 describes packet aggregation strategies at the ingress node of an OBS network. Section 20.5 presents a method of simulating optical network nodes employing the method of decomposition on different wavelengths. The capabilities of a GPU in support of the simulation process are highlighted in Section 20.6. The simulation of the Ingress node and associated numerical results are included in Section 20.7. Concluding remarks follow in Section 20.8.

20.2 Dense Wavelength Division Multiplexing (DWDM)

WDM is a favorite multiplexing technology in optical communication networks because it supports a cost-effective method to provide concurrency among multiple transmissions in the wavelength domain. Several, communication channels, each of which is carried by a different wavelength, are multiplexed into a single fiber strand at one end and demultiplexed at the other end, thereby enabling multiple simultaneous transmissions. Each communication channel (wavelength) can operate at any electronic processing speed (e.g., OC-192 or OC-768).

In DWDM, a large number of wavelengths are packed densely into the fiber with small channel spacing. DWDM amplifies all the wavelengths at once, without first converting them to electrical signals and it has the ability to carry signals of different speeds and types simultaneously and transparently over the fiber (protocol and bit rate independence, see [5]).

WDM allows the aggregation of several parallel wavelength channels, each of which may operate at a much slower speed. WDM and DWDM enable the optical network topology to be viewed as a collection of virtual *all-optical lightpaths*, which symbolize channels that optical packets traverse from source to destination, over the same wavelength if there is no wavelength conversion [6].

Wavelength routing and lightpath topology of a WDM/DWDM optical network have received widespread attention in the literature. Associated results indicate that it is extremely difficult to find the optimal topology for the various lightpath source-destination pairs in order to minimize the number of wavelengths required for an effective network operation, especially as today's network topologies become more and more complex. The absence or existence of wavelength converters in the network core increases or decreases the complexity of this problem, respectively [6].

20.2.1 Optical Packet Switching (OPS)

In OPS, the data is transmitted in optical packets which do not go through an optical-electrical-optical conversion but rather stay in the optical domain [7]. In contrast to traditional packet switching however, OPS faces the problem of the lack of an optical version of memory, relying mostly on fiber delay lines, which offer simply a delay to the optical packet and storage. An added issue is the recognition of the header from its payload by the intermediate nodes. Several suggestions have been made in order to tackle this problem, such as the sending of the headers on different channels (i.e., wavelengths), so that the node may "store" the payload in a fiber-optical delay line (FDLs) whilst the header is being processed [8].

20.2.2 Optical Burst Switching (OBS)

Today's technology is not yet mature enough to be able to tackle all the challenges imposed by an all-optical backbone Optical Packet Switching Network [9]. A burst is considered of an intermediate "granularity" switching entity in comparison to a call (or session) and an optical packet [8]. An adaptation of the ITU-T standard for burst switching in ATM, which is also known as ATM block transfer (ABT, OBS) is considered to be the next step towards the optical Internet in the near future [7].

In OBS, a control packet is sent first to set up a connection (by reserving an appropriate amount of bandwidth and configuring the switches along a path), followed by a burst of data without waiting for an acknowledgement for the connection establishment. OBS allows the switching of the data channels to remain in the optical domain, whilst doing any resource allocation in the electronic domain [8, 9]. This distinguishes OBS from circuit-switching as well as from other burst-switching approaches using protocols such as Reservation/scheduling with Just-In-Time switching (JIT) and tell-and-wait (TAW), also known in ATM as ABT-DT (Delayed Transmissions), all of which are two-way reservation protocols [7].

OBS can switch a burst whose length can range from one to several packets to a (short) session using one control packet, thus resulting in a lower control overhead per data unit. In addition, OBS uses out-of-band signaling, but more importantly, the control packet and the data burst are more loosely coupled (in time) than in packet/cell switching. In fact, they may be separated at the source as well as subsequent intermediate nodes by an offset time as in the Just-Enough-Time (JET) protocol. Alternatively, an OBS protocol may choose not to use any offset time at the source, but instead, require that the data burst go through, at each intermediate node, a fixed delay that is no shorter than the maximal time needed to process a control packet at the intermediate node.

Due to the limited "opaqueness" of the control packet, OBS can achieve a high degree of adaptivity to congestion or faults (e.g., by using deflection-routing) and support priority-based routing as in optical cell/packet switching. In OBS, the wavelength on a link used by the burst will be released as soon as the burst passes through the link, either automatically according to the reservation made or by an explicit release packet. In this way, bursts from different sources to different destinations can effectively utilize the bandwidth of the same wavelength on a link in a time-shared, statistical multiplexed fashion [9]. Note that FDLs providing limited delays at intermediate nodes, which are not mandatory in OBS when using the JET protocol, would help reduce the bandwidth waste due to blocked and dropped bursts and improve performance in OBS [8].

Several schemes have been proposed as variants of OBS, such as Tell-and-Go (TAG), Tell-and-Wait (TAW), Just-Enough-Time (JET) and Just-In-Time (JIT) [7]. In the TAG scheme, the source transmits the SETUP message and immediately after it transmits the optical burst. The TAW scheme is the opposite of the TAG scheme where the SETUP message is propagated all the way to the receiving end-device and each OBS node along the path processes the SETUP message and allocates resources within its switch fabric. A positive acknowledgment is returned to the transmitting end-device, upon receipt of which the end-device transmits its bursts and the burst will go through without been dropped at any OBS node. The offset can be seen as being equal to the round trip propagation delay plus the sum of the processing delays of the SETUP message at each OBS node along the path. In the case of JET and JIT, there is a delay called offset between the transmission of the control packet and the transmission of the burst. The JIT scheme uses immediate configuration in which the OBS node allocates resources to the incoming burst immediately after it processes the SETUP message and is simpler to implement [7, 10].

JET is based on TAG and the latency is calculated by $(t + p) \cdot L + T + d$, where t is the transmission time of the control packet, L is the total number of hops, p is the propagation delay for a single hop, T is an offset time and d is the transmission time of the data burst [3]. However, JET is considered in [8] to be based on reserve-a-fixed-duration (RFD) instead. In RFD, similarly to TAG, a source node transmits a control packet to reserve bandwidth, and transmits the burst after offset time, T. The difference here is that the bandwidth is reserved only for a certain period of time, specified by the control packet itself. This of course adds the constraint that the upcoming burst will have a certain maximum size [8].

20.3 Traffic Flow Models for WANs

Some of the main bursty and correlated traffic models for wide area networks (WANs) are reviewed below.

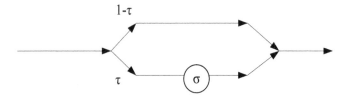

Figure 20.1 The GE distribution two stages.

20.3.1 The Generalized Exponential (GE) as an Inter-Arrival Time Distribution

The GE-type distribution [11] is uniquely determined by

$$F(t) = P(W \leq t) = 1 - \tau e^{-\sigma t}, \quad t \geq 0 \qquad (20.1)$$

where

$$\tau = \frac{2}{C_a^2 + 1}, \quad \sigma = \tau \lambda,$$

W is the random variable of an inter-arrival time and $(1/\lambda, C_a^2)$ are respectively, the corresponding mean and squared coefficient of variation (SCV). The GE distribution is a limited case of the Hyperexponential-2 (H_2) distribution and it is composed of two parallel imaginary stages ($i = 1, 2$), one with an exponential inter-arrival time with stage selection probability τ and mean $1/\tau\lambda$ and one with 0 inter-arrival time with stage selection probability $1 - \tau$ (see Figure 20.1).

According to the definition of the GE-type distribution, an individual arrival that selects the exponential interarrival time stage with probability τ is considered to be the head of an incoming data burst (batch). Upon completion of this non-zero inter-arrival time, all the following individual arrivals that enter the zero interarrival time branch of the GE distribution with probability $1 - \tau$ are considered to be the remaining members of the batch (see Figure 20.3). Note that the GE distribution has a counting compound Poisson process (CPP) of geometrically distributed batch sizes with mean $1/\tau$ [11]. Thus, it may be meaningfully used to model the inter-arrival times of bursty traffic represented by incoming data bursts (or, batches) (see Figure 20.2).

The choice of the GE distribution is further motivated by the fact that measurements of actual inter-arrival or service times may be generally limited and so only few parameters can be computed reliably. Typically, only the mean and variance may be relied upon, thus a choice of a distribution which

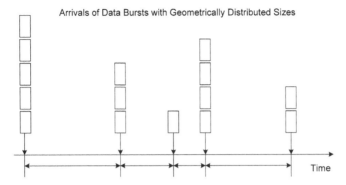

Figure 20.2 A schema of arriving of data bursts (batches).

implies least bias (for example introduction of arbitrary and therefore false assumptions) is that of the GE type distribution.

In the context of optical networks the GE-type distribution is particularly applicable for modelling bursty traffic [12]. Lightpath session arrivals in an optical circuit switching network, may be modeled by the GE representing batches of lightpath session arrivals leaving an edge node.

The GE-type distribution is versatile due to its pseudo-memoryless properties, which makes the solution of GE-type queues analytically tractable [11]. Note that, when SCV $C_a^2 = 1$, the GE distribution reduces to an Exponential distribution [13].

20.3.2 The Self-Similar Nature of Internet Traffic

Although the GE distribution captures traffic burstiness, nevertheless it does not actually reflect the true statistical characteristics of IP packet arrivals at the optical ingress node, which resides at the edge of an OBS network. Note that an ingress node receives packets in the electronic domain (usually IP Packets), aggregates them into optical bursts, which it forwards into the core optical network. In contrast, an egress node, which also resides at the edge of an optical network, receives optical bursts from the core optical network, converts them into the electronic domain, splits them into their individual packets and then forwards them to the various connecting networks.

For over a decade, it is known that traffic patterns in a diverse variety of packet switched networks show self-similar characteristics in both local as well as wide area networks [14]. As it was illustrated in [15], WAN traffic has self-similar properties and that although File Transfer Protocol (FTP) and

Network Virtual Terminal Protocol (TELNET) session arrivals could potentially be characterized by Poisson [15], the actual packet flows arrive in a very
bursty manner whilst their sizes appear to follow a heavy-tailed distribution.

In the literature self-similarity and LRD are used often interchangeably.
However, there is a number of self-similar processes with self-similarity degree H, called the Hurst parameter, that are not LRD. To avoid such confusion
asymptotic second-order self-similar processes are employed which imply
LRD and vice-versa, under the restriction of $1/2 < H < 1$ [16].

Self-similarity and LRD processes, the Pareto distribution and self-
similar traffic generators based on *on/off sources* and the $M/G/\infty$ delay
system are defined below.

• Second-Order Self-Similarity [16]
The process $X(t)$ is *exactly second-order self-similar* with Hurst parameter
$H(1/2 < H < 1)$ if

$$\gamma(k) = \frac{\sigma^2}{2}((k+1)^{2H} - 2k^{2H} + (k-1)^{2H}) \qquad (20.2)$$

for all $k \geq 1$, where, $\sigma^2 = E[(X(t) - \mu)^2]$, $\mu = E[X(t)]$ for all $t \in Z$, Z
being the set of integers and $\gamma(k)$ is the autocovariance function.

The process $X(t)$ is *asymptotically second-order self-similar* if

$$\lim_{k \to \infty} \gamma^{(m)}(k) = \frac{\sigma^2}{2}((k+1)^{2H} - 2k^{2H} + (k-1)^{2H}) \qquad (20.3)$$

where $\gamma^{(m)}(k)$ is the autocovariance function of the *aggregated process* $X^{(m)}$
of X at aggregation level m.

• Long-Range Dependence (LRD) [16]
By definition, $r(k) = \gamma(k)/\sigma^2$ denotes the *autocorrelation function*. For
$0 < H < 1, H \neq 1/2$,

$$r(k) \approx H(2H-1)k^{2H-2}, \quad k \to \infty \qquad (20.4)$$

In particular, if $1/2 < H < 1, r(k)$ behaves asymptotically according to
a power law as $ck^{-\beta}$ for $0 < \beta < 1$, where $c > 0$ is a constant, $\beta = 2 - 2H$
and

$$\sum_{k=-\infty}^{\infty} r(k) = \infty \qquad (20.5)$$

When $r(k)$ decays hyperbolically such that condition (20.5) holds, the corresponding stationary process $X(t)$ is called *long-range dependent*.

• On-Off Traffic Sources [16]

The 0/1 reward renewal process consists of N independent traffic sources represented by the stationary process $X_i(t), i \in [1, N]$, each with independently and identically distributed (i.i.d) *on* and *off* periods. Being in the *on* period, it means that the source is transmitting packets (usually modeled by a constant rate).

For the On/Off model, the total traffic up to time Tt, where $T > 0$ is an explicitly incorporated scale factor, is denoted by the cumulative process $Y_N(Tt)$ defined as

$$Y_N(Tt) = \int_0^{Tt} \left(\sum_{i=1}^N X_i(s) \right) ds,$$

where $\sum_{i=1}^N X_i(t)$ denotes the aggregate traffic at time t. For large T and large number of sources N, $Y_N(Tt)$ *behaves statistically as*

$$\frac{E(\tau_{on})}{E(\tau_{on}) + E(\tau_{off})} NTt + CN^{1/2}T^H B_H(t),$$

where τ_{on} and τ_{off} are the on and off periods, $H = (3 - a)/2$, and $B_H(t)$ *with parameter H and $C > 0$* is a real valued function, which depends only on the distributions of τ_{on} and τ_{off}. If the latter are heavy-tailed, then the total traffic is long-range-dependent [16].

• $M/G/\infty$ Delay System [16]

The $M/G/\infty$ queue, also known as the *busy server process*, describes a system with Poisson distributed connection arrivals, which immediately enter service, due to the infinite number of servers with a general service time. If the i.i.d service times are given by $\tau = \tau_\iota, \iota \in Z_+$, where Z_+ is the set of positive integers, then the counting process of the busy servers, describes the aggregate traffic process $X(t)$ in the Poisson source model with a single on period. If the service time distribution τ is heavy-tailed with a tail index α $(1 < \alpha < 2)$, then $X(t), t \in Z_+$, *is asymptotically second-order self-similar process with parameter $H = (3 - a)/2$* [16].

It is shown in [16] that this aggregate traffic is long-range dependent. Essentially, this model provides an attractive approach in network traffic modelling, based on Poisson distributed session arrivals with heavy-tailed sizes and/or durations with infinite variance. The aggregate traffic is used often in simulations to generate self-similar traces, also called synthetic traces. This model results in fractional Brownian motion with the proper scaling of the arrival and service distributions [17].

● **Pareto Distribution [18]**

When generating synthetic self-similar traces with the $M/G/\infty$ method, the Pareto distribution is widely used as it is a heavy tailed distribution. Its Cumulative Distribution Function (CDF) of Pareto random variable X is given by

$$F(x) = P[X \leq x] = 1 - \left(\frac{\beta}{x}\right)^a \tag{20.6}$$

where β is the minimum value of the random variable and α the so-called shape parameter indicating the tail index. When $1 \leq a \leq 2$, the Pareto random variable X has finite mean and infinite variance and the shape parameter's relation with the Hurst Parameter H, which measures the degree of self-similarity is given by $H = (3 - a)/2$ [18].

20.4 Optical Burst Aggregation Strategies

Incoming traffic to an optical backbone network, is consisted of IP packets coming from several inter-connecting networks. This aggregated traffic, causes the IP packets to arrive at the optical edge node in batches. In addition, due to the aggregated traffic's LRD properties, self-similarity persists even after the aggregation of traffic over longer timescales.

20.4.1 OBS Ingress Node

OBS is thought to be a valid, plausible IP-over-WDM implementation in the near future [1, 2, 19, 20]. Switching overhead in the core node inside the optical network can be quite large, which makes large-sized optical bursts a requirement for its alleviation. Consequently, IP packets arriving in the OBS ingress nodes are assembled into optical bursts based on destination or priority [21].

20.4.2 Burst Assembly Traffic Shaping

Many investigations have been conducted on the kind of impact of the burst assembly procedure with respect to inter-departure time and sizes of bursts

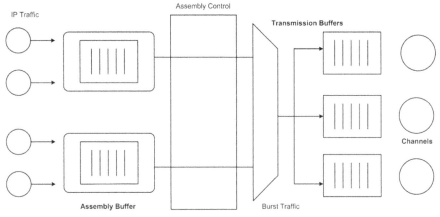

Figure 20.3 An ingress node to an optical network.

on the traffic characteristics [22–24]. It was shown in [25–28] that LRD of the incoming IP Traffic [29] persists after the burst assembly procedure. Nevertheless, a "shaping effect" is applied in the form of self-similarity degree reduction [30–32].

An ingress node model with a single buffer was proposed in [33], where arriving IP packets are classified based on their destination egress nodes and their priority class. The traffic of bursts, which is produced by the assembly procedure, is multiplexed into their respective unbounded transmission buffers. This makes the modelling of the ingress node easier to analyse. Note that a more realistic system design would be to have dedicated transmission buffers to each wavelength frequency (see Figure 20.3). However, such would require intense computational power.

20.5 Simulating Optical Networks with Self-Similar Traffic

20.5.1 Traditional Simulation Problems

Due to the high complexity of the self-similarity and the associated analytical problem, simulation became the most favorable tool to evaluate the performance of networks. Consequently, many self-similar trace generators were invented towards the creation of synthetic traces of network traffic such as *On/Off Sources* and $M/G/\infty$ (cf., Section 20.3).

The inherent complexity of optical networks is an additional obstacle to the speed and duration of the simulations. Due to WDM/DWDM, current

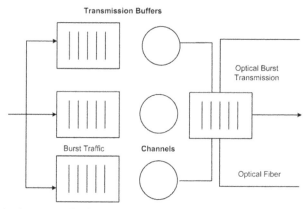

Figure 20.4 An ingress node that transmits optical bursts into the core optical network, multiplexed from several wavelengths.

optical fiber speeds in WANs exceed several Tbps, when each wavelength's speed is 10–40 Gbps. A simulation would practically need millions of simulated optical bursts and for each optical burst a large number of IP Packets. Therefore, a Uniform Random Number Generator with an extended period is necessary. One such huge-period generator was proposed in [34], called Mersenne Twister (MT), which has a massive prime period of $2^{19937} - 1$.

However, the computational requirements remain extremely taxing when running an optical network simulation that is fed by self-similar LRD traffic flows on an average home computer. Thus, supercomputers and other parallel systems are usually chosen for these types of simulations that potentially have very high costs and are not always readily available.

20.5.2 Simulating an Optical Network Edge Node on a Parallel System

When simulating optical circuit switching lightpath session arrivals, the task translates into generating session inter-arrival times. When the inter-arrival distribution is known or selected, this can be a relatively easy problem. Due to WDM, however, there is a large number of wavelengths that act as generators of lightpath session arrivals. Each wavelength could be modeled as an i.i.d. lightpath generator. It would therefore be beneficial if all of these generators could be simultaneously simulated (see Figure 20.4).

When dealing with LRD and self similarity, it is shown in [15, 33] that after aggregating IP packets into optical bursts, the degree of self-similarity decreases slightly due to the shaping of the incoming traffic. Nevertheless,

Transmission Buffers

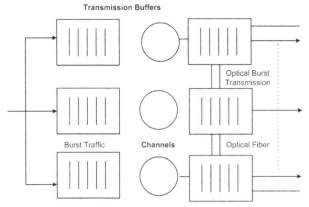

Figure 20.5 Ingress node that transmits optical bursts into the core optical network, multiplexing several wavelengths (decomposed).

LRD properties persist with different Hurst parameter and/or mean. Each wavelength transmission buffer would have its own traffic load with similar LRD characteristics to the incoming IP traffic at the node. Moreover, optical packets and bursts are transmitted into the optical network core, according to their routing, destination and priority (see Figure 20.4). Until now, many models used only one transmission buffer containing all optical packets and bursts as this would make writing a simulation easier, especially since for many years, simulations run on single core processor machines [33]. When incorporating several transmission buffers, however, it is possible for the simulation of each wavelength to decompose the overall transmission of the node into several smaller transmission generators, provided enough processors or system cores exist to simulate each wavelength concurrently and separately (see Figure 20.5).

Let the aggregate incoming traffic from various access networks (internet, LANs, etc.) arrive at the optical edge node with an initial Hurst parameter H. After the traffic shaping is conducted at the edge node, the optical bursts leaving the node would also have self-similar properties with a Hurst Parameter H', $H > H'$ [27, 33]. Similarly, each transmission rate for each channel would have self-similar properties and samples could be generated concurrently for each one, assuming the exact parameter values have been estimated (see Figure 20.5). Each generator could run on a different processor, thus speeding up the simulation significantly limited only by the number of processors available.

20.6 Simulating on a GPU

Modern CPUs have multiple cores, which means in essence that more than one processor exists within the same chipset. Today, CPUs with four cores have increased performance by allowing more processes to be handled simultaneously. However, even if the first steps towards multiple processors in one chipset have been made, it is hardly enough.

Alternatively, graphics cards created by NVidia [27] may be employed for the simulations to run on, instead of the actual CPU of the system. These Graphics Processing Units (GPUs) have proven to be a good cost-effective solution to the lack of processing power problem.

20.6.1 Differences between CPU and GPU

In the past five years, a lot of progress has been made in the field of graphics processing, giving birth to graphics cards that have a large amount of processors as well as memory sufficient to allow their use to other fields than simply processing and depicting graphics. The processing power (cf., floating-point operations per second) superiority of modern graphics cards (GPUs) to that of the CPU with NVidia's GeForce 8800 GTX reaches almost 340 GFlops whilst newer models like the GeForce GTX 280, can reach almost 900 GFlops [35]. Nevertheless, even if the GPU appears to have significantly more power than the conventional CPU, it is worth noting that the CPU is capable of handling different kinds of processes quickly, while the GPU is only capable of processing a specific task very fast. This latter task needs to be in the form of a problem composed of independent elements, due to the large parallelization of GPUs [35].

Note that the GeForce 8800 GTX is equipped with 128 scalar processors divided into 16 groups – called multiprocessors – of 8. The calculation power of such GPU was shown in [36] to be clearly superior to the CPU. However, the difference will not be that large unless this power is efficiently utilized.

20.6.2 GPU Architecture

Via the CUDA framework, the GPU is exposed as a parallel data streaming processor, which consists of several processing units. CUDA applications have two segments. One segment is called "kernel" and is executed on the GPU. The other segment is executed on the host CPU and controls the execution of kernels and transfer of data between the CPU and GPU [35].

Several threads that run on the GPU run a kernel. These threads belong to a group called block. Threads within the same block may communicate with each other using shared memory and may not communicate with threads of another block. There is a hierarchical memory structure, where each memory level has different size, restrictions and speed [35].

A disadvantage of the GPU was that it adopted a 32-bit IEEE floating-point numbers, which is well lower than the one supported by a general-purpose CPU. However, the recently released NVIDIA 280 series supports double precision floating point numbers.

20.7 Numerical Results

Simulating a decomposed optical edge (see Figure 20.5) node would require separate self-similar generators for each transmission rate. This transforms optical burst generation into a problem that can be parallelized into multiple processor cores. However, running each of these self-similar generators on a single CPU, would introduce a significant overhead as the self-similar generators can be extremely time-consuming. Assuming the time required for these generators can be minimized, this would greatly increase the efficiency of optical packet/burst generation simulations.

The aim of the experiments of this section is to investigate whether the use of a GPU for traffic generators can improve their efficiency and minimize their time requirements. Three types of generators were chosen, a generator based on the *GE-type distribution* for lightpath session arrivals, an LRD generator based on the $M/G/\infty$ *delay system* and an LRD generator based on *On/Off sources*. The length of the generated samples was limited specifically to reveal potential bottlenecks when running these generators on the GPU versus the CPU.

Simulations of a decomposed optical edge (see Figure 20.5) node were conducted on an NVidia 8800 GTX. For this investigation many different scenarios were considered, in order to provide a deeper insight on the GPU's capabilities on various test cases.

Lightpath session arrivals were generated based on the GE-type distribution, so that batches of sessions are taken into consideration. A total of 24,002,560 lightpath sessions were created, initially on one block on the GPU and on three types of general purpose CPUs, single, dual and quad core (Figure 20.6). Initially, measurements were taken by increasing the number of threads that simultaneously ran on the same block.

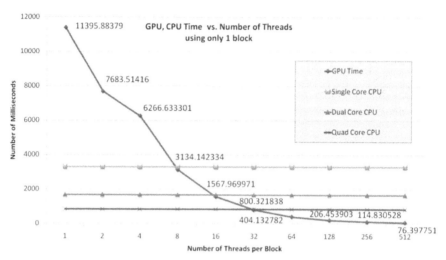

Figure 20.6 Simulation duration in milliseconds, of GPU and CPU simulations versus number of simultaneous threads per block.

The results showed that by using only one block (i.e., one multiprocessor) on the GPU, the overall simulation duration was significantly longer when the thread number was lower than 8. As the threads were increasing, the performance improvement was significant, surpassing even a Quad core CPU.

The same experiment was conducted, while increasing the number of blocks processing the kernel at the same time. This essentially increased the parallelization of the generator, increasing dramatically the performance. To this end, a parallel version of Mersenne Twister was employed to allow for simultaneous random number generators on each of the 190 wavelengths [37].

As the number of blocks is increasing, significant performance improvement is observed, as more blocks use more multiprocessors (see Figure 20.7). Note that a total of 16 blocks are required for all multiprocessors of the NVidia 8800 GTX to be used and up to 32 to be properly utilized for maximum performance.

However, as mentioned in Sections 20.3 and 20.4, the Poisson arrival process and the compound Poisson arrival process are not suitable to simulate network traffic for an OBS Network. Therefore, self-similarity generators need to be implemented since most are computational intensive. The $M/G/\infty$ generator was chosen specifically because it is not easily imple-

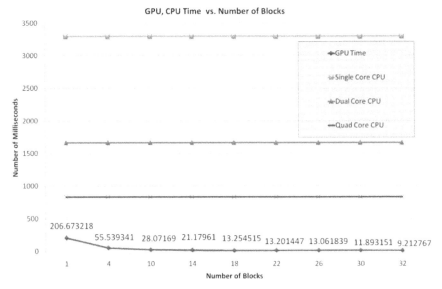

Figure 20.7 Performance comparisons between the GPU and different types of CPUs. Graph shows duration in milliseconds versus number of blocks.

mented on a parallel system, to allow increased complexity of calculations for each wavelength.

To this end, an optical edge node is being simulated having 190 wavelengths. Each wavelength produces streams of optical bursts, with a traffic load that exhibits self-similar properties. For demonstration purposes the generated samples per wavelength were limited to 302084, for a total of 58000128 bursts for the entire node. For each wavelength a series of self-similar traces is generated based on simulating individual $M/G/\infty$ *delay systems*. The results shown in Figure 20.8 indicate that the GPU performs even better when the complexity of the problem is high in comparison to the CPU, provided it has been parallelized enough to efficiently utilize the GPU's resources. This observation was also validated in [36].

An experiment with *On/Off Sources* was also conducted. This method is inherently easy to implement on a parallel system. It involves generating traffic from each wavelength of the fiber independently on a parallel system and then aggregating them into a multiplexed stream. Aggregation, however, needs to run on a single core since it requires access to all generated traffic streams. Essentially this creates a bottleneck, which limits the overall performance of the GPU. To illustrate this, the amount of traffic for

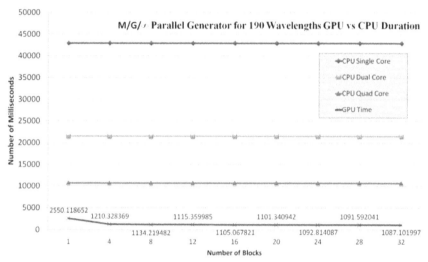

Figure 20.8 Performance comparisons between the GPU and different types of CPUs. Graph shows duration in Milliseconds versus number of blocks. Number of blocks start with 1 block using 16 Threads and continues increasing the blocks employing 128 threads per block.

each wavelength was limited. As shown in Figure 20.9, by increasing the number of threads per block and by using only one block, the GPU eventually surpasses the quad core CPU once the number of threads exceeds 128. Nevertheless, the performance difference is not that great. This is due to the aggregation overhead.

Furthermore, it is shown in Figure 20.10 that by increasing the number of blocks, no real difference in performance, is observed since the number of samples is too small whereas the aggregation overhead too great.

Increasing the number of samples would actually increase the performance difference between the CPU and the GPU, while at the same time diminishing the aggregation's overhead effect. This leads to the conclusion that without proper parallelization, the GPU remains significantly underutilized, thus making the general-purpose CPU the main choice for processes that were programmed without considering a parallel system.

20.8 Conclusions

The method of decomposition of an optical node and the simulation of the various segments in parallel are presented. These simulations run on an NVidia 8800 GTX graphics card acting as a parallel system. Results showed

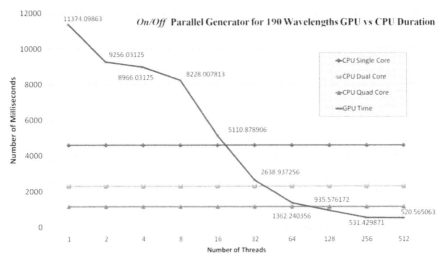

Figure 20.9 Performance comparisons between the GPU and different types of CPUs. Graph shows duration in Milliseconds versus number of threads per block.

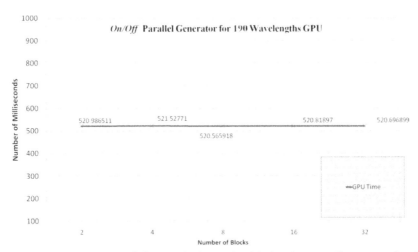

Figure 20.10 Performance of GPU while increasing blocks, for a small amount of traffic traces.

a GPU can provide significantly faster results in comparison to the CPU, provided the optical node simulation is decomposed properly and the simulation is designed for a parallel system. This allows independent systems based on different wavelengths to be simulated simultaneously. However, it is required that simulation designers develop their simulations efficiently;

otherwise worse performance may be experienced. Simulations, need to have high complexity and duration in order to justify the use of a GPU. In the future, more and more processing cores are expected to appear and the research community needs to create cost-effective algorithms that take advantage of their potential.

In the meantime, a hybrid simulation approach is advised, to allow multithreaded segments to be calculated on the GPU and single core bottlenecks on the CPU, in order to achieve maximum performance from the current technology.

20.9 Future Work

An entire new framework needs to be created around NVIDIA's CUDA [35] to allow speeding up self-similar traffic generators for realistic network traffic. In addition the self-similarity estimators like the Whittle Estimator and the Rescaled Range Analysis [38] need to be developed under CUDA to quicken processing of traces. A future investigation will focus on the optical burst assembly, under all possible aggregation strategies, on an optical edge node with incoming self-similar IP traffic. The GPU's processing capacity can allow simulation of a more detailed system without making approximations, which can lead to more accurate results. In addition analysis of the measured data (like distribution fitting) on a GPU could reveal statistical characteristics of the inter-arrival times of optical bursts and burst size distributions in considerably less time. This would reveal the amount of traffic shaping that actually occurs at the edge node. Similarly a comparison between CPU and GPU durations will be performed and possible bottlenecks will be revealed, both during the actual assembly as well as during the Hurst parameter estimation.

References

[1] S. R. Amstutz, Burst switching – An introduction, *IEEE Communications Magazine*, vol. 21, pp. 36–42, November 1983.

[2] M. Listanti, V. Eramo and R. Sabella, Architectural and technological issues for future optical Internet networks, *IEEE Communications Magazine*, vol. 38, no. 9, pp. 82–92, September 2000.

[3] M. Yoo and C. Qiao, Just-enough-time (JET): A high speed protocol for bursty traffic in optical networks, *Proceedings of the IEEE/LEOS Conference on Technologies for a Global Information Infrastructure*, pp. 26–27, August 1997.

[4] K. Dolzer, C. Gauger, J. Späth and S. Bodamer, Evaluation of reservation mechanisms for optical burst switching, *AEÜ International Journal of Electronics and Communications*, vol. 55, no. 1, pp. 1–8, January 2001.

[5] Introduction to DWDM for Metropolitan Networks, Corporate Headquarters Cisco Systems Inc., San Jose USA, Available at http://www.cisco.com/univercd/cc/td/doc/product/mels/dwdm/dwdm.pdf

[6] B. Mukherjee, D. Banerjee and S. Ramamurthy, Some principles for designing a wide-area WDM optical betwork, *IEEE/ACM Transactions on Networking*, vol. 4, no. 5, pp. 684–696, 1996.

[7] L. Xu, H. G. Perros and G. Rouskas, Techniques for optical packet switching and optical burst switching, *IEEE Communications Magazine*, vol. 39, pp. 136–142, January 2001.

[8] C. Qiao and M. Yoo, Optical burst switching (OBS) – A new paradigm for an optical internet, *Journal of High Speed Networks*, vol. 8, pp. 69–84, 1999.

[9] S. Verma, H. Chaskar and R. Ravikanth, Optical burst switching: A viable solution for terabit IP backbone, *IEEE Network*, vol. 14, no. 6, pp. 48–53, November/December 2000.

[10] G. Rouskas and H.G. Perros, A tutorial on optical networks, in *Advanced Lectures on Networking*, E. Gregori, G. Anastasi and S. Basagni (Eds.), Springer, Berlin/Heidelberg, pp. 155–193, 2002.

[11] D. D. Kouvatsos, Entropy maximization and queueing network models, *Annals of Operation Research*, vol. 48, pp. 63–12, 1994.

[12] K. Dolzer and C. Gauger, On burst assembly in optical burst switching networks – A performance evaluation of Just-Enough-Time, in *Proceedings of the 17th International Teletraffic Congress (ITC 17)*, Salvador, pp. 149–160, 2001.

[13] D. D. Kouvatsos, Maximum entropy and the G/G/1 queue, *Acta Informatica*, vol. 23, pp. 545–565, September 1986.

[14] V. Paxson and S. Floyd, Wide area traffic: The failure of Poisson modelling, *IEEE/ACM Transactions on Networking*, vol. 3, pp. 226–244, 1995.

[15] W. Leland, M. Taqqu, W. Willinger and D. Wilson, On the self-similar nature of ethernet traffic, in *IEEE/ACM Transactions on Networking*, vol. 2, pp. 1–15, 1994.

[16] K. Park and W. Willinger, Self-similar network traffic: An overview, in *Self-Similar Network Traffic and Performance Evaluation*, K. Park and W. Willinger (Eds.), Wiley-Interscience, pp. 1–38, 2000.

[17] F. Brichet, A. Simonian, L. Massoulié and D. Veitch, Heavy load queueing analysis with LRD on/off sources, in *Self-Similar Network Traffic and Performance Evaluation*, K. Park and W. Willinger (Eds.), Wiley-Interscience, pp. 115–142, 2000.

[18] K. M. Rezaul and V. Grout, A comparison of methods for estimating the tail index of heavy-tailed internet traffic, in *Innovative Algorithms and Techniques in Automation, Industrial Electronics and Telecommunications*, T. Sobh, K. Elleithy, A. Mahmood and M. Karim (Eds.), Springer, Dordrecht, pp. 219–222, 2007.

[19] C. Qiao and M. Yoo, Choices, Features and issues in optical burst switching, *Optical Networks Magazine*, vol. 1, no. 2, pp. 36–44, 2000.

[20] M. Yoo, C. Qiao and S. Dixit, Optical burst switching for service differentiation in the next generation optical internet, *IEEE Communications Magazine*, vol. 39, no. 2, pp. 98–104, 2001.

[21] D. Zhemin and M. Hamdi, Optical network resource management and allocation, in *The Handbook of Optical Communication Networks*, M. Ilyas and H. T. Mouftah (Eds.), CRC Press, 2003.

[22] K. Laevens, Traffic characteristics inside optical burst switched networks, in *Proceedings, Optical Networking and Communication Conference (OptiComm)*, July–August 2002, Boston, MA, vol. 4874, pp. 137–148, 2002.

[23] X. Yu, Y. Chen and C. Qiao, A study of traffic statistics of assembled burst traffic in optical burst switched networks, in *Proceedings, SPIE Optical Networking and Communication Conference (OptiComm)* Boston, MA, July–Aug 2002, vol. 4874, pp. 149–159, 2002.

[24] M. de Vega Rodrigo and J. Goetz, An analytical study of optical burst switching aggregation strategies, in *Proceedings of the Third International Workshop on Optical Burst Switching (WOBS)*, 2004.

[25] A. Ge, F. Callegati and L.S. Tamil, On optical burst switching and self-similar traffic, *IEEE Communications Letters*, vol. 4, no. 3, pp. 98–100, March 2000.

[26] F. Xue and S. J. Ben Yoo, Self-similar traffic shaping at the edge router in optical packet-switched networks, in *IEEE International Conference on Communications, ICC 2002*, 28 April–2 May 2002, vol. 4, pp. 2449–2453, 2002.

[27] G. Hu, K. Dolzer and C. Gauger, Does burst assembly really reduce the self-similarity, in *Proceedings OFC 2003*, vol. 86, pp. 124–126, 2003.

[28] X. Yu, J. Li, X. Cao, Y. Chen and C. Qiao, Traffic statistics and performance evaluation in optical burst switched networks, *J. Lightwave Technol.*, vol. 22, p. 2722, 2004.

[29] W. Willinger, V. Paxson, R. H. Riedi and M. S. Taqqu, Long-range dependence and data network traffic, in *Theory and Applications of Long-Range Dependence*, P. Doukhan, G. Oppenheim and M. S. Taqqu (Eds.), Birkhäuser, pp. 373–408, 2002.

[30] M. Izal and J. Aracil, On the influence of self-similarity on optical burst switching traffic, in *Proceedings of IEEE Globecom 2002*, Taipei, Taiwan, November, vol. 3, nos. 17–21, pp. 2308–2312, 2002.

[31] X. Yu, Y. Chen and C. Qiao, Performance evaluation of optical burst switching with assembled burst traffic input, in *Proceedings of IEEE Globecom 2002*, Taipei, Taiwan, November, vol. 3, pp. 2318–2322, 2002.

[32] R. Rajaduray, S. Ovadia and D. J. Blumenthal, Analysis of an edge router for span-constrained optical burst switched (OBS) networks, *J. Lightwave Technol.*, vol. 22, p. 2693, 2004.

[33] G. Hu, Performance model for a lossless edge node of OBS networks, in *Proceedings of IEEE Globecom 2006*, San Francisco, CA, November, pp. 1–6, 2006.

[34] M. Matsumoto and T. Nishimura, Mersenne twister: A 623-dimensionally equidistributed uniform pseudo-random number generator, *ACM Transactions on Modelling and Computer Simulation (TOMACS)*, vol. 8, no. 1, pp. 3–30, January 1998.

[35] NVIDIA, *NVIDIA CUDA Compute Unified Device Architecture*, Programming Guide, November 2007.

[36] D. Triolet, NVIDIA CUDA: Preview – BeHardware, Available at
http://www.behardware.com/art/imprimer/659

[37] V. Podlozhnyuk, Parallel Mersenne Twister, NVIDIA. Available at
http://developer.download.nvidia.com/compute/cuda/1.1-Beta/x86_website/projects/
MersenneTwister/doc/MersenneTwister.pdf

[38] A. Feldmann, Characteristics of TCP connection arrivals, in *Self-Similar Network Traffic and Performance Evaluation*, K. Park and W. Willinger (Eds.), Wiley-Interscience, pp. 367–397, 200.

21

Blocking Analysis in Hybrid TDM-WDM Passive Optical Networks

John S. Vardakas,* Vassilios G. Vassilakis and Michael D. Logothetis

WCL, Department of Electrical & Computer Engineering, University of Patras, 26504 Patras, Greece; e-mail: jvardakas@wcl.ee.upatras.gr

Abstract

Optical access networks are the ultimate solution to the problem of the last mile bottleneck between high-capacity metro networks and customer premises, capable of delivering future integrated broadband services. Nowadays, passive optical networks are the most promising and cost-effective class of fiber access systems. This rapid evolvement of fiber in access networks motivates the study of their performance. In this paper we develop teletraffic loss models for calculating connection failure probabilities (due to unavailability of a wavelength) and call blocking probabilities (due to the restricted bandwidth capacity of a wavelength) in hybrid TDM-WDM passive optical networks with dynamic wavelength allocation. The springboard of our analysis is well-established teletraffic models. The proposed models are derived from one-dimensional Markov chains which describe the wavelength's occupancy in the optical access network. The optical access network accommodates multiple service-classes with infinite or finite number of traffic-sources. The accuracy of the proposed models is validated by simulation and was found to be quite satisfactory.

Keywords: Wavelength division multiplexing, passive optical networks, blocking probability, Markov chains, Poisson process, quasi-random traffic.

---

* Author for correspondence.

D. D. Kouvatsos (ed.), Performance Modelling and Analysis of Heterogeneous Networks, 441–465.

List of Abbreviations

ATM   Asynchronous Transfer Mode
BPON   Broadband Passive Optical Network
CBP   Call Blocking Probability
CFP   Connection Failure Probability
DSL   Digital Subscriber Line
EnMLM   Engset Multi-rate Loss model
EPON   Ethernet Passive Optical Network
FTTH   Fiber-To-The-Home
FTTP   Fiber-To-The Premises
GPON   Gigabit Passive Optical Network
HDTV   High Definition Television
ITU-T   International Telecommunications Union – Telecommunication
             Standardization Sector
OEO   Optical-Electrical-Optical
OLT   Optical Line Terminal
ONU   Optical Network Units
POC   Passive Optical Concentrator
PON   Passive Optical Network
TCBP   Total Call Blocking Probability
TDM   Time Division Multiplexing
WDM   Wavelength Division Multiplexing

21.1 Introduction

The increased demand for more bandwidth and the expansion of the telecommunication services offered to residential homes, have forced conventional access network infrastructures to their bandwidth limits [1]. End users are no longer only interested in voice telephony and broadcast television; there is a massive request for high-speed Internet connections and for access to high-definition multimedia services, such as High Definition Television (HDTV), video conferencing and on-line gaming. This increased demand for more bandwidth has forced telecommunication carriers to develop economical subscriber networks based on optical technology, since current access solutions, like Digital Subscriber Line (DSL) systems became the so-called last mile-bottleneck. The final option for access technology is optical fiber, due to its unique properties, providing huge bandwidth with very low loss.

Optical access networks can be classified into three general categories [2]. In the point-to-point architecture, a dedicated fiber is used to connect directly the customers to the central office. This approach is effective in terms of bandwidth, but it has the limitation of high cost of the network user's equipment, which must realize Optical-Electrical-Optical (OEO) conversions. A second access solution is the active star architecture, a shared fiber approach, where a fiber is used to connect the central office with an active node, placed at the location of the network users. This approach reduces the total fiber length that has to be installed, compared to the point-to-point architecture, since only one fiber is needed for the connection of the active node to the central office and a number of branching fibers to connect each customer to the active node. In a third solution, the active node is replaced with a passive splitter/combiner in passive star architectures. This topology is widely known as the Passive Optical Network (PON). The decision for a particular architecture is a result of economic and technical considerations. Since a single fiber has to be installed for each customer in point-to-point architectures, the active node design and the PON are advantageous, where only one termination is needed. Moreover, the passive nature of PONs reduces the cost of the power supply and maintenance in the active node. Therefore, PONs have gained prominence and are reported as the optimum solution for utilizing the advantages of optical fibers in access networks [3]. PONs are already deployed today, including their application to the Fiber-To-The-Home/Fiber-To-The Premises (FTTH/FTTP) networks [4].

A PON consists of an Optical Line Terminal (OLT), located at the central office and optical nodes which are located in the customers' premises and called Optical Network Units (ONUs) [5]. All ONUs are connected to the OLT through a combiner/splitter which we call Passive Optical Concentrator (POC); the communication between the ONUs is realized only through the OLT (Figure 21.1). The POC is a passive device, where signals from the ONUs are combined and transmitted to the OLT with one fiber; signals from the OLT are separated and then sent to ONUs. In other words, in the downstream direction (from the OLT to the ONUs) traffic is sent from one point to multiple points, while in the upstream direction (from the ONUs to the OLT), traffic from multiple points, reaches only one point. In both directions, the traffic multiplexing is realized in the optical domain; therefore collisions may occur only between optical data streams.

PONs come in different flavors, depending on the multiple access schemes they deploy, in order to provide high utilization of the fiber bandwidth capacity. In Time Division Multiplexing (TDM) PONs, the optical

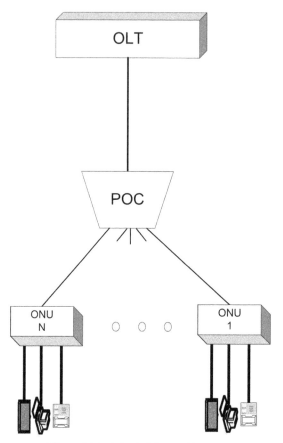

Figure 21.1 A basic PON architecture.

signals from the ONUs are transmitted in different time slots, in order to avoid collisions and are multiplexed at the passive element. Time-based architectures that have been already deployed are Asynchronous Transfer Mode PONs (ATM-PONs), Ethernet PONs (EPONs) and Broadband PONs (BPONs). These architectures use a 1550 nm wavelength for the downstream traffic and a 1310 nm for the upstream traffic [6]. Given the steadily increasing number of users and their bandwidth demands, current TDM-PONs must be upgraded in order to satisfy these growing traffic demands. One approach refers to the increase of the line rate of the existing TDM architectures. The Gigabit-capable PON (GPON) is specified by the International Telecommunications Union-Telecommunication Standardization Sector (ITU-T) G.984

[7]. GPON is capable of providing higher bandwidth, namely a line rate of 2.4 Gbps downstream and 1.2 Gbps upstream, supporting up to 64 users. GPONs support a framing format of 125 µs, which can host both Ethernet and ATM frames. GPON also supports quality of service, as it enables Service Level Agreement (SLA) negotiations between the OLT and the ONU.

Another upgrading approach for the TDM-PONs is to combine the Wavelength Division Multiplexing (WDM) technology together with the TDM technology. In the resulting hybrid TDM-WDM PON, multiple wavelengths are used in the upstream and downstream directions, so that the access network becomes flexible and efficient in providing the required bandwidth to the users [8]. In early TDM-WDM PONs systems, each ONU uses expensive optical transmitters, generating a unique wavelength. The assignment of the wavelength channels in these systems is static; each ONU is connected to the OLT by a specific pair of wavelengths, while wavelength reallocation is not possible. Under dynamic wavelength assignment, the wavelength routing is flexible and the performance of the PON is significantly enhanced. In these colorless-ONU architectures a wavelength-specific laser diode is necessary in the upstream direction, which increases the cost. Alternatively, other colorless-ONU solutions may reduce the total cost, such as the "spectral slicing" approach [9] and the installation of reflective modulators at each ONU [10]. These cost-effective approaches [1, 11] make TDM-WDM PONs the ideal solution for broadband access networks.

In this paper, we develop analytical loss models for calculating blocking probabilities in the upstream direction of a hybrid TDM-WDM PON, with dynamic wavelength allocation. In order for an ONU to be connected with the OLT, a free wavelength must be seized in the PON (actually the problem of finding a free wavelength is localized in the link between the POC and the OLT). We calculate the Connection Failure Probability (CFP), which occurs due to the unavailability of a free wavelength in the link between the POC and the OLT. We also calculate Call Blocking Probabilities (CBP); more precisely, the CBP that occurs after the establishment of the OLT-ONU connection, due to the restricted bandwidth capacity of the wavelength, as well as the Total CBP (TCBP) that occurs to a call either due to the inexistence of a free wavelength or due to the limited bandwidth capacity of the wavelength. We consider that calls belong to different service-classes with infinite or finite traffic-source population. The number of ONUs is finite, but large enough, so that they are more than the number of wavelengths in the PON. This consideration is done in order for our study to be meaningful (despite of the fact that it is also realistic). The proposed models are computationally efficient, because they are based on recursive formulas. Our analysis is validated through simulation; the accuracy of the proposed models was found to be quite satisfactory.

This paper is organized as follows. In Section 21.2 we describe the basic features of our service model, while in Sections 21.3 and 21.4 we present the analytical models for the infinite and finite traffic-sources cases, respectively. Section 21.5 is the evaluation section and we conclude with Section 21.6.

21.2 Service Model of the Hybrid WDM-TDM Passive Optical Network

We consider a hybrid TDM-WDM PON with multistage splitting, as shown in Figure 21.2. Multistage splitting is a technique that is used for increasing the number of the network's customers and also the utilization of the wavelength's bandwidth capacity [12, 13]. Thanks to dynamic wavelength allocation scheme that is applied to the PON, each ONU has the ability of connecting to the OLT by using any available wavelength; in contrast to the static wavelength allocation scheme where a certain wavelength is assigned to an ONU. The POC comprises a Passive Wavelength Router (PWR) which is responsible for routing multiple wavelengths in a single fiber toward the OLT.

The connection establishment procedure of an ONU to the OLT is as follows. When a call arrives to an ONU while no other calls either of the same ONU or the other ONUs in the group are in service, it searches for an available wavelength in the PON. This is actually done through connection setup signals between the OLT and the ONU. If a free wavelength is found by the OLT, it is assigned to the group of ONUs and to the call. Then, a connection is established and the call is serviced, otherwise (if the search of a free wavelength is unsuccessful) the call is blocked and lost (connection failure occurs). Since after the connection establishment, all calls from the same ONU group seize the same wavelength, while the bandwidth capacity of the wavelength is restricted, call blocking occurs. When all calls on a wavelength terminate, the connection also terminates and the wavelength releases its resources to become available to any new arriving call from any ONU group.

In the following sections we present analytical models for the CFP, the CBP and the TCBP in the upstream transmission, namely the direction from the ONUs to the OLT, while assuming that each ONU accommodates multiple service-classes with infinite or finite number of traffic-sources. The

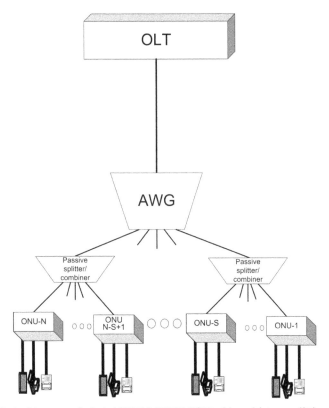

Figure 21.2 Architecture of a hybrid TDM-WDM PON with multistage splitting technique.

presented models involve the number of the ONUs per splitting device in a parametric way; therefore, they can also be applied to the case where the hybrid TDM-WDM PON is implemented with one-stage splitting as shown in Figure 21.1.

21.3 Model for Service-Classes with Infinite Number of Traffic-Sources

Let C be the number of different wavelengths and N be the number of ONUs in the PON of Figure 21.2. Each splitting device supports a group of S ONUs ($S<N$); therefore the number of the splitters in the access network is N/S (where N is a multiple of S). Each wavelength has a bandwidth capacity of T bandwidth units (b.u.). Calls arrive to the ONU according to a Poisson process (random call arrivals) and are groomed onto one wavelength. Each

ONU accommodates K service-classes. In order for each service-class k call to be serviced, it requires b_k ($k = 1, \ldots, K$) b.u. of the wavelength. The b.u. of each wavelength are commonly shared among the arriving calls. If the requested b.u. of a call are not available, the call is blocked and lost (Complete Sharing policy [14]). The arrival rate of service-class k calls is denoted by λ_k, while the service time is exponentially distributed with mean μ_k^{-1}.

In order to derive the CFP, we formulate a Markov chain with the state transition diagram of Figure 21.3, where the stage j represents the number of occupied wavelengths in the PON. The total arrival rate of calls from an ONU is denoted by $\lambda = \sum_{k=1}^{K} \lambda_k$. The establishment of a connection is realized with a rate that depends on the number of the ONUs, which have not established a connection yet. Having established a connection of one ONU, we have actually connected all ONUs of the same group. Therefore, after one connection establishment, the number of ONUs, which have not established a connection, is reduced by S (the number of ONUs in a group). Because of this, when the system is in state $[j-1]$, it will jump to the state $[j](N-(j-1)S)\lambda$ times per unit time (Figure 21.3), where $N-(j-1)S$ is the number of ONUs which have not been connected yet.

In order to define the rate by which a wavelength is released, we must consider not only the number of the occupied wavelengths j, but also the fact that the release of a wavelength coincides with the release of the last call, which seizes the wavelength and may belong to any service-class. The downward transition from state $[j]$ to state $[j-1]$ is realised jQ times per unit time, where Q corresponds to the mean service/seize rate of a wavelength. The rate Q can be determined by the product of (the conditional probability that b_k b.u. are occupied in the wavelength by only one call of service-class k, given that the wavelength is occupied) by (the corresponding service rate μ_k). Since the last call could belong to any service-class:

$$Q = \sum_{k=1}^{K} \mu_k y_k\,(b_k)\,\hat{q}(b_k) = \sum_{k=1}^{K} \mu_k y_k\,(b_k)\,\frac{q\,(b_k)}{\sum_{i=1}^{T} q(i)} \tag{21.1}$$

where $y_k(b_k)$ is the mean number of service-class k calls (with bandwidth requirement b_k) when b_k b.u. are occupied in the wavelength, and $q(i)$ is the occupancy distribution of the b.u. in a wavelength, which can be calculated by the well-known Kaufman–Roberts recursion for service-classes with infinite

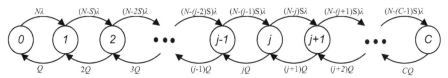

Figure 21.3 State transition diagram of a TDM-WDM PON with C wavelengths and dynamic wavelength assignment.

traffic-source population [15, 16]:

$$iq(i) = \sum_{k=1}^{K} a_k b_k q (i - b_k) \text{ for } i = 1, \dots, T \quad (21.2)$$

with $q(i) = 0$ for $i < 0$ and $\sum_{k=1}^{K} q(i) = 1$. The offered traffic-load a_k for service-class k comes from the group of S ONUs, therefore:

$$a_k = S \lambda_k \mu_k^{-1} \quad (21.3)$$

where the product $\lambda_K \mu_k^{-1}$ represents the offered traffic-load of one ONU. This is possible due to the Poisson process.

The calculation of the service rate Q of a wavelength is based on the fact that the release of a wavelength is realized through the end of the service of its last call that may belong to any service-class. The release of a wavelength, after the departure of the last service-class k call, occurs when the number of occupied b.u. in the wavelength is b_k. Therefore, the release of this last call occurs with a rate $\mu_K \cdot y_k(b_k) \cdot \hat{q}(b_k)$. The mean number of service-class k calls is given by [15]:

$$y_k (i) = \frac{a_k q (i - b_k)}{q (i)} \quad (21.4)$$

The probability $P(j)$ that j wavelengths are occupied in the link can be derived from the rate balance equations of the state transition diagram of Figure 21.3. A method for deriving the distribution $P(j)$ can be found in [14]. More specifically, from the rate-out = rate-in, we get the steady state equation:

$$(N - (j - 1)S)\lambda P(j - 1) + [(j + 1)Q]P(j + 1) = [(N - jS)\lambda + jQ]P(j) \quad (21.5)$$

where $P(j) = 0$ for $j < 0$ and $j > C$. By summing up side by side of (21.5) from $j = 0$ to $j - 1$, we get the recurrence formula:

$$P(j) = \frac{\lambda}{Q} \frac{N - (j - 1)S}{j} P(j - 1) \quad (21.6)$$

Consecutive applications of (21.6) yields to the equation:

$$P(j) = \left(\frac{\lambda}{Q}\right)^j \cdot \frac{\prod_{i=1}^{j}[N - (j-1)S]}{j!} \cdot P(0) \qquad (21.7)$$

where $P(0)$ is the probability that no wavelengths are in service. Using the normalization condition,

$$\sum_{j=0}^{C} P(j) = 1 \qquad (21.8)$$

it follows that:

$$P(0) = \left[\sum_{l=0}^{C} \left(\frac{\lambda}{Q}\right)^l \frac{\prod_{i=1}^{l}[N - (j-1)S]}{l!}\right]^{-1} \qquad (21.9)$$

Therefore, by using (21.7) and (21.9), we get the wavelength occupancy distribution in the PON:

$$P(j) = \left(\frac{\lambda}{Q}\right)^j \times \frac{\prod_{i=1}^{j}[N - (j-1)S]}{j!}$$
$$\times \left[\sum_{l=0}^{C} \left(\frac{\lambda}{Q}\right)^l \frac{\prod_{i=1}^{l}[N - (j-1)S]}{l!}\right]^{-1} \qquad (21.10)$$

The CFP is determined by $P(C)$, since a connection establishment is blocked and lost if and only if all the wavelengths are occupied.

To calculate the CBP $B_{k\mathrm{INF}}$ of service-class k calls of a particular ONU utilizing the access network, we rely on (21.2) while summing up the last b_k $q(i)$'s which correspond to the blocking states:

$$B_{k\mathrm{INF}} = \sum_{j=T-b_k+1}^{T} q(j) \qquad (21.11)$$

Note that the above model can be easily applied to the one-stage splitting case by setting $S = 1$.

We now proceed on the calculation of the TCBP, that is, the probability that a call is lost, either due to the restricted bandwidth capacity of a wavelength, or due to the unavailability of a free wavelength in the PON. To this end, we follow a reverse concept and we calculate the probability that a call is accepted for service by the PON. A call is accepted for service either when this call arrives at an ONU that has already established a connection while, at the same time, enough free b.u. are available in the wavelength, or this call is the first call that arrives at an ONU that has not established a connection yet, and a free wavelength can be found for the connection establishment. The probability P_{accept}^k that a service-class k call is accepted for service by the PON is given by

$$P_{\text{accept}}^k = P_s \cdot \frac{\sum_{i=1}^{T-b_k} q(i)}{\sum_{i=1}^{T} q(i)} + (1 - P_s) \cdot (1 - P(C)) \qquad (21.12)$$

where P_s is the probability that an ONU has already established a connection upon the call arrival.

The first term of the right part of (21.12) refers to the probability that a service-class k call has arrived to an ONU, which has already established a connection (with probability P_s) and there are at least b_k b.u. available in the wavelength (with probability that is given by the fraction of (21.12)). The second term of the second part of (21.12) signifies the probability, that a service-class k call arrives at the ONU, which has not established a connection (probability 1-P_s), and there is at least one available wavelength in the PON (probability 1-$P(C)$). The probability P_s is given by the summation of the probabilities $P(j)$ for $j = 1, \ldots, C$, multiplied by the probability that a specific ONU has established a connection:

$$P_s = \sum_{j=1}^{C} P(j) \frac{\binom{cN-1}{j-1}}{\binom{cN}{j}} = \sum_{j=1}^{C} P(j) \frac{j}{N} \qquad (21.13)$$

Thus, using (21.12) and (21.13), we can express TCBP of service-class k calls as

$$B_{k\text{INF}}^T = 1 - \left(P_s \cdot \frac{\sum_{i=1}^{T-b_k} q(i)}{\sum_{i=1}^{T} q(i)} + (1 - P_s) \cdot (1 - P(C)) \right) \qquad (21.14)$$

21.4 Model for Service-Classes with Finite Number of Traffic-Sources

In the afore-mentioned service model of the Hybrid TDM-WDM PON, we now consider that each service-class has a finite source population; let M_k be the number of traffic-sources of service-class k. To analyze the new model, we use the Markov chain of the corresponding infinite model (Figure 21.3), while properly modifying the parameters λ and Q.

In what follows, we show that the state transition diagram of Fig. 3 can also represent the Markov chain for the new service model with service-classes of finite population. Due to the finite traffic-source population the call arrival process in each ONU is not random (Poisson), but we consider it quasi-random. The call arrival rate from an idle traffic-source of service-class k is denoted by v_k. If no call is in service in each ONU, then the effective arrival rate to each ONU from service-class k is $M_k v_k$, and the total call arrival rate to each ONU becomes $\lambda = \sum_{k=1}^{K} \lambda_k M_k$.

The establishment of a connection can be realized by any call (from any service-class) that arrives from an ONU (belonging to an ONU group). Hence, the transition from state $[j-1]$ to state $[j]$, takes place $(N-(j-1)S)\lambda$ times per unit time; where $(N-(j-1)S)$ is the number of ONUs which have not been connected yet. The reverse transition, from state $[j]$ to state $[j-1]$, is realized jQ times per unit time, where the service rate of the wavelength Q is given as in (21.1), by

$$Q = \sum_{k=1}^{K} \mu_k y_{k_F}(b_k) \frac{q_F(b_k)}{\sum_{i=1}^{T} q_F(i)} \tag{21.15}$$

where $y_{k_F}(b_k)$ is the mean number of service-class k calls (with bandwidth requirement b_k) when b_k b.u. are occupied in the wavelength; $q_F(i)$ is the occupancy distribution of the b.u. in a wavelength, for finite number of traffic-sources per service-class.

The mean number of service-class k calls, $y_{k_F}(i)$, when i b.u. are occupied in the wavelength is given by

$$y_{k_F}(i) = \frac{a_{k_F} q_F(i - b_k)}{q_F(i)} \tag{21.16}$$

where $a_{k_F} = v_\kappa \mu_k^{-1}$ is the offered traffic-load per idle source of service-class k.

Since S ONUs utilize the same wavelength, the overall number of sources of the service-class k that demand service within one wavelength is $S \cdot M_k$. Therefore, the occupancy distribution of i b.u., $q_F(i)$, is given by the following recurrent formula, according to the Engset Multi-rate Loss model

(EnMLM) [17, 18]:

$$iq_F(i) = \sum_{k=1}^{K} a_{k_F} q_F(i - b_k)(S \cdot M_k - n_k(i) + 1) \text{ for } i = 1, \ldots, T$$

(21.17)

with $q_F(i) = 0$ for $i < 0$ and $\sum_{i=0}^{T} q_F(i) = 1$.

In (21.17) $n_k(i)$ is the number of service-class k calls which are in service when i out of T b.u. are seized in the wavelength. It has been proved [19] that this number can be approximated by $y_k(i)$, the average number of service-class k calls in the corresponding system, when infinite source population is assumed for each service-class. In the corresponding system with infinite number of traffic-sources, the offered traffic-load of service-class k is $S \cdot M_k a_{k_F}$; therefore, as in (21.4), we get:

$$n_k(i) \approx y_k(i) = \frac{SM_k a_{k_F} q(i - b_k)}{q(i)}$$

(21.18)

where $q(i)$ is the occupancy distribution of the b.u. of a wavelength, for the infinite traffic-sources case, and is given by (21.2). Note that the offered traffic-load in (21.2) is substituted with $S \cdot M_k a_{k_F}$.

The probability $P(j)$ that j wavelengths are occupied in the PON in steady state can be derived from the rate balance equations of the state transition diagram of Figure 21.3. Thereupon, the CFP is obtained by (21.10), as $P(C)$, when $\lambda = \sum_{k=0}^{K} M_k v_k$ and Q is determined by (21.15). For an ONU utilizing the PON, the CBP B_{k_F} of service-class k calls is calculated, as Time Congestion Probability [14], by

$$B_{k_F} = \sum_{j=T-b_k+1}^{T} q_F(j)$$

(21.19)

Following the same approach as in the infinite traffic-sources case, the TCBP can be calculated as in (21.14), by

$$B_{k_F}^{T} = 1 - \left(P_s \cdot \frac{\sum_{i=1}^{T-b_k} q_F(i)}{\sum_{i=1}^{T} q_F(i)} + (1 - P_s) \cdot (1 - P(C)) \right)$$

(21.20)

where $q_F(i)$ is the occupancy distribution of the wavelength and is given by (21.17) and P_s is the probability that an ONU has already established a connection upon the call arrival and is given by (21.13).

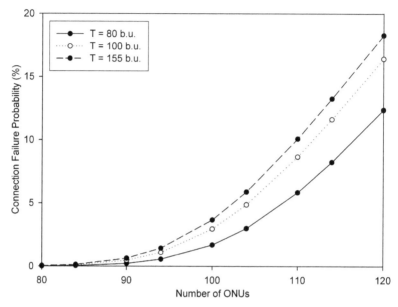

Figure 21.4 CFP versus the number of ONUs for different values of the wavelength bandwidth capacity T, for the infinite number of traffic-sources case.

21.5 Numerical Examples – Evaluation

In this section we evaluate the proposed models through simulation. To this end, we consider a hybrid TDM-WDM PON with $N = 100$ ONUs and capacity $C = 32$ wavelengths. The bit rate of the upstream direction is selected to be 155 Mbps, therefore, assuming that 1 b.u. $= 1$ Mbps, the wavelength capacity is $T = 155$ b.u. The ONUs form groups with $S = 2$, that is, one wavelength is shared by 2 ONUs. The PON accommodates three service-classes, $s1$, $s2$ and $s3$, with bandwidth requirements $b_1 = 48$ b.u, $b_2 = 36$ b.u and $b_3 = 24$ b.u., respectively. The simulation results are obtained as mean values from 8 runs with confidence interval of 95%.

We consider two cases of service-classes, the first case with infinite and the second case with finite source population. In the first case of infinite population, we comparatively present analytical and simulation results for the CFP versus the offered traffic-load in Table 21.1 and for the CBP and TCBP for the three service-classes, in Tables 21.2–21.4. The analytical CFP results are obtained through (21.1), (21.2), (21.3), (21.4) and (21.9), the analytical CBP results are obtained through (21.2), (21.3) and (21.11), whereas the analytical TCBP results are obtained through (21.2), (10, (13) and (21.14). The results

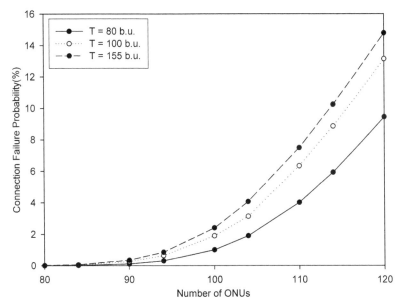

Figure 21.5 CFP versus the number of ONUs for different values of the wavelength bandwidth capacity T, for the finite number of traffic-sources case.

given in Tables 21.1–21.4 show a completely satisfactory accuracy of the proposed model.

Table 21.1 Analysis versus simulation for the CFP, for the infinite traffic-sources case.

Traffic-Load (erl)			Connection Failure Probability		
a_1	a_2	a_3	Analysis (%)	Simulation	
				Mean (%)	95% Conf.
0.04	0.060	0.090	0.00019	0.00125	0.000053
0.05	0.075	0.1125	0.010532	0.02303	0.00401
0.06	0.09	0.135	0.161948	0.17918	0.005699
0.07	0.105	0.1575	1.043973	1.07559	0.019481
0.08	0.120	0.180	3.672296	3.72838	0.040493
0.09	0.135	0.2025	8.540600	8.62377	0.075234
0.10	0.150	0.225	15.13064	15.1707	0.060469

Table 21.2 Analysis versus simulation for the CBP and TCBP of the first service-class, for the infinite traffic-sources case.

Traffic-Load (erl)			CBP 1st service-class			TCBP 1st service-class		
a_1	a_2	a_3	Analysis (%)	Simulation Mean (%)	95% Conf.	Analysis (%)	Simulation Mean (%)	95% Conf.
0.04	0.060	0.090	0.2966	0.2354	0.0129	0.1485	0.1290	5.3E-05
0.05	0.075	0.1125	0.5489	0.5383	0.0258	0.2831	0.2655	0.0041
0.06	0.09	0.135	0.8977	0.8989	0.0137	0.5754	0.5692	0.0057
0.07	0.105	0.1575	1.3479	1.3511	0.0292	1.4632	1.4600	0.0197
0.08	0.120	0.180	1.9013	1.9092	0.0346	3.6475	3.6424	0.0409
0.09	0.135	0.2025	2.5564	2.5456	0.0375	7.4061	7.3891	0.076
0.10	0.150	0.225	3.3097	3.311	0.0538	12.321	12.311	0.0611

Table 21.3 Analysis versus simulation for the CBP and TCBP of the second service-class, for the infinite traffic-sources case.

Traffic-Load (erl)			CBP 2nd service-class			TCBP 2nd service-class		
			Analysis (%)	Simulation		Analysis (%)	Simulation	
a_1	a_2	a_3		Mean (%)	95% Conf.		Mean (%)	95% Conf.
0.04	0.060	0.090	0.1508	0.1222	0.0076	0.0755	0.0659	6.1E-05
0.05	0.075	0.1125	0.2862	0.2851	0.011	0.1517	0.1424	0.0046
0.06	0.09	0.135	0.4792	0.4772	0.0066	0.3663	0.3502	0.0066
0.07	0.105	0.15575	0.7354	0.7323	0.014	1.1581	1.5652	0.0225
0.08	0.120	0.180	1.0582	1.0481	0.0241	3.2315	3.2122	0.0467
0.09	0.135	0.2025	1.4494	1.4386	0.0168	6.8695	6.8601	0.0868
0.10	0.150	0.225	1.9091	1.9061	0.0353	11.6587	11.631	0.0697

Table 21.4 Analysis versus simulation for the CBP and TCBP of the third service-class, for the infinite traffic-sources case.

Traffic-Load (erl)			CBP 3rd service-class			TCBP 3rd service-class		
				Simulation			Simulation	
a_1	a_2	a_3	Analysis (%)	Mean (%)	95% Conf.	Analysis (%)	Mean (%)	95% Conf.
0.04	0.060	0.090	0.0586	0.0507	0.0056	0.0295	0.0210	0.011
0.05	0.075	0.1125	00.1164	0.1181	0.0063	0.0668	0.0623	0.0162
0.06	0.09	0.135	0.2029	0.2038	0.004	0.2282	0.2099	0.0096
0.07	0.105	0.1575	0.3222	0.3224	0.0061	0.9522	0.9432	0.0204
0.08	0.120	0.180	0.4782	0.4784	0.007	2.9453	2.9212	0.035
0.09	0.135	0.2025	0.6732	0.6734	0.0164	6.4934	6.4792	0.0244
0.10	0.150	0.225	0.9088	0.9195	0.0147	11.1858	11.1643	0.0513

Table 21.5 Analysis versus simulation for the CFP, for the finite traffic-sources case.

Traffic-Load (erl)			Connection Failure Probability		
			Analysis (%)	Simulation	
a_{1_F}	a_{2_F}	a_{3_F}		Mean (%)	95% Conf.
0.004	0.006	0.009	0.00019	0.00118	0.00026
0.005	0.0075	0.01125	0.01091	0.08952	0.00569
0.006	0.009	0.0135	0.16706	0.15812	0.01861
0.007	0.0105	0.01575	1.07119	1.0305	0.04524
0.008	0.012	0.018	3.74699	3.54465	0.11311
0.009	0.0135	0.02025	8.67161	8.2953	0.13968
0.01	0.015	0.0225	15.3037	14.6961	0.11751

In the second case of finite population, the same number of traffic-sources is assumed for all service-classes: $M_1 = M_2 = M_3 = 10$. We comparatively present analytical and simulation results for the CFP versus the offered traffic-load in Table 21.5 and for the CBP and TCBP for the three service-classes, in Tables 21.6–21.8, respectively. The analytical CFP results are obtained through (21.2), (21.10), (21.15), (21.16), (21.17) and (21.18), the analytical CBP results are obtained through (21.17) and (21.19), whereas the analytical TCBP results are obtained through (21.10), (21.13), (21.17) and (21.20). As the results of Tables 21.5–21.8 reveal, the accuracy of the proposed models is satisfactory. The small divergence between the analytical and simulation results is an outcome of the approximation that is introduced in the models for the calculation of $n_k(i)$. Nevertheless, this discrepancy is minor and in a well-acceptable level.

The results of the infinite traffic-sources case (Tables 21.1–21.4) can be compared to the corresponding results of the finite traffic-sources case (Tables 21.5–21.8), because we have considered an equal offered traffic-load each time; i.e. $a_1 = M_1 a_{1_F}$, $a_2 = M_2 a_{2_F}$ and $a_3 = M_3 a_{3_F}$. Comparing the above-mentioned results, one can observe that the CFP of the finite population case are higher than the corresponding results of the infinite population case. The reverse situation happens for the CBP results; higher CBP are obtained for infinite population. Both CFP and CBP results are reasonable. When lower CBP results are obtained for finite traffic-sources, it means that more calls are serviced by a wavelength and therefore the mean service/seize

Table 21.6 Analysis versus simulation for the CBP and TCBP of the first service-class, for the finite traffic-sources case.

Traffic-Load (erl)			CBP 1st service-class			TCBP 1st service-class		
a_{1_F}	a_{2_F}	a_{3_F}	Analysis (%)	Simulation		Analysis (%)	Simulation	
				Mean (%)	95% Conf.		Mean (%)	95% Conf.
0.004	0.006	0.009	0.2978	0.2698	0.0118	0.1489	0.1286	0.0002
0.005	0.0075	0.01125	0.5515	0.4969	0.0147	0.2842	0.2489	0.004
0.006	0.009	0.0135	0.9023	0.813	0.0253	0.5811	0.5301	0.014
0.007	0.0105	0.01575	1.3549	1.2462	0.0319	1.4858	1.3975	0.0345
0.008	0.012	0.018	1.9108	1.7161	0.0291	3.7043	3.4901	0.0863
0.009	0.0135	0.02025	2.5681	2.394	0.0404	7.5014	7.346	0.1066
0.01	0.015	0.0225	3.2331	2.9733	0.0371	12.4437	11.799	0.0897

Table 21.7 Analysis versus simulation for the CBP and TCBP of the second service-class, for the finite traffic-sources case.

Traffic-Load (erl)			CBP 2nd service-class			TCBP 2nd service-class		
a_{1_F}	a_{2_F}	a_{3_F}	Analysis (%)	Simulation		Analysis (%)	Simulation	
				Mean (%)	95% Conf.		Mean (%)	95% Conf.
0.004	0.006	0.009	0.1508	0.1304	0.0059	0.7548	0.7088	0.0208
0.005	0.0075	0.01125	0.2868	0.2531	0.011	0.1521	0.136	0.0259
0.006	0.009	0.0135	0.4808	0.4186	0.0152	0.3707	0.3437	0.0447
0.007	0.0105	0.01575	0.7381	0.6428	0.0231	1.179	1.1461	0.056
0.008	0.012	0.018	1.0621	0.917	0.017	3.2863	3.199	0.0513
0.009	0.0135	0.02025	1.454	1.3227	0.022	6.9627	6.8157	0.0712
0.01	0.015	0.0225	1.9147	1.7045	0.062	11.779	11.099	0.0654

Table 21.8 Analysis versus simulation for the CBP and TCBP of the third service-class, for the finite traffic-sources case.

Traffic-Load (erl)			CBP 3rd service-class			TCBP 3rd service-class		
				Simulation			Simulation	
a_{1_F}	a_{2_F}	a_{3_F}	Analysis (%)	Mean (%)	95% Conf.	Analysis (%)	Mean (%)	95% Conf.
0.004	0.006	0.009	0.0581	0.0488	0.0045	0.0291	0.0199	0.0064
0.005	0.0075	0.01125	0.1157	0.102	0.0047	0.0666	0.0615	0.0118
0.006	0.009	0.0135	0.2021	0.1749	0.0106	0.2316	0.1919	0.0163
0.007	0.0105	0.01575	0.3213	0.2805	0.00984	0.9717	0.9144	0.0249
0.008	0.012	0.018	0.47697	0.4064	0.0121	2.9981	2.8854	0.0178
0.009	0.0135	0.02025	0.631424	0.5811	0.0091	6.5839	6.5001	0.0241
0.01	0.015	0.0225	0.9061	0.8146	0.0457	11.303	10.773	0.0668

time of the wavelength is increased, or, in other words, the CFP is increased. The consistence of the results validates our models.

Having validated the models and evaluated their accuracy through simulation, we proceed to investigate the effect of the number of ONUs, N, and the wavelength capacity T to the CFP. Figures 21.4 and 21.5 correspond to infinite and finite traffic-sources cases, respectively, and show the variation of the CFP versus the number of ONUs, N, for three wavelength capacities: $T = 80$, 100 and 155 b.u. In the infinite traffic-sources case, the offered traffic-load for $s1$, $s2$ and $s3$, is $a_1 = 0.08$ erl, $a_2 = 0.12$ erl and $a_3 = 0.18$ erl, respectively. In the case of finite population, the number of traffic-sources of $s1$, $s2$ and $s3$, is $M_1 = M_2 = M_3 = 5$, while the offered traffic-load per idle source is $a_{1_F} = 0.016$ erl, $a_{2_F} = 0.024$ erl and $a_{3_F} = 0.032$ erl, respectively. As it was anticipated (Figures 21.4 and 21.5) the increase of the number of ONUs strongly increases the CFP, especially when the wavelength bandwidth capacity T is large. This is due to the fact that the increase of the ONUs leads to higher demand for connections in the PON and therefore to higher CFP. The decrease of T leads to smaller CFP, since the release of a wavelength is more frequent when the values of T are small, given that the other traffic parameters are the same.

21.6 Conclusion

In summary, we propose teletraffic loss models, for hybrid TDM-WDM PONs with dynamic wavelength allocation that accommodate service-classes of infinite or finite number of traffic-sources. The proposed models are derived from one-dimensional Markov chains, which describe the wavelength's occupancy in the PON. The analytical calculations of CFP, CBP and TCBP are based on the systems state distributions. The models are validated and evaluated through extensive simulation, based on the results. Their accuracy was found to be absolutely satisfactory. Based on the proposed models, several investigations could be carried out, as for example, we showed the impact of the number of ONUs and the wavelength bandwidth capacity on the CFP. In our future work we shall extend our research by considering queues in each ONU while including in our study not only the upstream but also the downstream direction (from OLT to the ONUs).

Acknowledgement

The authors would like to thank Dr Ioannis D. Moscholios (University of Patras, Greece) for his support.

References

[1] F. Effenberger, D. Cleary, O. Haran, G. Kramer, R. Ding Li, M. Oron and T. Pfeiffer, An introduction to PON technologies, *IEEE Communications Magazine*, vol. 45, no. 3, pp. S17–S25, March 2007.

[2] B. Mukherjee, *Optical WDM Networks*, Springer, 2006.

[3] T. Koonen, Fiber to the Home/Fiber to the premises: What, where, and when?, *Proceedings of the IEEE*, vol. 94, no. 5, pp. 911–934, May 2006.

[4] M. Abrams, P. C. Becker, Y. Fujimoto, V. O'Byrne and D. Piehler, FTTP deployments in the United States and Japan – Equipment choices and service provider imperatives, *IEEE/OSA Journal of Lightwave Technology* , vol. 23, no. 1, pp. 236–246, January 2005.

[5] R. Davey, F. Bourgart and K. McCammon, Options for future optical access networks, *IEEE Communications Magazine*, vol. 44, no. 5, pp. 50–56, October 2006.

[6] M. Ma, Y. Zhu and T. H. Cheng, A systematic scheme for multiple access in ethernet passive optical networks, *IEEE/OSA Journal of Lightwave Technology*, vol. 23, no. 11, pp. 3671–3682, November 2005.

[7] ITU-T G.984.1, SG 15, Gigabit-capable passive optical networks (G-PON): General characteristics, March 2003.

[8] A. R. Dhaini, C. M. Assi, M. Maier and A. Shami, Dynamic wavelength and bandwidth allocation in hybrid TDM/WDM EPON networks, *IEEE/OSA Journal of Lightwave Technology*, vol. 25, no. 1, pp. 277–286, January 2007.

[9] N. J. Frigo, P. P. Iannone and K. C. Reichmann, Spectral slicing in WDM passive optical networks for local access, in *Proceedings of the 24th European Conference on Optical Communication, ECOC 1998*, Madrid, Spain, September 1998, pp. 119–120, 1998.

[10] N. J. Frigo, P. P. Iannone, K. C. Reichmann, J. A. Walker, K. W. Goossen, S. C. Arney, E. J. Murphy, Y. Ota and R. G. Swartz, Demonstration of performance-tiered modulators in a WDM PON with a single shared source, in *Proceedings of the 19th European Conference on Optical Communication, ECOC 1995*, Brussels, Belgium, 17–21 September 1995, pp. 441–444, 1995.

[11] D. Gutierrez, W.-T. Shaw, F.-T. An, K. S. Kim, Y.-L. Hsueh, M. Rogge, G. Wong and L. G. Kazovsky, Next generation optical access networks, Invited Paper in *Proceedings of IEEE/Create-Net BroadNets 2006*, pp. 1–10, October 2006.

[12] G. Maier, M. Martinelli, E. Salvadori and M. CoreCom, Multistage WDM passive access networks: Design and cost issues, in *Proceedings of IEEE International Conference on Communications, ICC 99*, Vancouver, Canada, 6–10 June, pp. 1707–1713, 1999.

[13] K. E. Han, W. H. Yang, K. M. Yo and Y. C. Kim, Design of AWG-based WDM-PON architecture with multicast capability, in *Proceedings of IEEE INFOCOM 2008*, Phoenix, USA, 13–19 April, pp. 1–6, 2008.

[14] H. Akimaru and K. Kawashima, *Teletraffic – Theory and Applications*, Springer Verlag, Berlin, 1993.

[15] J. S. Kaufman, "Blocking in a shared resource environment", *IEEE Transactions on Communications*, vol. 29, no. 10, pp. 1474–1481, 1981.

[16] I. Moscholios, M. Logothetis and G. Kokkinakis, Connection dependent threshold model: A generalization of the Erlang multiple rate loss model, *Performance Evaluation*, Special Issue on Performance Modeling and Analysis of ATM and IP Networks, vol. 48, nos. 1–4, pp. 177–200, May 2002.

[17] G. Stamatelos and J. Hayes, Admission control techniques with application to broadband networks, *Computer Communications*, vol. 17, no. 9, pp. 663–673, 1994.

[18] I. D. Moscholios, M. D. Logothetis and P. I. Nikolaropoulos, Engset multi-rate state-dependent loss models, *Performance Evaluation*, vol. 59, nos. 2–3, pp. 243–277, February 2005.

[19] M. Glabowski and M. Stasiak, An approximate model of the full-availability group with multi-rate traffic and a finite source population, in *Proceedings of 12th GI/ITG Conference on Measuring, Modeling and Evaluation of Computer and Communication Systems (MMB) and 3rd Polish-German Teletraffic Symposium (PGTS), MMB and PGTS 2004*, Dresden, Germany, 12–15 September, pp. 195–204, 2004.

Author Index

Subject Index